DIADROMY IN FISHES
Migrations between Freshwater and Marine Environments

DIADROMY IN FISHES

Migrations between Freshwater and Marine Environments

ROBERT M. McDOWALL

CROOM HELM
London Sydney

TIMBER PRESS
Portland, Oregon

First published in 1988 by Croom Helm
11 New Fetter Lane, London EC4P 4EE

© 1988 Robert M. McDowall

Printed in Great Britain at the
University Press, Cambridge

ISBN 0–7099–5503–0

British Library Cataloguing in Publication Data

McDowall, R.M.
  Diadromy in fishes: migrations between
  freshwater and marine environments.
  1. Fishes — Migration
  I. Title
  597′0525   GL639.5

  ISBN 0–7099–5503–0

First published in the USA 1988 by Timber Press
9999 S.W. Wilshire, Portland, Oregon 97225

Library of Congress Cataloging in Publication Data

McDowall, R.M. (Robert Montgomery), 1939–
  Diadromy in fishes.

  Bibliography: p.
  1. Diadromous fishes.   I. Title.
QL639.5.M33   1988   597′.0525   88-4888
ISBN 0–88192–114–9

Contents

Preface

Diadromy as a phenomenon was, for all practical purposes, introduced to ichthyology and fisheries biology in the late 1940s by George Myers at Stanford University. He clarified definitions of existing terminology and introduced some new and useful terms, that have stood for nearly 40 years. My own introduction to the terminology came in the late 1960s while studying with Giles Mead at the Museum of Comparative Zoology at Harvard University (Mead had been a student of Myers). Since that time I have worked extensively on the southern hemisphere freshwater salomoniform fishes — the glaxiids and relatives — many of which are diadromous. It was, I imagine, because of my interest in these fish, (giving me a distinctive southern perspective,) that I was invited by the Northeastern Division of the American Fisheries Society to present a paper reviewing Massachusetts. The paper was entitled Common Strategies of Anadromous and Catadromous Fishes. As I pursued the task of a broad review of diadromy, it rapidly became apparent than no-one had before attempted such a review, and that the literature of the subject was large in volume, broad in scope, and scattered very widely through diverse journals and books. Much was known about a limited group of diadromous fishes, especially the salmonides, anguillid eels, lampreys, and sturgeons, but that little was recorded in an accessible way about the many other groups that are also diadromous. The paper presented at the Boston AFS symposium eventually focused on a few issues, in which I attempted to draw attention to the taxonomic, geographical and ecological diversity represented by a diadromous fishes. The results of the literature review were far too voluminous to allow verbal presentation and have become the basic material of this book. It is offerred as a resource that can from a basis for understanding the different types of diadromy (anadromy, catadromy, amphidromy), how widely diadromy occurs in diverse groups of fishes, and where it occurs geographically. It looks at the significance of diadromy in fish behaviour and ecology.

The literature grows very rapidly, in this as in other fields of ichthyology, and much of that referred to here dates from 1970 or later. It is my hope that this review will assist biologists in focussing more clearly on the various aspects of fisheries related to diadromy and form a basis for a growing understanding of the phenomenon.

Acknowledgements

My thanks go, first of all, to Michael Dadswell, Christine Moffit and Richard Saunders, of AFS, who initially invited me to prepare a review of diadromy for their Boston symposium. This early stimulus, their helpful suggestions, as well as funding provided to attend the meeting, are together responsible for prompting the review. The symposium provided me with invaluable opportunities for discussions with fisheries biologists from many parts of the world, who are interested in varied aspects of diadromy. I had helpful and stimulating discussions with Mart Gross, with whom I shared the opening session of the symposium, and also with Stephen Blaber, Michael Bruton, Ronald Griffin, Gene Helfman, James McCleave, Christoper Moriarty, Geoff Power, Roger Rulifson, John Thorpe, Clement Walton, and many others too numerous to name.

During preparation of the volume, I corresponded with and obtained advice from a large number of ichthyologists from many parts of the world: C.E. Dawson (pipefishes); P.R. Dano (European fishes); D.A. Erdman (Central America); W.Fulton (Tasmania); M.T.D. Giamas (South American anchovies); D.A. Hensley (Central America); B. Jonsson (Scandinavia); R.A. Kinzie (Hawaii); L. Lesack (Africa); D.A. McAllister (osmerid smelts); J.A. Maciolek (Pacific islands); N. Menezes (South America); P.J. Miller (gobies and eleotrids); N.Mizuno (Japanese gobies); J.S. Nelson (sticklebacks); I.C. Potter (lampreys); J. Reynolds (African cluepeids); G.G. Teugels (Indian Ocean); T.B. Thorson (elasmorbranchs); and many others whose literature or correspondence contributed to the substance of the text. To all these, and to Brenda Bonnett, Fisheries Research Centre librarian, who pursued a lot of difficult library interloans, go my thanks.

1

Introduction

Animal migrations have always stirred the imaginations and interests of man, both in relation to his obtaining food, and also to his basic curiosities as a naturalist. They have probably played an important role as a protein source for man throughout history, first, when he was a hunter – gatherer seeking food for himself and his family, later as civilisation developed in obtaining food both for personal consumption and for the wider community group, and later again for barter and sale. This interest in migratory animals is likely to have developed largely because, during migrations, animals were observed to be highly concentrated geographically and temporally, and also were known to follow well-defined and predictable pathways. These characteristics made migratory animals highly accessible as food resources (Figure 1.1), and enabled primitive man to plan his food gatherering and food storage/preservation activities.

The migrations of fishes are among those animal migrations that have become well known to man and which he has exploited for food purposes, and many writers have described them in colourful terms. Roule's (1933) account of migration in fishes talks of how:

The imagination is stirred at the idea of these vast assemblies, their concentration, their setting out, their gradual breaking up. It follows them along their way, sometimes in deep water, sometimes on the surface, swimming in eager cohorts. It goes with them in their invading ardour, in their irruption into spaces which had hitherto been unoccupied, upon which their troops suddenly and invincibly impose their domination and the power of their numbers.

Tchernavin (1939) wrote of how the return of salmon from the sea is:

one of the most fascinating phenomena in nature. This huge, glittering marine fish exposes itself to the most perilous adventure, rushing into small streams too shallow to cover its enlarged body, and becoming an easy prey to every creature with leisure to hunt it: man, eagle, even the bear and dog.

Figure 1.1 North American Indians netting salmon (*Oncorhynchus* sp.) at Celico Falls, Columbia River, in a way they have done for generations; this photo about 1942 Source: Oregan Historical Society

... The effort which some Salmonidae make in overcoming obstacles to reach the really inland region for spawning is one of the most arresting facts in natural history.

According to Talbot and Sykes (1958):

The study of fish migrations has long intrigued many persons. Much of the early work was inspired by scientific curiosity or undertaken as a hobby by wealthy owners of riparian rights. More recently, however, knowledge of fish migrations has been necessary for the intelligent management of some species since many fisheries depend on congregations of migrating fish.

Nowhere is interest in the migrations of fishes more strongly represented than in those species that migrate regularly between the sea and fresh water—species that have become known as 'diadromous' fishes. There is a mystique

Figure 1.2: Runs and catches of diadromous fish vary greatly from day to day as illustrated by daily catch records of whitebait (*Galaxias* spp.) in the Waiatoto River, New Zealand

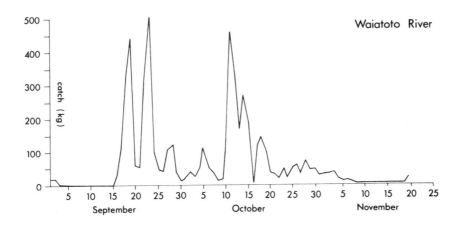

about the way diadromous fishes 'turn up' at river mouths on their way up-stream or downstream, doing so predictably with regard to season of the year, yet often totally unpredictably from day to day or hour to hour (Figure 1.2). They may attract huge concentrations of fishermen of all varieties. Haslar (1966) captured the sense of excitement and wonder generated by watching fish migrating up stream from the sea, when he wrote:

> No-one who has seen a 20 kg salmon fling itself into the air again and again until it is exhausted in a vain effort to surmount a waterfall, can fail to marvel at the strength of the instinct that draws the salmon up river to the place where it was born (Figure 1.3).

Salmonids have unquestionably played a major role in studies of migratory fishes. Tchernavin (1939) considered that this is because:

> The life history of the salmon is one of the most amazing phenomena in nature. Why does the young salmon (the parr) suddenly change its spotted, trout-like livery for a bright silvery one, transform itself to a smolt, and venture from its parent stream down to the vast, alien spaces of the sea? Why does it undergo strange changes on its return to the rivers; and why ... does it die when it comes back to its native home? Whence derive these disastrous habits?

3

Figure 1.3 Atlantic salmon (*Salmo - salar*) jumping at falls on its way upstream to spawn in a Swedish river (R. Bognarsson)

Salmonids have been the subject of extensive detailed studies

of a quantity and diversity unparalleled elsewhere in the fish literature. In part this is due to the importance of salmonids as food and game fishes, but in part it is due to the diversity of the life history patterns found among species and populations of salmonids. (Miller and Brannon, 1982.)

Old history shows that fish migrations have always been important to man as a food gatherer. There is mention in the French literature of a picture of an Atlantic salmon carved on a reindeer bone that dates back about 14,000 years (Anderson and Brimer, 1976); the name salar was evidently given to the species at the time the Roman armies of Julius Caesar invaded western Europe, about 2,100 years ago. Hall (1985) mentioned exploitation of two diadromous African fishes — the mullet, *Myxus capensis*, and the eel, *Anguilla mossambica*, as long as 6,000 years ago; this is probably not unique in archaeology. Netboy (1980) cited the mention of salmon in the writings of the Roman scholar Pliny in the first century A.D., and referred to descriptions of salmon in the fisheries literature of England and Scotland as early as the eighth century. Regan (1911) cited reports dating from the thirteenth century discussing the management of

English salmon fisheries, while exploitation of lampreys is documented as early as the fifteenth century in Finland and Sweden (Sjoberg, 1980; Tuunainen, Ikonen and Auvinen, 1980), being mentioned in sixteenth century Swedish taxation returns, and apparently being discussed by the pioneering taxonomist Linnaeus, following his journey to northern Sweden in 1732 (Sjoberg, 1980). Svardson (1979) published drawings of two coregonid whitefishes on the door of an old Swedish church that dated back to the sixteenth century. Some towns in Europe were so proud of their wealth of salmon that they had a salmon on their coat of arms (Roule, 1933). This is a question discussed at length by Moule (1842), who described how both civic authorities and the nobility in the United Kingdom, Ireland, Germany, France, and other European countries had various fish, including salmon, as prominent parts of crests, seals and coats of arms (Figure 1.4). In eighteenth century England, salmon were obviously common and important as food, and were:

> sold for a penny or two pence a pound, whilst it was a regular condition in indentures of apprenticeships or in agreements between master and servant that Salmon should not be given for dinner more than so many times a week (Regan, 1911).

Figure 1.4 Salmon were included in the coats of arms of the nobility and civic authorities in Europe, a recognition of their importance to the people — in this instance the crest of the Princes of Upper Salm in Lorraine, France (Moule)—1842

Old paintings in art galleries indicate that salmon was on the tables of the poorest people in the Netherlands, while archaeologists have discovered numerous sturgeon scales in excavations of a prehistoric village near Rotterdam, showing that this fish must also have played a significant part in the diet of the local people; sturgeon fishing must have been an important occupation for them (Korringa, 1967). There are bound to be many other indicators of such early use by man of diadromous fishes like salmons and whitefishes, and there can be no doubt that in some countries primitive man knew where and when food resources would be present and available in large quantities, and made good use of them. Reading Roule (1933) again, we are told how in France:

> Once, ... it was understood that labourers should have salmon two or three times a week, at least, so plentifully stocked were the neighbouring streams. ... Our prehistoric ancestors carved on the walls of their caves the outlines of this fish ... up to the most recent times there was a profusion of salmon which stands out in striking contrast to the shortage of that fish today. ... also with sturgeons, at those state banquets at which the guests sparkled with precious stones and wore magnificent clothes ... one of the principal dishes often consisted of a huge sturgeon cooked whole, served up on a great pewter or silver dish ... its practical disappearance is of quite recent date.

At the advent of the spread of European culture into other parts of the world during the sixteenth to nineteenth centuries aboriginal peoples in many countries were routinely observed to be harvesting shads, alewives, salmons, smelts, sturgeons, galaxiids, eels, gobies, and the like. This was certainly true in New Zealand, where the Polynesian Maoris were harvesting many of the native diadromous fish species at various life history phases — adult lampreys, anguillid eels at all stages from glass eel onwards, retropinnid smelts, the so-called southern graylings *(Prototroctes)*, various galaxiids, and others. It was also happening in North America, where the Indian people were exploiting the massive runs of the various species of shad (Cluepidae), smelt (Osmeridae) and salmon, char and whitefish (Salmonidae). In many of these former colonial areas commercial harvesting by the Europeans developed rapidly once the values of the fish species involved were recognised, and some of these fisheries have come to constitute very important commercial fishery resources.

The question of why diadromous fishes have had such historical/culinary importance is an interesting one. Harden-Jones (1968) argued that if migration is an adaptation towards abundance, this would explain why many of the important commercial species are migratory; the hypothesis is that they are of commercial importance because they are abundant, and are abundant because they are migratory. This argument depends, in part, on demonstrating that migratory species actually are more abundant, and while what Harden-Jones said may be true, I am unaware that this has been examined rigorously. Of the 10 most productive fish species in world fisheries catches (FAO, 1978), the first nine, were migratory fish (Harden-Jones, 1981). These nine migratory species contributed about one-third of total world fisheries catches in 1977, but of the nine none was diadromous. They were all marine migratory fishes. Although this shows a compelling relationship between *high fish catches* and the *occurrence of migration* it does not establish a causal connection between *abundance* and *migration*, and that seems, to me to be the critical issue. Harden-Jones (1981) noted that Cushing (1969) has asked the question concerning 'whether fish are abundant because they migrate or *vice-versa*', and confessed to having no answer. I venture to suggest that the commercial importance of migratory fishes may be due also, at least in substantial measure, to their spatial and temporal concentration during migration, rather than or in addition to their actual abundance. This is a suggestion also offered by Smith (1985) who considered that migration brings fish into situations that allow easy harvest as they concentrate along migration routes.

Many migratory fish species move between fresh and marine waters and such species are amongst the best-known of fish migrations. As they enter river mouths from the sea, they tend to pass through a very narrow 'bottle-neck', at least spatially, and to a lesser extent temporally. This brings together even more massive concentrations of harvestable animals in both space and time, than is true of many other types of migrant animals, and makes possible their capture in vast numbers (Figure 1.5). A wide diversity of fishes in many lands is represented. Included among these fishes are the sturgeons (*Acipenser* spp., family Acipenseridae) of Europe and North America, although absolute numbers of these species were probably never high. Migrations of Atlantic salmon (*Salmo salar*, family Salmonidae) in rivers entering all parts of the cool and cold North Atlantic Ocean are widely renowned, as are those of the Pacific salmons (*Oncorhynchus* spp.) which migrate into the rivers of western North America and northeastern Asia. Their abundance was enormous, salmon catches in Washington State (USA), primarily from the Fraser River, amounting to an average of more than nine million fish per year between 1935 and 1955 (Foerster, 1968). Netboy (1980) listed total catches of all salmons as exceeding 400,000 tonnes in several years between 1967 and 1975, nearly one-half being pink salmon (*Oncorhynchus gorbuscha*). Migrations of the smelts (family Osmeridae) enter into northern, cool-temperate, river systems. McKenzie (1964) discussed a stock of *Osmerus mordax* amounting to 350 million fish, and an annual catch of 14.5 million in the Miramichi River in New Brunswick, Canada. This species occurs widely along the Atlantic coast of North America. There are immense migrations of northern temperate alewives, shads and other clupeids (family Clupeidae); Bigelow and Schroeder (1953) discussed annual catches of shad (*Alosa sapidissima*) reaching as high as 800,000 fish in the Merrimac River in New England, and of alewive (*Alosa pseudoharengus*) populations in the Gulf of Maine (1896) of 22 million fish, these being only a fraction of the alewife runs into the St Johns River, nearby. Leggett (1973) said that 'the "shad run" of the Atlantic coast of North America is a well-known phenomenon' and

Figure 1.5 Large accumulations of chinook salmon (*Oncorhynchus tshawytscha*) on the spawning grounds, Rakaia River, New Zealand.

'within the space of a few weeks, shad by the tens of thousands come in from the sea and move up river to spawn'.

The migrations of hilsa (*Hilsa* spp., family Clupeidae) are well known in India and parts of Southeast Asia. Total catch of all *Hilsa* species in 1970 was more than 13,000 tonnes (FAO, 1974), the catch in the Hooghly River (India) alone reaching 1,731 tonnes in 1970–71 (Jhingram, 1975). Freshwater eels (*Anguilla* spp., family Anguillidae) are best known in the North Atlantic and western North Pacific, but also occur and are probably more abundant in other temperate and tropical lands of the Indo-Pacific, and have for long attracted attention from fishermen and gourmets in many areas. Cairns (1941) recounted watching anguillid elvers migrating up a New Zealand river, comprising a shoal about 5 m wide and 2–3 m deep, which continued past his observation point for more than eight hours. The widespread sea bass or barramundi, *Lates calcarifer* (family Centropomidae), is widely taken as a food fish, catch exceeding 2,000 tonnes in northern Australia (Morrisey, 1985) and in India amounting to 283 tonnes in 1968–70 (Jhingram, 1975).

Wootton (1976) drew attention to some old reports of extraordinarily large catches of the three-spined stickleback, *Gasterosteus aculeatus*; in particular there was a report by a Mr Pennant dating from the eighteenth century, and published by Regan (1911):

One in seven or eight years amazing shoals appear in the Welland [river in England] and come up the river in the form of a vast column. The quantity is so great that they are used to manure the land, and trials have been made to get oil from them. A notion may be had of this vast shoal by saying that a man employed by the farmer to take them has got for a considerable time four shillings a day by selling them at a halfpence a bushell.

Wootton's (1976) response to this was to exclaim that this man was collecting 96 bushells a day! (A bushell is a rather antiquated volume measure formerly used for grain, that equals eight gallons, so the catch was 768 gallons a day for a considerable time.) Wootton referred to similar occurrences in recent times, and also to reports by Baggerman (1957) who described how the Dutch took them during the war to produce fish meal and to feed to ducks. The scope of these stickleback migrations is difficult to imagine.

Amongst the most astonishing migrations are those of the gobiid *Sycidium plumieri* (family Gobiidae) into streams of the islands of the Caribbean. Erdman (1961) described how he had observed a run of these tiny fish about 25 mm long, moving upstream in a column about 15 cm wide, 2.5 rows of fish deep, passing his observation point at a rate of 300 per second. He calculated that during an average migration lasting two days, 90 million fish would have moved past. This little fish was the basis of a significant fishery (the Jamaican 'whitebait' or equivalent local names) in several Caribbean islands — Puerto Rico, Martinique, Dominica, Haiti, Jamaica, and others. Migrations of the diminutive juveniles of closely related species of *Sicyopterus* are known in several Pacific island and Asian rivers, the tiny fish about 20 mm long once being taken in huge quantities in a fishery in the Philippines (Manacop, 1953).

Much less widely known, but equally spectacular (at least with respect to the numbers of fish migrating) are some of the peculiarly Southern Hemisphere fish groups. Notable among these are the southern pouched lamprey (*Geotria australis*, family Geotriidae), one observer describing the capture of 'tonnes' in one night from a New Zealand river (Phillipps and Hodgkinson, 1922). The Australian whitebait (*Lovettia sealii*, family Aplochitonidae), once ran in Tasmanian rivers in large quantities, with a peak seasonal catch of 484 tonnes of these tiny fish 40–45 mm long (Blackburn, 1950). Similarly the New Zealand whitebait (*Galaxias* spp., family Galaxiidae) once ran in vast numbers, catch for the peak year of records being 320 tonnes along a coastline of about 500 km, a total catch of more than 500 million fish. Catches by one fishermen in a single day of 1.5–2 million fish are known in recent years (McDowall and Eldon, 1980). Massive runs of the southern smelts (*Stokellia* and *Retropinna* spp., family Retropinnidae) occur in New Zealand rivers (McDowall, 1984c). And there are, no doubt, other fish groups in which migrations of these types and magnitudes take place. Apart from these well-known and spectacular migrations, in which the fish are of economic or food importance, there is a highly diverse array of fishes that migrate in large numbers in equally diverse countries of the globe, but whose migrations are little reported, often poorly understood, and which do not attract the attention of man as a fisherman. There is little doubt that many more instances of migrations between marine and fresh waters remain undiscovered, or are still undescribed in the literature.

2

Terminology and Some Definitions

It has become customary to classify fishes, particularly those that live in fresh water, according to their capacity to cope with waters of differing salinities. Many fishes are marine stenohaline and are strictly limited to marine waters of high salinity. Similarly, many are stenohaline freshwater species which are strictly confined to low salinity, fresh waters. However, many fish species are much more adaptable and occupy waters of diverse salinities. Various authors have classified such euryhaline fishes according to their ability to tolerate differing salinities; usually, for no obvious reason, these classifications have applied primarily to freshwater fishes. Boulenger (1905) distinguished between:

1. Those freshwater fishes that live a part of the year in the sea;
2. Those living normally in the sea but of which certain colonies have become landlocked;
3. Those which, although entirely confined to fresh waters, have as nearest allies species living in the sea; and
4. Those belonging to families entirely or chiefly restricted to fresh water.

Later, Nicholls (1928) distinguished between continental freshwater fishes, which were equivalent to Boulenger's group 4 — those restricted to fresh water, and peripheral species, these being freshwater fishes which have 'better marked affinities with salt water groups'.

Myers (1949a) further analysed the categories of fishes that variously occupy fresh waters, listing the following:

1. Primary — those strictly intolerant of salt water;
2. Secondary — those rather strictly confined to fresh water, but relatively salt tolerant, at least for short periods;
3. Vicarious — presumably non-diadromous, freshwater representatives of primarily marine groups;
4. Complementary — freshwater forms often or usually diadromous;
5. Diadromous — fish which regularly migrate between fresh and salt water

at a definite stage of the life cycle;

6. Sporadic — fishes which live and breed indifferently in salt or fresh water, or which enter fresh water only sporadically and not as part of a true migration.

Darlington (1957) has combined categories 3 to 6 in Myers' listing and labelled them with Nicholls' term peripheral to cover species that are regarded as having — or being closely related to or derived from — species that have high salinity tolerances. Myers (1963) subsequently adopted this same use of peripheral.

McHugh (1967) presented a classification of estuarine-living fishes distinguishing between

1. Freshwater fish that occasionally enter brackish waters;
2. Truly estuarine species which spend their entire lives in the estuary;
3. Anadromous and catadromous species;
4. Marine species which pay regular, seasonal visits to the estuary, usually as adults;
5. Marine species which use the estuary primarily as a nursery ground usually spawning and spending much of the adult life at sea, but others returning seasonally to the estuary;
6. Adventitious visitors which appear irregularly and have no apparent estuarine requirements.

Haedrich (1983) considered this classification to be applicable to most estuarine conditions, but considered that the 'greatest problem ... is not in the definitions themselves, but in having sufficient information to assign individual species to a category [how true this is!]'

In the recent literature amongst fish species that occur in fresh waters, three groups are generally recognised (Myers, 1949a; Darlington, 1957).

Primary — species that are confined strictly to fresh waters (these terms seem usually to be applied, for some reason, at the family level).

Secondary — freshwater species that are found almost exclusively in fresh water; they seem to have sufficient tolerance of higher salinities to be found occasionally in brackish/saline waters, and to be able to disperse through the sea, at least for short distances.

Peripheral — these appear able to live readily in both fresh and salt water. Some of them live primarily in one or the other, but venture into different salinity waters from time to time, others move freely between fresh and salt water on a regular or irregular, short-term basis, and yet others have explicit life history stages that are predictably and alternately fresh water and/or marine.

Fishes that are diadromous — the primary concern of this book — are classified as peripheral under the above categories.

None of the above sets of definitions specifically addresses the question of diadromy. It was Myers (1949b) who was instrumental in bringing diadromy and related terms into modern usage in fisheries biology, and he used the term diadromy to describe fish species that alternate various life stages between marine and fresh waters. The question of how and from where within the diverse array of differing types of fish movements the diadromous fishes are determined is an important one, if the term diadromy is to have any useful meaning. Myers (1949b) defined diadromous fishes as: 'Truly migratory fishes which migrate between the sea and fresh water'. Diadromy can thus properly be regarded as a specialised form of migration, so that critical to the use of diadromy, and the determination of which groups are included in the category diadromous, is the question of how migration is defined. However, if we turn to Myers' (1949b) discussion and look for his definition of migration, we are frustrated. Essentially all that Myers tells us is that, according to Heape (1931), migration is different from emigration (a one way moving out with no return), and nomadism (more or less aimless and homeless wandering). It is clear from Myers' (1949b) review of Heape (1931), and from his subsequent discussion, that he takes migration to mean a regular, ordered movement that is representative of a species, and involves outward and return movements between defined areas at predictable stages of the life cycle. The existence of return movements, and the occurrence of predictable, ordered movements, are two of the features critical to a definition of migration. Northcote (1979), recognising that his was a rather restricted definition, defined migration as:

> movement resulting in an alternation between two or more habitats (i.e. a movement away from one habitat followed by a return again), occurring with regular periodicity (usually seasonal or annual), but certainly within the life span of an individual and involving a large fraction of the breeding population. Movement at some stage in this cycle is *directed* rather than a random wandering or a passive drift, although these may form part, or one leg of a migration.

The existence of synchrony of movement, both in space and in time, is often a distinctive feature of animal migrations, but is not an obligatory aspect of the phenomenon. Some authors have emphasised the bi-directional character of true migration, referring to an initial removal migration (away from the spawning site) followed by a later 'return migration' (Baker, 1978) — there is a 'coming and going with the seasons — even if not on an annual basis — and the traveller needs a return ticket' (Harden- Jones, 1981). The concept of a return movement is an important and fundamental one that was recognised long ago (Landsborough- Thomson, 1926, Heape, 1931). Logically, migration should probably be seen as having its source in the place of hatching/birth, and a return migration that brings it back to that place. Mottley (1934), writing of salmonids, regarded the term migration as:

> the tendency for all the young to disperse downstream away from the place of hatching and may involve an extended journey to sea. It also implies a return migration on the part of mature fish if the population is to maintain its place in a particular range of water.

The accuracy of the return migration is widely variable and may involve a return to the actual birth place known in ichthyology as homing and defined by Gerking (1959) as 'the return to a place formerly occupied instead of going to other, equally probable places'. This seems a definition preferable to that of Frost (1963) viz. 'the habit of sexually mature fishes repeatedly to return to spawn at one particular place' which would preclude the description of species that spawn only once (semelparous fishes like species of *Oncorhynchus*) as engaging in homing. Homing is a phenomenon that has received high emphasis in the salmonid fishes, although Melvin, Dadswell and Martin (1986) considered it to be associated with most anadromous fishes. I doubt that this claim can be substantiated. The alternative to homing is a return migration to fresh water that may just involve a movement to the habitat type of birth, though not necessarily in the same geographical location — the latter probably being the more usual, although homing is a well-known phenomenon. This is not only in salmonids where it is best known but is reported in other groups like some of the alosid shads (family Clupeidae). The question of the presence or absence of homing is not fundamental to the definition of either migration or diadromy. Whether or not homing takes place, the return migration necessarily involves complex physiological and behavioural characteristics that enable the fish to find their way back to suitable habitats. Orientation, navigation, and homing have been discussed at length by Haslar (1966, 1971), and by various authors in McCleave, Arnold, Dodson and Neill (1984).

The length of the migration is also not of much significance to the definitions of migration and diadromy. A diadromous migration may in many instances be only a few kilometres — to take the fish through the estuarine/tidal zone of varying salinities. Or yet may involve many thousands of kilometres as in some of the Pacific salmons (*Oncorhynchus* spp.) and the freshwater eels (*Anguilla* spp.); total migration distances approaching 8,000 km are suggested for some salmons. Vladykov (1963) described out-migration distances of 2,400 and 2,700 km in the Atlantic salmon (*Salmo salar*), so that with a return migration still to take place, total sea distances of at least 4,800 and 5,400 km must be travelled. Differences of vast magnitude are known even within species, especially as regards their penetration of fresh waters. This is true of the chinook salmon, *O. tshawytscha*, some populations of which may spawn in coastal streams less than 100 km from the sea, while others reach the farthest extremities of vast river systems like the Fraser and Columbia, up to 4,000 km from the sea. Podlesnyy (1968) found that the omul of the Kara Sea, *Coregonus autumnalis*, may spawn 1,000 km up the Yenisei River, but only a few kilometres up other rivers in the same vicinity.

The migration itself may involve anything between, on one hand, an outward and return migration along well-defined routes and between precise sources and destinations (as is typically true of many birds but is not as characteristic of fishes — Baker's, 1978, 'to and fro' migrations), or on the other hand, it may include a rather undirected wandering movement which forms a large loop that brings the migrants back to their source, or back to locations in which the migrants can begin a more clearly directed return migration to the source habitat type. This is true of many fishes which have dual phases in their migrations — often a well-defined, down-river migration that may be largely passive (induced by river currents), then a further, often somewhat passive, loop wandering at sea that eventually brings them back towards their source, and is followed by an active purposeful upstream migration to the original habitat. Thus the degree of activity in migration of diadromous fishes varies from entirely passive, as in the downstream migrations of newly hatched larval fishes, to highly active as in all instances of upstream migrations.

Taking account of the fact that diadromy is a special type of migration, how, then, are diadromy and diadromous fishes to be defined?

Looking at the diverse behavioural patterns of the peripheral fishes mentioned earlier (i.e. those that move between waters of widely varying salinities) a series of very informal and ill-defined categories can be recognised, as follows:

1. In some species there are occasional and erratic movements of isolated individuals, or small aggregations, usually from the sea into low elevation river estuaries and lakes, often penetrating only as far as tidal/brackish waters, but sometimes further. Many families of fishes, from across a broad spectrum of higher fish taxa, would be included here.
2. In other species movements of significant numbers of fish take place, often diurnally or tidally, fish moving up-river to feed on a rising tide and moving out again on the falling tide. Some clupeids, many mullets, various flatfishes, and many other families behave in this way.
3. In yet other families, the fish may take up seasonal residence in such habitats, remaining in fresh or brackish water for weeks or even months at a time, but without a majority of the population behaving in this way. Again, diverse families may be represented here, including further flatfishes, mullets, gobies, blennies, clupeids, and many others.

It seems to me that in none of the above categories are the movements of a type that fall within definitions of migrations such as that cited earlier from Northcote (1979). As such, these fishes cannot be regarded as diadromous either, and need to be regarded, instead, as facultative, euryhaline, marine wanderers. In most instances these fishes are of marine origin and enter fresh water for the purpose of feeding. Their spawning habitats are normally in the same medium as the predominant habitat occupied, i.e. usually in the sea, not fresh water. Very few fish seem to be facultative as regards spawning in waters of differing salinity — although some are.

Because the fishes that fall into the above three categories are not strictly migratory, they cannot be regarded as diadromous either.

4. In some families of fish the movements are much more explicitly ordered as regards timing and direction, involve the vast majority of the populations, and are generally obligatory. Those sections of populations that do not make these movements can be regarded as deviants. Reproductive habitats often routinely differ from feeding, rearing and maturing habitats, and spawning may take place in a habitat that is otherwise not occupied, or is occupied for only a brief part of the life cycle.

Fishes coming within this category are truly migratory, and because their migrations involve movements between fresh and saline waters, they are also diadromous, according to Myers' (1949b) definition. Although an explicit and precise distinction cannot be drawn between fishes that fall into categories 1–3 above, and those in category 4 in which migration is a more or less orderly and essential part of the life cycle, there is nevertheless value in attempting to make such a distinction for the purposes of this discussion. And while the way I have drawn this distinction is undoubtedly arguable, both in principal and in specific examples, such argument seems scarcely germane to the discussion that follows, in which I attempt to explore and describe the strategies that fish use in their migrations.

If we take migration to mean something like the definition above of Northcote (1979) (p.3), and regard diadromy as a special category of migration, then:

1. Some groups of fishes can clearly be designated diadromous;
2. Others can not; but
3. For some groups it is not so clear; and
4. For many groups not enough is known to decide.

My point here is to narrow down the definitions somewhat, to exclude from a listing of diadromous fishes, those groups that undertake facultative and opportunistic movements between fresh and salt water and/or that may not involve a significant proportion of the populations. I see diadromy to refer to fish species that have a migration between fresh and salt water that is, at least, carried out at predictable times, in predictable directions, at predictable stages of the life cycle, which is characteristic of the species, which involves the majority of the members of the populations and which is, in general, essential to the completion of the life cycle/reproduction, i.e. the marine– fresh water migratory movements are more or less predictable, universal, and obligatory. An important aspect of this definition, and of the various subcategories, is that they apply to the species, rather than to their migrations.

Thus Myers' (1949b) term diadromy can, I believe, most usefully and appropriately be restricted to apply to those fish that *normally, as a routine phase of their life cycle, and for the vast majority of the population, migrate between marine and fresh waters.* They do this twice (i.e. sea to fresh water and back to sea again — or the reverse); obviously dual migrations are mandatory or the entire population would end up either in fresh water or in the sea! Migration occurs at regular and predictable phases of the life cycle. Under such a definition, groups like petromyzontids, geotriids, mordaciids (lampreys), acipenserids (sturgeons), clupeids (shads, etc.), salmonids (trouts and salmons), osmerids (smelts), galaxiids (southern whitebaits), retropinnids (southern smelts) gasterosteids (sticklebacks), and many other groups may be described as diadromous. Such a definition excludes some groups of fishes that a looser definition might include.

Having defined diadromy in this somewhat restrictive fashion, it is important to recognise that in a great many of the species in which diadromy is recognised, there are populations, or sections of populations, that deviate from the pattern. This deviation takes place in a variety of ways (see Chapter 13). This is often referred to as 'landlocking', but Rounsefell (1958) makes the point that:

> The word 'landlocked' as it has been used is, of course, a misnomer in the majority of cases; only rarely is a population denied access to the sea. ... The more usual situation is for so-called landlocked races or species to occupy streams or lakes with access to the sea freely traversable in both directions. Furthermore, in many cases, both resident and anadromous salmonids, even of the same species, may occupy the same water for at least part of the year.

Non-diadromous deviants of otherwise diadromous species may comprise some individuals in the deme that fail to go to sea with their siblings; or there are reported instances in which some individuals may alternate between being diadromous and non-diadromous, in an irregular fashion; or deviation may comprise populations of a species that fail to migrate, either because they 'choose' not to or are prevented from doing so by barriers to migration, either natural barriers or man-made barriers; or in yet other groups there are species related to or often evidently derived from diadromous species that are non- diadromous.

There is a wide range of variation, as becomes evident in the detailed analysis of the occurrence and characteristics of diadromy through the groups of fishes which follows.

Cohen (1970), following the definitions of diadromy suggested by Myers

(1949b), calculated that there were about 115 species of fish that can be designated diadromous, these constituting 0.6 per cent of all fish species (about
20,000 by Cohen's estimation); he hypothesised that as knowledge of the life
histories of marine and shelf fishes increased, the number shown to be diadromous would also rise. The present analysis indicates that the number of diadromous species at present recognised is nearer 170 species, and this figure can
be regarded as a minimum, likely to rise, perhaps quite substantially.

Although it has long been accepted that some fish groups, like lampreys and
salmons are diadromous, acceptance has been much less forthcoming for other
groups. This applies particularly to some of the southern diadromous fishes,
like retropinnids and galaxiids, and some rather surprising statements have
been made in this regard. Breder and Rosen (1966), describing the life history
of the common smelt, *Retropinna retropinna* (from New Zealand), were clearly dubious about diadromy. Acceptance of the fact that diadromy is a feature
of life histories in the Galaxiidae has been especially difficult to obtain. Campos (1984) said that the inanga, *Galaxias maculatus,* does not have a marine
phase in the strict sense, but rather an estuarine phase that reproduces in salt
water with the juveniles feeding near the coast before entering the rivers.

Benzie (1968) expressed similar uncertainties about the marine phase of
other diadromous galaxiids. Rosen (1974) expressed doubts about diadromy
in the inanga, considering that there was no evidence that any galaxiid is capable of undertaking any major transoceanic migration and that it is still a question as to what exactly occurs in the sea and where.

Nevertheless there seems to be little doubt that at least five New Zealand
species, at least two others in Australia, and perhaps a further species in South
America do go to sea and are fully diadromous. The inanga, *G. maculatus* has
been reported at sea at distances up to 700 km from land in the southwestern
Pacific Ocean (McDowall, Robertson and Saito, 1975), and it occurs in all three
areas.

One of the conflicts in determining which species should be regarded as diadromous, and which not, involves a quite diverse array of species which migrate up or down rivers associated with freshwater lakes, such as the large lakes
of the African rift valleys, the lakes of Arctic Siberia, or the Great Lakes of
North America. Migrations of fishes that occupy these lakes and their tributary rivers are often structurally similar to movements described between salt
water and rivers. Fish species may live and feed in lakes and migrate into
tributaries to spawn, and either die or return to the lakes to resume feeding and
recover condition again before a repeat spawning migration. Such cycles resemble marine-migratory species that are designated diadromous but differ
from marine-migratory strategies in one important feature — the osmoregulatory aspect of the migration (to which Myers (1949b) drew attention) is mis-
sing. For this reason, I have excluded such lake-migratory species from this
review of diadromy, which I envisage as covering those fish species that migrate between fresh water and the sea, this being the sense in which Myers

(1949b) originally coined the term. Even this apparently simple and clear-cut (if arguable) decision has its difficulties when examining the fish faunas of large, semi-saline lakes/seas like the Caspian, Azov, Black and Baltic, in which salinity may be quite low, highly variable, and with strong horizontal salinity geographical gradients. Apparently diadromous species in these seas can easily be divided into two categories:

1. Those species which belong to families or genera that are clearly recognised as being diadromous in other parts of the world — such as Caspian shads (Clupeidae), lampreys (Petromyzontidae), sturgeons (Acipenseridae), etc. and Baltic whitefishes (Salmonidae); these families are generally described as 'peripheral' (Darlington, 1948, 1957, Myers, 1949b).

2. Those species which belong to families or genera that are otherwise generally regarded as exclusively fresh water in habit, but which venture into the reduced salinity waters of the Caspian etc. — such as the rudd (*Scardinius erythrophthalmus*), roach (*Rutilus rutilus*), and other cyprinids; these families are generally designated as primary freshwater fishes (Darlington, 1948, 1957, Myers, 1949b).

In this review of diadromy I have chosen, perhaps arbitrarily, to include those fishes in category 1 above and exclude those in category 2. I recognise that this, too, is arguable, but bounds have to be set in any review, and whether a group of fishes is included or excluded has little impact on an attempt to synthesise the phenomenon of diadromy and identify principles and common features in the behaviour of diadromous fishes. This latter is of more importance than arguing about what groups should or should not be included, or about how definitions should be made.

Looking, now, explicity at strictly diadromous fishes, as defined above, a series of distinct subcategories can be identified. The terms anadromous and catadromous have been used widely in the fish literature for many years; the allied term amphidromous is less commonly used (although the phenomenon is probably no less frequent in occurrence in fishes). Shubnikov (1976) credits Kessler (1877) with first introducing the terms anadromous and catadromous, but Meek (1916) was probably responsible for their introduction to and wide use in the English literature. However, as Schmidt (1936 — and following him Myers, 1949b) explained Meek was also responsible for so broadening their meanings that anadromous and catadromous became virtually equivalent to contranatant (upstream) and denatant (downstream). Meek (1916) regarded anadromous fishes as those involved in any upward or shoreward migration even in the sea, while catadromous was used to refer to any seaward offshore or downward migration. With regard to freshwater fishes, the result of such definitions was that all diadromous fishes (ones that move between fresh and salt water) became both and alternately anadromous (contranatant) and catadromous (denatant), at various stages of their lives. This meant that the terms anadromous and catadromous lost most of their heuristic value as well as duplicating other well-understood terminology in the fish literature. Thus, the Zoological Record has adopted this usage of the terminology, and refers to anadromous and catadromous migrations of the same species. Anadromous is sometimes used to refer to upstream migrations of fishes *within* river systems, or from rivers into lakes. I regard this as an unnecessary distortion of Myers' nomenclature; furthermore Myers coined the term potamodromous for migrations within fresh water — although this term does not indicate the direction of any spawning migration. Lagler, Bardach and Miller (1962) described amphidromy as completely free movement between fresh and marine waters, not for the purpose of breeding. They cited *Sicydium* and considered that the irregular movements of the Asiatic milkfish, *Chanos*, into and out of fresh water makes it amphidromous. I think not, and regard the definition of Myers (1949b) as being much more explicit and restrictive than this. In my view *Chanos* is a euryhaline, facultative, marine wanderer.

Myers (1949b) is frequently, and I think properly, credited with establishing the terms as they are now most frequently used in the ichthyological literature. Certainly it is Myers who is most frequently referred to, and whose definitions are most often quoted (McDowall, 1968a); I see no reason to be any different. Shubnikov (1976) attempted a redefinition of the terms, but he essentially took us back to those of Meek (1916), with the attendant difficulties discussed above of equating them with contranatant and denatant, and I see little point in this. Myers' (1949b) definitions of the various terms were as follows (see Figure 2.1):

Diadromous — truly migratory fishes which migrate between the sea and fresh water. This is a general and inclusive term that seems to have been use first by Myers.

Anadromous — diadromous fishes which spend most of their lives in the sea and which *migrate to fresh water to breed* (my italics). In most anadromous species most of the life cycle is spent in the sea and the key point is a return migration to fresh water for breeding, by fully grown, mature to ripe adults.

Catadromous — diadromous fishes which spend most of their lives in fresh water and *which migrate to the sea to breed* (my italics). In most catadromous species most of the life cycle is spent in fresh water and the key point is a return, seawards migration of fully grown, mature to ripe adults, for the purpose of breeding.

Amphidromous — diadromous fishes whose migration from fresh water to the sea or vice-versa *is not for the purpose of breeding but occurs regularly at some other stage of the life cycle* (my italics). In amphidromous species occupation of fresh and marine waters varies widely, and the key point is that the migration is not a breeding one, but is typified by a return migration

Figure 2.1: Outlines of life history patterns in anadromous, catadromous and amphidromous fishes
Source: Adapted from Gross

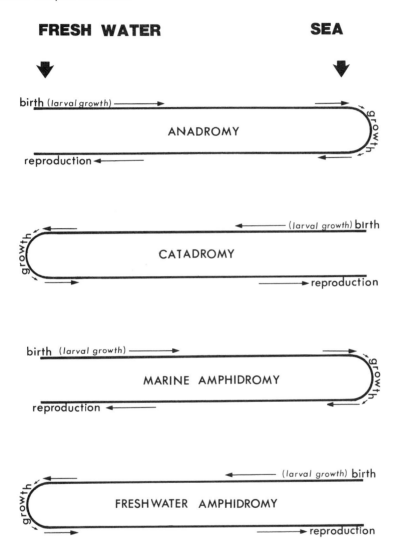

of well-grown juveniles, which continue to feed and grow for months or even years prior to maturation and breeding.

Amphidromous, like diadromous, is a new term suggested by Myers, for what he described as being 'a small known (but undoubtedly really large) number of species which appear to *need* to visit salt, or conversely, fresh water, *at some period of the life cycle other than the breeding period* (my italics). Amphidromy occurs in two distinct forms: marine amphidromy, in which spawning is marine and the larvae/juveniles are temporarily in fresh water before returning to the sea to grow to maturity; and freshwater amphidromy, in which spawning is in fresh water and the larvae/juveniles are temporarily marine before returning to fresh water to grow to maturity. (This distinction does not seem to have been made before.) The terms anadromy, catadromy and amphidromy are exclusive, specialised forms of diadromy that seem, to me, to cover all possibilities.

Myers (1949b) drew attention to Heape's (1931) classification of migration as either gametic, climatic, or alimental, and added a fourth type of migration that is of great importance to diadromous fishes, namely osmoregulatory. These four categories should not be seen as alternatives to the various forms of diadromy, or as being mutually exclusive, and it should be recognised that all the various forms of diadromy are, by definition, osmoregulatory. The osmoregulatory aspect of diadromous migrations is the fundamental difference between diadromous and non-diadromous migrations (i.e. migrations that are undertaken entirely in fresh or marine waters), and otherwise the general patterns of migration are not strikingly different. Fontaine (1975) also drew attention to different categories of migration, recognising trophic (= alimental of Heape, 1931), reproductive (= gametic), and amphibiotic (roughly equivalent to Heape's climatic though Fontaine's term, in the extreme, encompasses the deposition of fish eggs out of water, as is known for the non-diadromous, marine, Californian grunion (*Leuresthes tenuis*, family Atherinidae, Walker, 1952), or the New Zealand inanga (*Galaxias maculatus*, family Galaxiidae, McDowall, 1978a). Fontaine's amphibiotic refers to change of habitat undergone during or as a result of migration.

Diadromous fishes are routinely described as being highly euryhaline. Thus Evans (1984) in a recent review of gill physiology in fishes, tabulated the 'euryhaline' families and included catadromous and anadromous fishes. However, several authors have drawn attention to an aspect of diadromy that seems, to me, to have had far too little attention. As Norman and Greenwood (1963) put it:

Fishes like the salmon, which leave the sea to spawn in fresh water (anadromous species), or those like the eel which live in fresh water as adults but spawn in the sea (catadromous species) *do not necessarily have a wide salinity tolerance since the ability to adapt from one medium to another is confined to certain phases in their life histories.* In both types the change, once made, is not readily reversed.' (My italics)

Black (1957) recognised the same issue:

It is significant to realise that diadromous (anadromous and catadromous) fishes are not necessarily euryhaline, but are able to migrate because of hormonal activity that results in physiological changes required for survival in the new environment.

Fontaine (1975) labelled species of this sort as amphihaline, defining them as species that are able to 'pass periodically, at well-defined stages of their life histories, from salt to fresh water and vice-versa', the important point being that they are not facultatively euryhaline at all times. Their ability to move between salt and fresh water is periodic and transient, and is related to carefully mediated and timed physiological and hormonal changes connected with migration. At other times, such species are unable to make such a transition. The well-studied smoltification process in various of the salmons (*Salmo* and *Oncorhynchus* species, Hoar, 1976) falls into this category. Several studies have shown that the ability of such salmonids to adjust to rapidly varying salinities is a transient ability that increases with growth and varies seasonally. Somewhat surprisingly, Nordlie, Szelistowski and Nordlie (1982) have found the same feature to be true of the grey mullet, *Mugil cephalus*, which is commonly described and generally regarded as euryhaline and facultative. They showed that at high environmental salinities, the osmoregulatory abilities of this species increased with growth.

Fontaine saw euryhalinity to be the general ability of fish to withstand the salinity/osmoregulatory changes involved in a movement between salt and fresh water, an ability that many diadromous and even more non-diadromous fishes possess — but an ability that some diadromous species may not possess. Fontaine suggested that freshwater spawning amphihaline species be described as potamotocous and marine spawning amphihaline species as thalassotocous. These names are probably not as important as the distinction drawn by Black (1957,) Norman and Greenwood (1963) and Fontaine (1975), a distinction that has been little recognised but which seems, to me, to be of fundamental importance in understanding the migrations of diadromous fishes. It should not be expected that amphihalinity and euryhalinity are discrete conditions, but rather that there is a broad continuum of conditions between the extremes of total euryhalinity — with facultative movement between fresh water and sea water, and strict amphihalinity — in species in which movement is highly restricted and possibly obligatory. The question of the physiology of osmoregulation is of fundamental importance to all fish that migrate between fresh water and the sea — whether diadromous or just euryhaline wanderers. This topic has had extensive discussion , e.g. Fontaine (1975), Evans (1984), and has been reviewed at some length for migratory fishes by McKeown (1984).

Anadromy and catadromy are also, by definition, essentially gametic, i.e. they are defined, at least in part, around breeding requirements, while amphidromy is equally explicitly non-gametic. Components of many migrations are probably also alimental and climatic, although many of the actual spawning migrations of some (but not all) diadromous fishes are non-alimental (e.g. salmons, lampreys, eels, etc., which cease feeding at the onset of migration and never resume it before spawning and dying). I suspect that it is the relative importance of these four components of migration to survival and the successful transferral of genetic material to subsequent generations, that are of particular relevance to the strategic importance of diadromy in fishes (see Chapter 10, p. 146).

3

The Taxonomic Distribution of Diadromy

Knowledge of the occurrence of diadromy is very uneven, both as regards its geographical distribution and its occurrence in the various taxa of fishes. Attention tends to be focused almost exclusively on Northern Hemisphere fishes, especially from the Arctic-Boreal regions, with those from the Southern Hemisphere and the tropics getting much less, if any, mention (Dingle, 1980). The fisheries literature contains quite frequent accounts of the penetration into some of the great rivers of the tropics by marine fish species. Welcomme (1979) described how movement of fishes upstream into fresh waters is a common feature of many tropical river systems. Fish of marine origin regularly move many hundreds of kilometres up such rivers as the Niger in Africa (Reed, 1967), the Mekong in Asia (Shiraishi, 1970), or the Magdalena in South America (Dahl, 1971). However, there is very little detail about these movements, or of others into other great tropical rivers, like the Nile, the Amazon, or the Congo. Knowledge might be typified by Shiraishi's (1970) account of the Mekong — the largest river in Asia and, according to Shiraishi, the tenth largest in the world. About the most definitive thing he was able to say about the fauna of that river was that it is generally believed that many fishes start migrating upstream at the beginning of the wet season, but practically nothing has been clarified about their migratory behaviour.

Diadromy is not a well-recognised phenomenon in the tropical rivers of South America, either. Gery (1969), in reviewing the fisheries ecology of that region, did not actually mention the occurrence of diadromy in any South American fishes (although this has been reported for a few far southern families like the Geotriidae, Galaxiidae and Aplochitonidae). However, apart from the vast array of ostariophysans and some osteoglossids, all of which are essentially fresh water, there are a few peripheral clupeoids, cyprinodontids, and a few perciforms, some of which might prove to be diadromous.

A recent review of the fish faunas of some of the great rivers of the world (Davies and Walker, 1986) reveals little about diadromy. Welcomme (1986) made no mention of the occurrence of diadromy in his review of the fishes of the Niger River, in Africa, and the same is true of Banister's (1986) discussion of the Zaire, Skelton (1986) of the Orange-Vaal, Jackson (1986) of the Zambesi, Lowe-McConnell (1986) of the Amazon, and Stanford and Ward (1986) of the Colorado. It is difficult to believe that these great river systems have no diadromous fishes, but their faunas are not well known and the rivers involved are huge. A clue to the difficulty of observing diadromy may be gained from Lowe-McConnell's (1986) comment that the marine/fresh water interface moves 200 km up or down the Amazon, a distance greater than the distances that some diadromous fish move. Even those accounts that do mention diadromy are far from explicit, e.g. Pantulu (1986) defined diadromy as movement from marine to estuarine and/or fresh water, or vice versa; this is a rather broader and looser definition that that of Myers (1949b). With this in mind it is worth noting that Pantulu (1986) did mention a few species that he regarded as diadromous in the Mekong River, and Di Persia and Neiff (1986) reported a few in the Uruguay River. But no details are provided by these authors, and it seems that little is known.

Of some interest is Roberts' (1978) survey of the fishes in the Fly River of Papua New Guinea. He recorded such species as *Datnoides quadrifasciatus* (family Lobotidae — tripletails) 870 km up the river, *Nibea semifasciata* (family Sciaenidae — drums) 836 km, *Acanthopagrus barda* (family Sparidae — porgies) 525 km, and *Brachiurus villosus* (family Soleidae — soles) 930 km upstream. Similarly, Gunter (1938) listed *Dasyatis sabina* (family Dasyatidae — rays) 320 km up the Red River in North America, *Elops atlanticus* (family Megalopidae — tarpons) 480 km up the Black River, and *Trinectes maculatus* (family Pleuronectidae — flounders) 160 km up the Mississippi. And there are numerous other examples, and studies, that show extensive penetration of fresh water by species regarded as marine. A very large array of fish families may be involved. Herre (1958) listed a wide diversity of Philippines fishes that move regularly in and out of rivers, noting 234 species in 56 families. But he made the point that for most of these species only a small proportion of the population enters fresh waters; the bulk of the populations never enter fresh water at all.

> Those which find a stream or lake at hand certainly thrive there very well ... That such a journey is obligatory is disproven by the great numbers of the same species which never leave salt water.

Such fish cannot properly be described as either diadromous, or migratory using a definition such as that of Northcote (1979) discussed above. Miller's (1966) analysis of the freshwater fish fauna of Central America is also illustrative. Miller identified 456 species in fresh and brackish waters, of which 269 were primary or secondary species (i.e. are largely confined to fresh waters, Myers, 1949a, Darlington, 1957), with 187 being peripheral (i.e. they belong to groups with high salinity tolerances). These 187 species belonged to 30 different families and 73 genera. Dominant amongst these were Ariidae (sea catfishes), Atherinidae (silversides), Gerridae (mojarras), Gobiidae (gobies), and Eleotridae (sleepers), but they also included Pristidae (sawfishes), Hemirhamphidae (halfbeaks), Syngnathidae (pipefishes), Tetraodontidae (puffers), and others. Miller considered that only 31 of these peripheral fishes were 'permanent residents' and 26 were thought to spend much of their lives inland from the sea while most were sporadically or temporarily in fresh water (commonly in the early life history stages. Miller did not explicitly identify any of these species as diadromous, although some of them undoubtedly are, like *Anguilla* species and possibly some of the clupeids and mullets.

Analyses of freshwater faunas in other areas, like Africa (Roberts, 1975) and Pakistan (Mirza, 1975) produce similar sorts of results, with little emphasis on diadromy. However, the situation in most tropical lands is still very poorly defined. Bruton, Bok and Davies (1987) concluded that in southern Africa there are no anadromous species and only five that are catadromous — a mugilid mullet (Mugilidae) and four freshwater eels (Anguillidae); he did not mention the occurrence of amphidromy there. Dr G.G. Teugels (personal communication) advised that he knew of no really diadromous fishes in west (Atlantic) Africa, although there are possibly some clupeids. Again, they are very poorly understood.

There are numerous published discussions of the nature and occurrence of diadromy in fishes, but mostly these are brief and superficial. Textbook accounts of diadromy typically discuss anadromy and mention salmons and lampreys, and they list catadromy and cite eels. The more erudite discussions may also cite groups such as galaxiids as catadromous, but these are rarely accurate. One recent example is Moyle and Cech (1982), who wrote:

Anadromous fishes are those that spend the adult phase of the life cycle in salt water (or large bodies of fresh water) but move up stream in streams and rivers to spawn (for example Pacific salmons [*Oncorhynchus* spp.]). Catadromous fishes such as the eels of the family Anguillidae and some ... of the family Galaxiidae ... spend most of their life in fresh water but spawn in salt water. Catadromy is much less common than anadromy.

Occasionally other families, with which individual authors may have had some personal involvement, like sticklebacks or mullets, are included in the discussion. Thus although diadromy in fishes is well known and widely mentioned in the literature, it seems a poorly understood and often erratically reported phenomenon. 'Everyone' knows that some salmonids are anadromous, and that eels are catadromous; many realise that lampreys, sticklebacks, and some clupeids are anadromous. Beyond that there is confusion and much misunderstanding. There is sometimes mention of apparently little-known southern temperate groups usually referred to as 'southern trouts, smelts or graylings'.

Haedrich (1983) observed that catadromy is a rare phenomenon, and by comparison with anadromy, it is.   Berg (1959) regarded catadromy as so rare that, apart from the anguillid eels, all other diadromous fishes leave salt water and go into fresh water for spawning (i.e. are anadromous).   Catadromy is particularly poorly recognised and understood, and a typical example of the way it is reported can be found in Nikolsky (1963): 'Some migratory fishes feed in the rivers but pass out into the sea to spawn, performing a catadromous migration like the eel, etc.' and such migrations 'are also performed by certain members of the families Galaxiidae and Gobiidae'.

Ommanney (1964) considered that:

There are not nearly so many of them [catadromous fishes] and the best known is the common eel.   In the Southern Hemisphere there are some small fish called smelts in the Falkland Islands and when young, whitebait in New Zealand, which run down the streams and spawn in schools in the surf.

Apart from the fact that catadromy is not as common as anadromy and Ommanney's reference to eels, there is scarcely one fact in his statement that is both comprehensible and correct.   Norman and Greenwood (1963), discussing catadromy, mentioned only anguillid eels and *Galaxias maculatus* which they described as returning 'to its original home in the sea to spawn'. McKeown (1984) in briefly reviewing diadromy, mentioned groups like salmonids, lampreys, eels and sticklebacks, but discussed diadromy somewhat more widely, adding:

Fish from the southern hemisphere such as some of the southern trout (*Galaxias* spp), as well as certain smelt and grayling species also display catadromous behaviour.   Other examples are the Japanese ayu *Plecoglossus altivelis* and the Hawaiian climbing goby *Awaous guamensis*.   Myers (1949[b]) describes a third type of diadromous fish termed *amphidromous* ... Migrations of some gobiid fishes are of this type.

Again, in this statement the truth is elusive — one galaxiid is 'marginally' catadromous, southern smelts are anadromous, while most diadromous galaxiids, the southern graylings and the ayu are, in fact, all amphidromous! And as must by now be obvious a distinctly wider diversity of fishes than 'some gobioid fishes' is amphidromous. Hynes (1970) generalised that almost without exception prespawning migration is upstream — on which basis almost all diadromous fishes must be anadromous, which of course is untrue.

Confusion has long been rife as regards the galaxiids, prompting ichthyologists to ask: 'Why are the southern families of salmoniform fishes (Galaxiidae, Aplochitonidae, Retropinnidae, Prototroctidae) catadromous, when their northern counterparts (such as the Salmonidae and Osmeridae) are primarily anadromous?' But when we review these southern families we find that they are not catadromous; only one species, *Galaxias maculatus*, can be described as marginally catadromous, and the problem has been that this fish has incorrectly been described as migrating to sea to spawn as it was in Ommanney (1964) (quoted above) and has been perennially by numerous other authors. Uninformed authors have accepted this incorrect description and extrapolated unduly from erroneous accounts, classifying all the southern families as catadromous. Some of the southern families are anadromous and others amphidromous (see p.33). Perhaps the most astonishing comment about diadromous fishes is one by Kendeigh (1961) who wrote of anguillid eels that 'The male, however, remains in brackish water of the bays and estuaries. *It is here that mating takes place as the female returns to the sea to spawn*' (my italics). How this happens and what happens to the male afterwards, is not revealed. In addition to such errors, knowledge of the diadromous fishes generally in the ichthyological literature is highly deficient, so that understanding of the phenomenon leaves a lot to be desired. Thus there is obviously a need to examine fish life histories and clearly describe and define them according to the various categories of diadromy as defined by Myers.

I would like now to explore the patterns in the occurrence and distribution of diadromy, through the families of fishes (i.e. phylogenetically), in part because this is an interesting question in its own right, but primarily because it may help us to elucidate some of the 'why' questions related to the occurrence of diadromy.

I have explored widely in the recent literature on fish life histories, seeking to identify the families and species in which diadromy is present. The results of this enquiry undoubtedly underestimate the numbers of species that are diadromous, as in some geographical areas the life cycles of many species are poorly understood, especially in the tropical waters of Africa, South America and Asia. I think this is also true of some groups of fish. In particular the life histories of the gobioid fishes are poorly known, but where known they have proved to be diadromous in quite a few instances. I make no claim for an exhaustive coverage of the literature; however, I have made an extensive search, and this has certainly revealed a great many more groups than are commonly discussed in the general ichthyological literature. Limitations in the completeness of the coverage of diadromy here seem unlikely to impede an analysis of the distribution of diadromy through the taxa of fishes (although it may distort an analysis of the geographical distribution of diadromy in its various forms). In the following discussion the taxonomic arrangement of the families in the various groups is, in general, taken from Nelson (1976).

Diadromy is clearly a well-established habit in the Agnatha, occurring in the three families of lampreys (Petromyzontidae, Northern Hemisphere lampreys; Geotriidae and Mordaciidae, Southern Hemisphere lampreys), although not in the hagfishes (Myxinidae) which are entirely marine. Amongst the Chondrichthyes (sharks, skates and rays, etc.), this primarily marine group has some, although few euryhaline representatives (e.g. *Carcharinas leucas, Pristis microdon*) and some that are entirely freshwater (e.g. the Argentinean freshwater ray *Potamotrygon*, Castex and Castello, n.d.), but none of these seems to be strictly diadromous. *C. leucas* has been described as diadromous by some commentators, but Thorson (1976) is not of this view, finding no evidence of an obligatory freshwater period. Nor is there evidence of any general seasonal migration up or down river, or of any movement related to reproduction. *C. leucas* should therefore be regarded as a euryhaline, facultative wanderer. The same is probably true of *Pristis perotteti*, which even reproduces routinely in fresh water (Thorson, 1982). Neither of these elasmobranchs should be regarded as diadromous. Nelson (1976) listed 10 species of the Rajiformes as fresh water but no other elasmobranch.

Diadromy is widely present in the Osteichthyes (bony fishes) but its occurrence is erratic. In the diverse array of primitive and ancient relict groups — the lungfishes (Ceratodidae, Lepidosirenidae, Protopteridae), coelocanth (Latimeriidae), bichirs (Polypteridae), paddlefishes (Polydontidae), gars (Lepisosteidae), bowfins (Amiidae), mooneyes (Hiodontidae), notopterids (Notopteridae), osteoglossids (Osteoglossidae), mormyrids (Mormyridae), gymnarchids (Gymnarchidae), etc., only the sturgeons (Acipenseridae) are diadromous, and they are firmly so. It is of more than passing interest that, with the exception of the coelocanth and the sturgeons, these ancient groups are wholly confined to fresh water, and tend to be warm-temperate to tropical in distribution — though they may not always have been. There is little information on the fossil habitats of these groups, although Romer (1966) reported *Amia* from freshwater Eocene deposits.

The primarily marine Elopiformes — the tarpons (Elopidae, Megalopidae), bonefishes (Albulidae) etc., are well known to enter fresh water in some instances and have been somewhat loosely described as diadromous. The oxeye herring, *Megalops cyprinoides*, appears possibly to be diadromous, but as far as I can determine no other elopiform is, as their migrations are not predictable, nor do they include a majority of the populations, nor are they obligatory; apart from the inclusion of *M. cyprinoides*, and this is open to debate, I exclude this group from the listing of diadromous fishes.

The Anguilliformes (eels) is a very large and diverse group, with about 20 families, of which only one, the Anguillidae, is really diadromous; all anguillids are diadromous, but all other eels are strictly marine, with the possible exception of one muraenid eel (family Muraenidae).

The diverse and primarily marine Clupeiformes (the herrings, Clupeidae; anchovies, Engraulidae, etc.) have some diadromous representatives in both these families, but most species are marine and a few fresh water.

The Salmoniformes (trouts, salmons, smelts, etc.) is also a diverse and large order of fishes. Amongst the salmoniforms, the pikes and their relatives (Esocidae, Umbridae) are restricted to fresh water. There are some that are oceanic, marine, pelagic fishes that are strictly marine (e.g. the argentines, family Argentinidae) and many that are bathypelagic or mesopelagic marine fishes (e.g. the bristlemouths, Gonostomatidae; slickheads, Alepocephalidae; viperfishes, Chauliodontidae and other families). But there is also a group of families, variously related and conveniently (for this discussion) grouped in the Salmonoidei, where diadromy is very widespread. Almost all of the salmonoid families — Salmonidae, Osmeridae, Plecoglossidae, Salangidae, Galaxiidae, Aplochitonidae, Retropinnidae and Prototroctidae — can be described as largely, or at least strongly, diadromous. Only the aberrant Sundasalangidae (Roberts, 1981) are not, and are entirely fresh water; they seem to be neotenic derivatives of the already aberrant and neotenic Salan

Thus far, this somewhat tedious procession through the orders and families of fishes reveals a fairly orderly pattern, with groups at the family level being fairly easily categorised as either diadromous, or not. But as we press on through the groups of higher fishes this suddenly becomes no longer possible. There are many families (the majority) that are easily classified as non-diadromous. The entire Ostariophysi, for instance, with about 60 families are, with only a couple of exceptions, strictly fresh water. Two families of catfishes (Plotosidae and Ariidae) are variously marine and fresh water, but are not shown to be generally diadromous. The 16 families of Myctophiformes (lizardfishes, Synodontidae; lanternfishes, Myctophidae; barracudinas, Paralepididae; lancetfishes, Alepisauridae and others) are confined strictly to marine waters. Many of them, like the marine salmoniformes, are mesopelagic or bathypelagic.

Throughout the Paracanthopterygii and Acanthoptergyii the pattern is distinctly different from that seen thus far. According to Nelson (1976), most families in these two large orders are either marine or freshwater; a few are both, and occasionally one or a few species in a family may be diadromous. I have found only one quite explicit instance of diadromy in the whole of the primarily marine Paracanthopterygii (about 31 families, including the cods, Gadidae; hakes, Merluciidae; eel-pouts, Zoarcidae; cusk-eels, Ophidiidae; toadfishes, Batrachoididae; anglerfishes, Ceratiidae etc.). There are also very few freshwater representatives in these families. Amongst the acanthopterygians (about 245 families with the greatest diversity of fishes), diadromy turns up intermittently, irregularly, and quite unpredictably. In most instances it is represented in any family by one or a few species, and nearly always by a very small minority of species.

Some acanthopterygian orders appear to be exclusively marine, like the Lampridiformes (ribbonfishes, Trachipteridae; oarfishes, Regalicidae etc.), Zeiformes (dories, Zeidae, Oreosomatidae), Dactylopteriformes (flying gurnards, Dactylopteridae), Beryciformes (alfonsinos, Berycidae; squirrelfishes, Holocentridae etc.), Pegasiformes (seamoths, Pegasidae), Gobiesociformes (clingfishes, Gobiesocidae). The other eight orders have minority freshwater occurrence, and in a significant proportion of these, it appears that the freshwater representatives are non-diadromous. This seems to be true of the Atherini-

31

formes (silversides, Atherinidae; rainbowfishes, Melanotaeniidae), Syngnathiformes (pipefishes and seahorses, Syngnathidae), Synbranchiformes (swampeels, Synbranchidae), Tetraodontiformes (pufferfishes, Tetraodontidae), so any diadromy in the acanthopterygians occurs in the Scorpaeniformes (scorpionfishes, Scorpaenidae), Gasterosteiformes (sticklebacks, Gasterosteidae), Perciformes (the vast bulk of the spiny-rayed fishes), and Pleuronectiformes (flatfishes — particularly the flounders, Pleuronectidae, and soles, Soleidae etc.).

The Scorpaeniformes is primarily a marine family, but there is one possibly diadromous species. Within the small Gasterosteiformes (two families), two species of Gasterosteidae are diadromous. Other species are variously marine, marine–euryhaline, freshwater–euryhaline, or freshwater, with a fairly even spread of all types.

In the very large order Perciformes (about 145 families) the great majority of families is exclusively marine (about 91 families) and a few are marine–euryhaline (22 families). A very small number is exclusively fresh water (about 11 - the sunfishes, Centrarchidae; perches, Percidae; climbing gouramies, Anabantidae; snakeheads, Channidae etc.). Even fewer are freshwater–euryhaline (cichlids, Cichlidae; nurseryfishes, Kurtidae; leaffishes, Nandidae), while a few are best described as marine–freshwater.

This leaves us with, by my analysis, only about 11 perciform families that include members that are quite clearly known to be diadromous. These are the temperate basses, Percichthyidae; the Bovichthyidae; flagtails, Kuhliidae; theraponids, Teraponidae; sculpins, Cottidae; mullets, Mugilidae; snooks, Centropomidae; sandperches, Mugiloididae; sleepers, Eleotridae; gobies, Gobiidae; and flounders, Pleuronectidae.

Returning to the earlier question of the phylogenetic distribution of diadromy in fishes, the answer is very clear. Diadromy tends to occur primarily, and is strongly represented in the more primitive and ancient fish groups — lampreys, sturgeons, eels, the various salmonoid families, and clupeids, and is otherwise a rare and highly intermittent phenomenon. Questions regarding the phylogenetic distribution of diadromy can be asked in two ways:

1. Are the diadromous fish primarily primitive/ancient? The answer is yes — the lampreys, eels, salmonoids and clupeids are derived from ancient, basal stocks of fishes that all date back well into the Mesozoic (Romer, 1966);

2. Are the ancient and primitive fish primarily diadromous? The answer here is no; few of the ancient and/or relict groups are diadromous. Most of them, probably, are fresh water, and there are interesting questions to explore about whether they are primitively fresh water or secondarily so, and whether freshwater habitats are refuges for unsuccessful groups that persist as relicts.

Before further dissecting the occurrence of diadromy, it is timely briefly to classify the various diadromous species according to the definitions of Myers (1949b), discussed above (p.20, see also Figure 2.1).

1. Anadromy — species that have a migration from the sea, into fresh water for the purpose of spawning. I take this to mean a migration of adult fish whose primary and more or less immediate goal is reproduction. The reciprocal migration from fresh water to the sea may involve newly hatched larvae through to well-grown juveniles a year or more old. Without any doubt the best known and most often cited diadromous fishes are anadromous.

Groups that can be classified as anadromous include: some lampreys, sturgeons, salmonids, osmerids, salangids, retropinnids, some aplochitonids, clupeids, an engraulid, gasterosteids, a gadid cod, some percichthyids, and some gobies.

2. Catadromy — species that have a migration from fresh water to the sea for the purpose of spawning there. Again, I take this to mean migration of adult fish, the primary and more or less immediate goal of which is reproduction. The reciprocal migration from the sea to fresh water is normally of well-grown juveniles, weeks to months old, occasionally older. It is little recognised that catadromy is not a common phenomenon, although Haedrich (1983) has previously pointed this out.

Groups that can, on this basis, be described as catadromous, include the anguillid eels, some of the clupeids, one of the megalopids, a centropomid, some percichthyids, a bovichthyid, a scorpaenid, some cottids and mugilids, eleotrids, pleuronectids, and possibly a galaxiid.

3. Amphidromy — species that migrate between the sea and fresh water, in which migration in neither direction is related to reproduction. It normally involves spawning in fresh water, the young moving to sea very soon after hatching, a period of feeding and growth in the sea often lasting several months, a return migration to fresh water of well-grown juveniles, and a further period of months to several years of feeding and growth in fresh water prior to maturation and spawning there. This category of migration is, by far, the least recognised of the types of diadromy, and yet it is quite widely present amongst diverse families of fishes.

In some measure, amphidromy is a catch-all for left-over groups. Families included here include prototroctids, most diadromous galaxiids, a plecoglossid, mugiloidid, and some eleotrids and gobiids.

4

Detailed Analysis of Diadromy: Anadromy

1. LAMPREYS — FAMILIES PETROMYZONTIDAE, GEOTRIIDAE AND MORDACIIDAE

Lampreys are highly characteristic and distinctive, primitive vertebrates. They are very elongate and slender in form, rather eel-like, and are of modest size, the adults of the largest anadromous species, like *Petromyzon marinus*, larger stocks of *Lampetra tridentata*, and *Geotria australis*, frequently reaching about 600 mm. Others, like *Lampetra fluviatilis* and *Mordacia mordax*, reach about 450 mm. Non-anadromous species are distinctly smaller, 200–300 mm long (Hardisty and Potter, 1971c; Potter, 1980). The mouth opening is surrounded by a circular sucking disc covered with rows of horny teeth. At the centre of the disc is a piston-like tongue, the tip of which is also covered with teeth.

Most lampreys are Northern Hemisphere petromyzontids, with about 35 species occurring widely in the Arctic to warm temperate waters of all the northern continents. They spread south as far as the Mediterranean, and the Florida Peninsula and Mexico. Six petromyzontids are anadromous (Potter, 1980). The single geotriid and two of the three mordaciids are also anadromous. Anadromy in the three families is of a comparable type.

Typical of the anadromous species is the Atlantic sea lamprey, *Petromyzon marinus*. In this species spawning usually takes place in fresh water well up stream. Migrations in excess of 300 km are known, although some spawn only a little above the reach of the tide. Spawning migrations occur in spring, and the eggs are laid in primitive nests in the gravels. Although most reports suggest that after spawning the adults all die, Nikolskii (1961) thought that they may spawn several times; he seems to be largely alone in holding this view. On hatching, the larval ammocoetes burrow into mud and silt along sluggish stream margins and live for several years (usually 3.5 to 6.5 years and sometimes more than 7) as filter feeders on aquatic micro-organisms. A metamorphosis takes place during the summer and autumn, resulting in a miniature of the adult. During metamorphosis feeding ceases temporarily — probably for

Figure 4.1: Life history patterns in a parasitic anadromous lamprey (*Lampetra fluviatilis* – outer circle) and a non-parasitic non-anadromous lamprey (*L. planer* – inner circle) family Petromyzontidae): black and coarse stippled areas are periods of starvation; T,H transformation; M, macrophthalmia; S, spawning
Source: Hardisty and Potter, 1971b

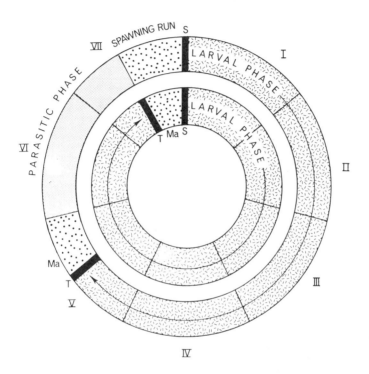

three or more months.

The small sub-adults make their way downstream during the autumn, some soon after metamorphosis, others staying in fresh water for several months. In general they do not feed before entering the sea, although there are a few instances of feeding in fresh water. Davis (1967) reported that 85 per cent of landlocked Atlantic salmon (*Salmo salar*) in a Maine lake had been attacked by newly metamorphosed sea lampreys (*Petromyzon marinus*). And Maitland (1980b) described the attacks by sea lampreys on coregonids in Loch Lomond, Scotland. There is some evidence that they may dwell in the brackish river estuaries for a time while they adjust to the changed salinities (Hardisty and Potter, 1971a). In the sea they live as fish parasites, for about 28 months. As maturity approaches, lampreys cease feeding, move inshore, and migrate back up river systems, but there is nothing to suggest that homing takes place. They do not feed again, and spend several months slowly migrating up river to the spawning grounds, the gonads developing and growing as they move (Beamish, 1980a; Maitland, 1980a).

The Arctic lamprey of the northern Pacific, *Lampetra japonica*, has an essentially similar life cycle, as do *Caspiomyzon wagneri* in the Caspian Sea, *Lampetra tridentata* and *L. ayresii* in rivers of western North America (Beamish, 1980a), and *L. fluviatilis* in western Europe and the Mediterranean (Maitland, 1980a). Beamish (1980b) described a summer metamorphosis in the American *Lampetra ayresi*, with a slow movement to sea over the following year. They feed at sea over the summer, and then return to fresh water, to spawn and die, about two years after metamorphosis. The life cycle of the European river lamprey, *L. fluviatilis*, is very similar (Figure 4.1). Nikolskii (1961) described the occurrence of both winter and spring migrating *L. fluviatilis*. Spring-run examples have much more mature gonads than winter ones as in the case of *C. wagneri* (Berg, 1962). *L. tridentata* may spend 3.5 years between metamorphosis and post-spawning death, and moves much more widely at sea.

Southern lampreys of the families Geotriidae and Mordaciidae are as widespread in the south as petromyzontids are in the north. The Geotriidae contains only the anadromous pouched lamprey, *Geotria australis* (Figure 4.2), that is found in southern Australia, New Zealand, the Chathams, Chile and Argentina. This lamprey is distinguished by the development of a baggy pouch below the head in mature males, the function of which is unknown. Very considerable distances are covered at sea by this species, it being found as much as 1500 km east of southern Argentina (Potter, Prince and Croxall, 1979). The Mordaciidae has two anadromous species, *M. mordax* in eastern Australia and *M. lapicida* in Chile. (No lampreys occur in southern Africa.) The life cycles of the southern species closely resemble those of the northern Petromyzontidae. Potter (1970) showed that *Mordacia mordax* has a larval life of about 3.5 years. Metamorphosis, which occurs in summer in Tasmania, and later on in autumn in mainland Australia, is followed by a downstream adult migration. *Geotria australis* behaves similarly, although the spawning migration may take longer

Figure 4.2: Pouched lamprey *Geotria australis* (family Geotriidae); (a) ammocoete; (b) pouched adult male

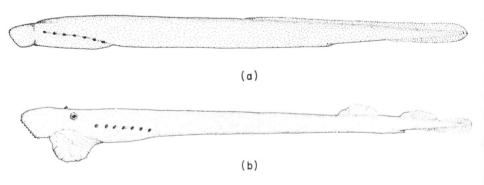

(a)

(b)

than in Northern Hemisphere species (11–12, perhaps up to 16 months, compared with 9–10 months (Hardisty and Potter, 1971b)). The return migration of mature adults occurs during the spring, from July onwards. Inland penetration by both *Mordacia mordax* and *Geotria australis* may be very extensive. Potter and Strahan (1968) reported that they can 'travel many hundreds of miles during the spawning run since they are recorded from many of the tributaries of the River Murray such as the Darling, Murrumbidgee, Wakool, Edward and Goulburn'. These are rather sluggish rivers. In New Zealand *Geotria australis* occurs at least 80 km up the Rakaia River, which is a very swift-flowing, bouldery river that has rapid fall to the sea and in which migration would be much more arduous. Lampreys (at least *G. australis*) have a considerable ability to climb, being able to make their way up rocky falls, fastening themselves to the rocky surface using their oral suckers.

Deviation from the typical lamprey life history pattern takes place in two ways. For example, at the intraspecific level, populations of *Lampetra tridentata* are known from at least one lake in Oregon, USA (McPhail and Lindsey, 1970). 'Landlocked' stocks of the sea lamprey have developed in the Great Lakes, in which the marine feeding phase has been replaced by a lacustrine feeding phase. I.C. Potter (personal communication) advised that the land-locking of *Petromyzon marinus* is believed to have occurred long before the influence of man on the Great Lakes, probably during the Pleistocene, although a movement from Lake Ontario into the upper lakes of the system did result from man's influence. This development seems to be a unique artefact of human disruption of habitats and migratory patterns, and does not occur elsewhere in the lampreys.

At the interspecific level, lampreys have proved capable of evolving towards a foreshortened life cycle. The outcome has been the evolution of distinct, non-parasitic species, in which there is a metamorphosis from juvenile to mature adult (as in *Lampetra planeri*, Figure 4.1). The sub-adult to adult parasitic feeding phase in the sea has been eliminated and the newly metamorphosed adults become sexually mature without further feeding or growing. The life cycle of such derived species is completed wholly in fresh water. This is true of most of the genera of northern petromyzontids (several species in most of the genera). In the Southern Hemisphere one of the mordaciids is a non-parasitic species, but this is not true of geotriids (Potter, 1980, Vladykov and Kott, 1979).

2. STURGEONS — FAMILY ACIPENSERIDAE

The sturgeons are rather bizarre fishes which may live for a very long time, some being aged at over 100 years (Anon., 1954), and which may grow to an immense size, an example over five metres being reported by Bigelow and Schroeder (1953); Vladykov and Greeley (1963) mention one weighing over

1200 kg (*Acipenser oxyrhynchus*), and Berg (1962) another of 1500 kg (*H. uso huso*).

Sturgeons are widely distributed on all the continents and in all oceans of the Northern Hemisphere, from the Arctic to the tropics, most species being temperate in range. The family is of great age, undoubtedly more than 200 million years and perhaps 400 million years (Nelson, 1976, McEnroe and Cech, 1985). The family contains about 27 species in four genera (Vladykov and Greeley, 1963), of which *Acipenser* and *Huso* contain anadromous species. The extent of anadromy varies widely from species to species. Doroshov (1985) classified sturgeons according to his interpretation of the extent of diadromy. He listed *Acipenser oxyrhynchus*, *A. sturio*, and *A. medirostris* as 'truly anadromous', and *Huso huso*, *A. transmontanus*, *A. guldenstaedtii*, *A. nudiventris*, and *A. fulvescens* as 'anadromous or semi-anadromous'. The last three of these he described as being able to establish non-anadromous populations.

The Atlantic sturgeon, *Acipenser oxyrhynchus* occurs along the Atlantic coast of North America and is one of the very large species. It is found from New Brunswick to Florida, and also in the Gulf of Mexico, northern coasts of South America and perhaps Bermuda (Huff, 1975). Spawning takes place in fresh water. The eggs are demersal and adhesive and hatch in about a week. The fry are tiny (about 11 mm) and they move gradually downstream and spend three to seven years in river estuaries before they move to sea, feeding as they move. The spent adults also make their way gradually downstream, eventually returning to the sea within months after spawning. Little is known of life in the sea. Atlantic sturgeons are late-maturing fish, females maturing at about 12 and males about nine. Spawning may take place only once in two to five years (Smith, 1985). Some of the spawning females leave the sea during the autumn, and move upstream to spend the winter in deep holes near the spawning grounds. These overwintering females are believed not to feed during the period in fresh water. Other spawners migrate in the spring and summer, the migration beginning earlier in the south (February in Georgia) than further north (June–July in the Gulf of St Lawrence, Vladykov and Greeley, 1963; Huff, 1975).

Another Atlantic species, the shortnose sturgeon, *Acipenser brevirostrum*, has a similar life history (Dadswell, 1979). This is one of the smaller sturgeons, reaching a length of only a metre; it occurs from New Brunswick to Florida. The shortnose sturgeon spawns in rivers sometimes not far upstream, though they are known to penetrate nearly 250 km up the Hudson River in New York state. Non-spawning adults accompanied by larger juveniles have spring upstream and autumn downstream migrations that are associated with the migrations of the current year's spawning fish. Some females undergo the autumn spawning migration to overwinter before a spring spawning. Dadswell (1979) suggested that they might home on to the summer foraging grounds and that there may be pair bonding. He regarded populations in the northern St Johns River in New Brunswick as being only marginally anadromous. They do not go far to sea, all captures at sea being taken within a few kilometres of the shore (Dadswell, Taubert, Squiers, Marchette and Buckley, 1984). 'This species occupies a somewhat intermediate position between the fully anadromous sturgeons ... and the freshwater lake sturgeon. [Whether *Acipenser brevirostrum*] can be considered anadromous depends on the limits of the definition of "anadromous" '. They considered that populations further south were more completely anadromous (Dadswell *et al.*, 1984), and suggested that this is related to the very cold water temperatures in the winter, at the northern limits of this species' range. McCleave, Fried and Towt (1977) described *A. brevirostrum* as being primarily in the estuarine and brackish waters of Maine.

Other North American sturgeons resemble the above in their life history patterns, some like the western green sturgeon, *A. medirostris* (occurring widely in the North Pacific in Siberia and western North America) being regarded as fully anadromous. Others like the white sturgeon, *A. transmontanus* (Pacific coast of North America from Alaska to California), are thought to move little beyond estuaries (Miller, 1972; Kohlhorst, 1976). McEnroe and Cech (1985) described *A. medirostris* as entering estuaries in the autumn and migrating slowly upstream to spawn, in winter. They did not know how long the juveniles stayed in fresh water.

The Arctic Lena sturgeon, *A. baieri*, in Siberian Russia resembles the Atlantic sturgeon, being fully anadromous, although Sokolov and Malyutin (1977) thought that the adults continued to feed during the spawning migration. Nikolskii (1961) recorded it migrating 1300 km up the Yenisei River, and said that the spent adults return to the sea, while juveniles may spend five or six years in fresh water before moving to sea. The European species, *A. sturio*, may remain in fresh waters for up to three years before going to sea (Wheeler, 1969). The ship sturgeon, *A. nudiventris*, occurs in the Caspian, Black and Aral Seas, from which it enters rivers to spawn, and may migrate long distances – it once penetrated the Danube River to beyond Budapest. It spawns in both spring and autumn (different stocks) and may spend many months in fresh water, after leaving the sea on the spawning migration and before spawning actually takes place (Berg, 1962, said not less than 10–11 months in the Syr-Darya River). Spawnings are at least two years apart, possibly more. The sterlet, *A. ruthensis*, is also a Caspian and Black Sea sturgeon, and seems to have a life history similar to that of *A. nudiventris*. The stellate sturgeon, *A. stellatus*, is yet another Caspian and Black Sea sturgeon, found also in the Adriatic Sea, and it, too, has autumn and spring spawning races. All of these Russian–eastern European sturgeons seem to have stocks that overwinter in fresh water. It appears that they become semi-dormant in deep holes during the winter period, and become active again after the ice-melt in spring.

Two Russian species of *Huso* are also anadromous, with a life history comparable to that of the Atlantic sturgeon (Berg, 1962). The beluga, *H. huso*

(Caspian, Black and Adriatic Seas) is reported to reach an enormous size, growing to more than four metres in length and 1500 kg in weight (Berg, 1962, Frank, 1969). Nikolskii (1961) described both autumn and spring migrating races of beluga, and said that the autumn-run fish go further upstream to spawn than spring-run fish. This species may move at least 1000 km up the Volga River to spawn. After hatching, the young of the beluga are said to go straight to sea, not lingering in the rivers (Nikolskii, 1961). The related kaluga, *H. dauricus*, occurs in eastern Pacific Siberia. It too grows to more than four metres in length but evidently does not move far out to sea (if at all). Some individuals may spawn only a few kilometres upstream but others may go 600–700 km upstream (Berg, 1962). Sturgeons may live for more than 50 years — very large kaluga must be a great age as Berg (1962) reported examples of 650 kg as 50–55 years old, a beluga of 640 g that was 58, and another of 1000 kg that was 75. How old, then, would they be at 1500 kg? They may spawn only every second year when young; the frequency of spawning may decrease to every three, four, or more years with increasing age (Berg, 1962). Little is known of the other anadromous sturgeons, the group as a whole being surprisingly poorly known for fish that have generated so much culinary interest.

Deviation in sturgeons from the anadromous life cycle discussed above occurs in various ways. Some genera are wholly fresh water (*Scaphirhynchus*); some species of otherwise anadromous genera are wholly freshwater in habit (*A. ruthensis* in Asia and *A. fulvescens* in North America). Berg (1962) reported that the anadromous *A. sturio* in Russia has landlocked populations, Nikolskii (1961) reported this for *A. baieri* in Siberia and *A. sturio* in western Europe and Haynes, Gray and Montgomery (1978) said the same for the North American *A. transmontanus*, which has become landlocked in hydro-lakes in the Columbia River. Taubert (1980) described a landlocked population of *Acipenser brevirostrum* from behind an impoundment in Massachusetts, USA and Dadswell *et al.* (1984) another from a lake in South Carolina. Berg (1962) wrote of both non-migratory/fluviatile and anadromous populations of *H. dauricus*.

3. GARS — FAMILY LEPISOSTEIDAE

The gars are elongated, predatory fishes of the fresh waters of North and Central America, with about seven species in one genus. Although regarded by some as diadromous, they are, in my view better regarded as euryhaline wanderers, some species moving into marine waters at times, some doing so quite freely, e.g. the longnose gar, *Lepisosteus osseus*, and the alligator gar, *L. platyrhynchus*. Hildebrand and Schroeder (1927) described *L. osseus* as a species that 'ventures' into salt water, while Suttkus (1963) considered that in neither species is there evidence of a regular pattern of migration to and from the sea.

4.TROUTS AND SALMONS — FAMILY SALMONIDAE

The trouts and salmons are probably the archetypical anadromous fishes and are amongst the most studied of fishes owing to their great importance for angling and as food. They are of medium to large size (reaching more than 1.5m in length). Salmonids are characteristic of cool and cold waters of all northern continents, occurring from the Arctic drainages of Europe, Asia and North America, south as far as the Mediterranean, northern Africa, some of the southern states of the United States of America, and northern Mexico. The family contains about 68 species in nine genera. All of these spawn in fresh water (Nelson, 1976) but anadromy is strongly and widely represented.

Anadromy in salmonids may be typified by the Atlantic salmon, *Salmo, salar*. This fish spawns during winter, in gravelly nests in swiftly flowing streams (as in almost all *Salmo* spp.) and it may take place a long distance upstream (up to 1000 km), though this is not always the case. Some weeks after hatching in the gravels, the young alevins emerge into the streams where they feed and grow. Freshwater life in juvenile Atlantic salmon may last only about a year, but is often longer, up to eight years in extreme cases (Power, 1969). Migration to sea follows a well-defined physiological change known as smoltification, at which time the fish become able to osmoregulate in sea water, and at which time their coloration changes. Not all the young actually go to sea, some maturing as 'precocious males' in fresh water, at an age of only a year. Atlantic salmon range very widely at sea, stocks from the east coast of North America, Britain, and Scandinavia all reaching Greenland, many hundreds of kilometres away (Figure 4.3). The return migration of mature adults to fresh water and the migration upstream to spawn may take place after only a year at sea (early maturing males known as 'grilse'), or may occur after about two to four years at sea. During the upstream spawning migration the adult fish do not feed, but as they migrate upstream their gonads enlarge at the expense of body and visceral fat deposits and other body sources of energy. After spawning many adults die but some, mostly females, return to the sea to feed, and recover condition again, before spawning a second time. A very few females may spawn a third time. In the very cold waters of Ungava Bay rivers in northern, Arctic Canada, Power, Dumas and Gordon (1987) found that some Atlantic salmon move downstream only into the estuaries and do not really go to sea at all.

The brown trout, *Salmo trutta*, is widely distributed in Europe from the White Sea in Arctic Russia, south to the Mediterranean and northern Africa, and west into Turkey. It is closely related to the Atlantic salmon. It has a more variable life cycle than the Atlantic salmon, but shares many of its features. The essential difference is that the migratory cycle is much more facultative, and it appears that not only do some fish in a population go to sea and others not, but also that some fish may go to sea in one year, return, and not go to sea the next. Females tend to be more migratory and to travel further at sea than

Figure 4.3: Migratory patterns of Atlantic salmon, *Salmo Salar*, in the North Atlantic
Source: After Netboy

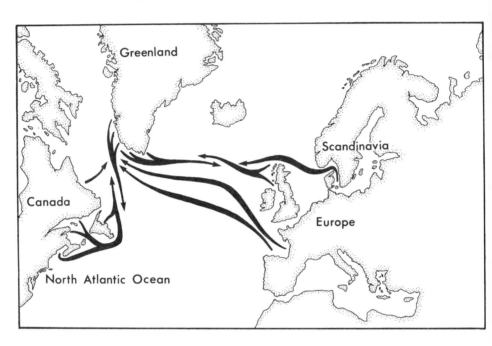

males. The cycle is so facultative and variable that it might be argued that the brown trout is scarcely anadromous. However, sea-migration of an essentially anadromous type in the brown trout is widely present in northern regions and persists as far south as the Bay of Biscay. Further south, populations are confined to fresh water (Trewevas, 1953). Sea-migratory and non-migratory/fluviatile stocks of brown trout were for a long time described as belonging to distinct species, but this view is no longer adhered to. Rounsefell (1958) described how 'sea trout' introduced into some rivers in Maine (USA) resulted in the occasional return of a few sea-run fish but also in the establishment of freshwater angling for brown trout; in other rivers, non-migratory brown trout were released resulting not only in freshwater angling but in the return of quite a number of fine sea trout. This is consistent with Nordeng's (1983) later reports that both resident and anadromous brown trout are sympatric in Scandinavia and that each type produced progeny which segregated into some of each type, while some individuals were found to pass through both life history types. Post-spawning survival in the brown trout is much higher than in the Atlantic salmon, and it may spawn repeatedly over five to ten or more years. Movement at sea, however, appears to be much more restricted and less - clearly defined, migrations coastwise are chiefly estuarine, and usually short (Rounsefell, 1958).

In the sea, brown trout probably remain largely close inshore, although they sometimes wander (rather than 'migrate'?) longer distances. Wheeler (1969) suggested distances of up to about 300 km as 'not uncommon' and distances of more than 600 km being on record. Jonsson(1985) described a population in a Norwegian lake in which there were sympatric anadromous and resident stocks, the anadromous fish smolting and leaving for the sea during the spring, the return migration of spawning adults occurring in autumn; resident and migrant fish were observed to spawn together.

Two eastern North American trouts, the rainbow trout, *Salmo gairdneri*, and the cutthroat, *S. clarki*, are also anadromous. The rainbow is known for having lower post-spawning survival than most other *Salmo* spp. *S. clarki* probably does not move far afield in the sea and may lead a characteristic estuarine existence moving in and out with the tide (Neave, 1949). In neither species are the gonads greatly enlarged at the time of leaving the sea, but they grow and mature during the upstream freshwater migration (Rounsefell,1958). Movement to sea of the rainbow trout occurs after smolting, Conte and Wagner (1965) showing that its smolting is related to size (about 150 mm) rather than to age. Behnke (1966) described the life history of the Kamchatka trout, *S. mykiss*, as being comparable with that of the rainbow trout (there appears to be some doubt about whether the two species are actually distinct, a question that seems to have been avoided by most salmonid ichthyologists — however, see Okazaki, 1984). Cutthroat migrate during summer—autumn into small headwater streams to spawn, seldom a great distance inland (mostly less than 160 km). The spent kelts return to sea and the young remain in fresh water for one to four years before they, too, go to sea. However, few cutthroat overwinter in the sea, whether mature or not, but will return to fresh water after a few months at sea (Johnston, 1981).

The Kamchatka salmon, *Salmo penshinensis*, occurs along the coasts of the Kamchatka Peninsula and the Okhotsk coastline in eastern Siberia. Berg (1962) described it as anadromous in these areas. It enters fresh water during autumn but does not spawn until the following spring, usually not far upstream from the sea; spent adults return to sea to feed after spawning. This cycle resembles other *Salmo*.

Some of the species of char (genus *Salvelinus*) are also anadromous, although Rounsefell (1958) described them as typically spending only a small portion of each season in the sea. Typical of these are the arctic char, *S. alpinus*, which is virtually circum-Arctic, the brook char, *S. fontinalis*, in eastern North America, the dolly varden, *S. malma*, in western North America and eastern Asia, and two rather more localised species - *S. leucomaenis* in northeastern Asia and possibly *S.confluentis* in Puget Sound, in western North America (Behnke, 1980).

The Arctic char is, with little doubt, the most widespread northerly distributed species of freshwater/anadromous fish; it occurs widely (as widely as there is land within the Arctic circle, reaching about 82.5°N (Johnson, 1980));

it spreads south into northern New England in the eastern United States; anadromy is well developed. Arctic char are slow growing, and do not reach maturity until a considerable age (4–8 years, sometimes up to 10). Their life in the sea seems highly variable between invividuals and between stocks. The juveniles may remain in fresh water, and not go to sea until they are 5–6 years old (Grainger, 1953). Moore (1975b) showed that in Baffin Island populations of northern Canada, the fish less than five years old did not go to sea at all, but that older fish live at sea during the brief far northern summer. Once they have begun moving to sea, Arctic char smolts then make annual migrations to (during spring) and from (autumn) the sea prior to their first spawning. At sea, they are generally found along the coasts and in bays, in shallow water, and perhaps do not move far from their source river. Moore (1975b) considered that they live extensively nearshore, in the intertidal zone, Sprules (1952) suggested that they spend some time in brackish water during the late summer, before moving further upstream, and Grainger (1953) showed that they may move in and out of estuaries prior to the upstream migration. Moore (1975b) described their movements from fresh water to the sea during spring, at about the time the ice thaws, they move some distance at sea (up to 40 km). The entire population leaves the sea in late summer–autumn, returning to fresh water. Dempson and Green (1985) said that the return migration of mature fish began in mid-July and continued until late September in the Fraser River, Labrador, the larger fish migrating earlier than smaller ones. Not all the return migrants spawn; the immature fish leave the sea with, or a little later than, the spawners and accompany them onto the spawning grounds. Arctic char home accurately onto their natal streams, Glova and McCart (1974) suggesting that they actually return to the same spawning ground throughout their lives. Spawning takes place in river rapids or around lake shores, on the latter (according to Johnson, 1980) because the rivers may totally freeze during the winter. Both post-spawning adults and non-spawners overwinter in fresh water, being largely inactive in places where unfrozen water remains. However, some fish may remain in fresh water for several years without returning to the sea. The composition of the spawning aggregations is highly variable and Johnson (1980) observed that the spawning population may comprise some new sea-migrants, some fish that had left the sea a year or two previously, and some that had never been to sea; all three types appear to intermingle on the spawning grounds and to spawn at about the same time. The fish survive spawning, and return to the sea to feed in the spring, when the ice melts. They spawn several, perhaps many times, and reach a great age — Johnson wrote of reports of Arctic char up to 33 years old and having been involved in six or seven spawnings with gaps of three or four years; however, annual spawning is known (Nordeng, 1961).

The eggs develop during the winter, often under ice, and hatch in the spring, up to three months after spawning. The young feed for a considerable time in fresh water, probably several years (four to five commonly but up to nine) before going to sea, with a much less marked smoltification than is known for species of *Oncorhynchus*.

Nordeng (1983) identified three sympatric stocks of Arctic char in Norwegian rivers and lakes – 'anadromous, small, and large freshwater residents'. He found that the progeny of each type segregated to produce some of each sort, regardless of the parentage, and that an individual char may pass through more than one of the above types of stock during its life. In the south (south of about 65°N), he found that anadromy failed to occur, but when southern, non-anadromous fish were transplanted north to where there were anadromous stocks, the transplanted fish became anadromous.

Other species of *Salvelinus* resemble *S. alpinus* in details of the life history. Dolly varden, *S. malma*, have a sub-Arctic and Arctic distribution in eastern North America and western Siberia, and have a life history similar to Arctic char, i.e. autumn spawning, spring hatching, life in fresh water for several years before a sea migration by smolts at about three years of age. The whole population leaves the sea to overwinter in fresh water, some spawning annually, others every two years. Some males remain in fresh water as 'residuals', mature early at a small size and participate in spawning amongst the larger adults. The dolly varden is said to spend only a month or two feeding in the sea. Rounsefell (1958) believed that it was less than 60 days in the Karluk River on Kodiak Island, Alaska. Spawning survival is said to be lower in the dolly varden than in the Arctic char, few surviving more than two spawnings (Armstrong and Morrow, 1980).

Brook char, *S. fontinalis*, occur in eastern Canada, from the sub-Arctic and south into the eastern United States as far as the Appalachian Mountains of Georgia. They also have a comparable life history strategy in the more northern parts of their range, but further south they mature earlier and may spawn annually. In some areas populations include both anadromous and non-migratory individuals (Dunbar and Hildebrand, 1952). Wilder (1952) studied a population in which faster-growing, larger individuals were anadromous but slower-growing, smaller ones, were not as likely to be migratory. He thought that anadromous brook char may not go to sea every year. McCormick, Naiman and Montgomery (1985) and Power (1980) described brook char in northern latitudes as going to sea in spring and summer. They spend a few months at sea and return upstream in autumn, to spawn. Brook char probably do not range very widely when at sea. Males are believed to spawn annually, once mature, but females may spawn only once every second year, or only two out of three years (Power, 1980). Whereas it is suggested that in the north this species, and other chars, leave the sea and move into fresh water to escape the very cold temperatures of the Arctic winter, others (e.g. Mullan, 1958) suggest that further south, brook char leave the sea in the spring and move to fresh water to escape the warmer sea temperatures of the summer.

Salvelinus leucomanis occurs in eastern Asia (Sakhalin to Korea and Japan). The young of this species go to sea at a year or less, and grow there for several years, before there is an adult return to fresh water to spawn. Some authors classify the related lake char, *Salvelinus namaycush*, as diadromous, but this seems quite indefinite. It has been reported from salt water by some observers (Walters, 1955); Norden (1970) considered that it ventures into sea water in the northern parts of its range, but Dunbar and Hildebrand (1952) and Scott and Crossman (1973) both described the lake trout as the least tolerant of salt water of all the chars. Morin, Dodson and Power (1980) found that it tolerates salinities up to 19.2 ppt, but this shows only that it is to some extent euryhaline, and not that it is diadromous. Boulva and Simard (1968), however, said that its upper salinity limit was between 11 and 13 ppt, and that it does not enter oceanic salt water with a salinity around 35 ppt. There seems little evidence of regular and semi-obligatory fresh water–sea water migration in the lake trout, and therefore little cause to classify it as diadromous.

In the Asian genus *Hucho* one species, *H. perryi*, is anadromous. It occurs along coasts and in bays, returns to the sea after spawning (Okada, 1960) and does not move very far upstream into fresh waters (Berg, 1962).

Anadromy is widely present (though not well described) in the whitefishes and ciscoes (genus *Coregonus*). The extent of their anadromy seems to vary widely. Some, like the Siberian omul, or Arctic cisco (North America), *C. autumnalis*, are described as fully anadromous in Arctic Canada, Asia and Europe, running into fresh water in early spring in Canada (Wynne-Edwards, 1952). The omul may move long distances (at least 1500 km) upstream in river systems, although it does not occur in the upper reaches of rivers (Berg, 1962). Burkov and Solovkina (1976) found that it forages widely in coastal seas, and that with the onset of the cold temperatures of autumn (October), the immature fish move into river mouths, penetrating no further up than the tidal zone, where they live in depressions in the river bed. After the ice melts and breaks up, they return to the sea. The sexually mature fish migrate into fresh water much earlier, during summer (August–September), to spawn there in autumn (October). Feeding does not occur during the migration (Nikolsky and Reshetnikov, 1970). There is considerable spawning survival, and individuals may spawn two or three times, but not necessarily annually. In this species a distinct, downstream, post-spawning migration of spent fish is reported during late autumn–winter. The spent adults join the immature fish to overwinter in the tidal/estuarine areas of the rivers. The young are believed to emigrate from the rivers after hatching, during the spring.

The Siberian cisco, *Coregonus albula*, also occurs very widely from Europe to North America. Ustyugov (1972) showed that it migrates from the sea during the autumn (September to November) and spawns in spring (May–June). Spawners are between four and nine years old and do not feed during migration. Siberian cisco are repeat spawners, some spawning up to five times. The young stay in fresh water for several years, eventually migrating to sea during the summer. The Bering cisco, *C. laurettae*, possibly has a similar life history strategy in the Bering Sea drainages of Alaska (McPhail and Lindsey, 1970), and possibly also of Siberia, but its upstream migration is thought to be

Figure 4.4: Life history of the lake whitefish, *Coregonus clupeaformis*
Source: Morin Dodson and Power 1981

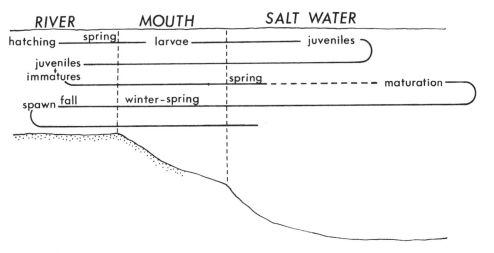

in autumn. The lake whitefish, *C. clupeaformis*, in spite of its common name, is anadromous (Figure 4.4). It occurs very widely across Arctic and temperate Canada. Adults migrate from the sea and upstream, to spawn in autumn. The eggs develop beneath the ice over the winter and the young stay in the river gravels for some time before moving to sea. They return as juveniles/immatures to overwinter in fresh water in the lower reaches of the rivers, moving back to sea in spring to mature over the summer, and later return to spawn in fresh water (Morin, Dodson and Power, 1981). The lake herring, *C. artedii*, occurs in northern (Hudson Bay) drainages of Canada. It migrates from coastal seas during the summer, moving gradually upstream in fresh waters over a prolonged period. Spawning is in autumn and the spent adults return to the estuaries, to overwinter and then go back to sea. The young hatch in the spring, but do not remain in the gravels, moving quickly to sea, where they feed and grow over the summer. Juveniles and immature fish move into the estuaries for the winter, and go back to sea in spring as the ice melts. They mature at sea and return to fresh water to spawn (Morin *et al.*, 1981).

The broad whitefish, *C. nasus* (in nearly all Arctic drainages of North America, Asia and Europe), has been described as 'evidently anadromous' (McPhail and Lindsey, 1970), moving downstream at least into brackish water and having a summer return migration. Spawning migrants are said to cease feeding during migration, but I have found no information on the downstream movement of the young of this species of *Coregonus*, or the age at which they enter the sea. Svardson (1979) recognised two species of *Coregonus* in Scandinavia/Baltic Sea drainages — *C. lavaretus* and *C. wildgreni*, of which the former seems to be anadromous, although both appear to be poorly understood.

47

C. lavaretus (which is northern circumpolar in the Arctic seas) was described by Svardson as spawning every two years, and during the spring, but Berg (1962) and Lindroth (1957) considered that it is in the autumn. Spawning occurs over gravelly rapids, the post-spawners returning to the Baltic before the winter. The eggs are non-adhesive and semi-bouyant, and Jager, Nellen, Schoffer and Shodjai (1981) considered it likely that young larvae or even eggs get transported into the estuary and are exposed to brackish water. In the Russian/Siberian Arctic this species leaves the sea and moves into rivers in the late summer, both mature and immature fish migrating as the sea and rivers ice up; they do not move far upstream. It seems that there may be two seasonal migratory forms (Berg, 1962). Another Arctic species, *C. muskun*, was regarded as anadromous by Berg (1962). Lindroth (1957) showed that in Sweden some eggs and most fry are carried passively downstream and into coastal waters of the Baltic Sea, but this species does not migrate far upstream by comparison with other anadromous *Coregonus*. It leaves the sea in autumn for a November spawning, although, again, there seem to be both spring and autumn migrant races. The English houting, *C. oxyrinchus*, is described by Wheeler (1969) as being typically estuarine, only rarely moving far to sea. The same appears to be true of the Canadian whitefish, *C. canadensis* (Scott and Crossman, 1973). *Coregonus artedii* is said, by Scott and Crossman (1973), to enter 'coastal salt water' but is not described as diadromous. The ussuri whitefish, *Coregonus ussuriensis*, is found in the Amur River (Nikolskii, 1961) where it is primarily riverine in habit; anadromy seems possible but unlikely.

The inconnu, *Stenodus leucichthys*, is a further coregonine species, that is anadromous. It occurs widely in northwestern North America and in the Arctic drainages of Siberia and Russia, as far east as the White Sea (Alt, 1969). Its range extends south into the Bering Sea (Korf Bay). Smith (1957) reported it as having 'an inclination to undergo extensive migrations' — more than in other coregonines. Berg (1962) regarded it as partly riverine, some fish spending all their lives in rivers, and partly anadromous. Growth in the inconnu is very slow and maturity is reached at a considerable age (11–14 years, Berg, 1962); some live to 20 years or more. McPhail and Lindsey (1970) described it as probably not straying far from river mouths. It evidently has an upstream migration lasting several months over the summer period following ice melt, with no highly concentrated or well-defined movements. Feeding ceases during migration. Spawning is during the autumn (October), Berg, (1962). However, McPhail and Lindsey (1970) referred to a 'tremendous rush' of spawned-out adults, moving downstream during the autumn, after the late summer–autumn spawning; thus there is apparently substantial post-spawning survival. The young are described as taking several years to move downstream to the sea (Berg, 1962). Another related genus, *Prosopium*, is essentially fresh water in habit though *P. cylindraceum* may venture into brackish waters in the northern part of its range (Norden, 1970). Morin *et al.* (1980) showed that it will tolerate salinities up to 19.2 ppt but, as with some other salmonids already discussed, this demonstrates euryhalinity rather than diadromy.

Pacific salmons (genus *Oncorhynchus*) comprise six species. The essential structure of their life cycles resembles that of the Atlantic salmon — spawning in fresh water in gravelly streams, a movement to sea of the juveniles, growth to maturity in the sea, and a return migration of spawning adults to the stream of birth. Returning adults cease feeding on or before entry to fresh water. Pacific salmons are characterised by virtually complete post-spawning mortality, though Robertson (1957) showed that a very few precocious males of the chinook salmon (*O. tshawytscha*) may survive spawning. However, although the basic cycle is the same in all species, details vary very widely.

At one extreme, in the pink salmon, *O. gorbuscha*, and the chum salmon, *O. keta* (both present in western North America and in north-eastern Asia from northern Japan, northwards into Siberia), spawning may take place close to the sea (it may be in tidal waters in the pink salmon (Hanavan and Skud, 1954)). In these species the gonads tend to be nearly mature before entry into fresh water (Rounsefell, 1958). However, spawning may be well upstream in the chum — upstream at least 200 km in Asian chum (Sano, 1966), and nearly 500 km in Canadian populations of both pink and chum salmon (Neave, 1966a,b). There are reports of chum spawning over 3,000 km up the Yukon River (Gilbert and O'Malley, 1921), but this is undoubtedly an extreme. Pink salmon are distinctive in having a very rigid two-year life cycle — there are separate odd-year and even-year gene pools (though three-year-old populations have appeared recently in the lacustrine populations in the North American Great Lakes). Chum are known to leave the sea on their spawning migration in both spring and autumn, and a similar dual migration is known in some other *Oncorhynchus*, e.g. the chinook, *O. tshawytscha*. On hatching and emerging from the gravels, the young of both pink and chum salmon may go almost immediately to sea, without feeding beforehand, though in the chum feeding in fresh water can be quite prolonged prior to the migration to sea (Miller and Brannon, 1982). In the pink salmon some feeding may take place (Rounsefell, 1958). These species spend two (pink) or more years in the sea, before returning to spawn.

The coho salmon, *O. kisutch*, is very widely distributed in lands bordering the northern Pacific, occurring from the northern islands of Japan (Hokkaido) northwards along the Siberian coastline, east to Alaska and south along the coastline of North America to central California. Coho migrate during autumn, and at the time of leaving the sea their gonads are small; they mature as the fish migrate upstream to spawn, usually only moderate distances up from the sea. The young, when they emerge from the spawning gravels during the spring, may either go very rapidly to sea (Foerster, 1955), or more usually, may remain in fresh water streams, for one year, sometimes two, rarely three years (Miller and Brannon, 1982). The chinook has a similar geographical distribution to that of the coho. The life cycle of the chinook is also similar to that of the coho, though, as noted above, well-defined spring and autumn spawning runs can be identified. It spawns very widely, from streams not far upstream from the sea (autumn run fish which leave the sea with their gonads in an advanced stage of maturation), to other spawning grounds vast distances into the inland, perhaps 2,000 km upstream in extreme examples (spring run fish in which the gonads are small when the fish leave the sea, but mature during migration, Miller and Brannon, 1982). Most of the young chinooks move out to sea during the first year of life, some after only a few weeks feeding in fresh water, many after about three months, some staying for up to a year, and a few for two years in streams of the far north. Some of the outmigrants may dwell in brackish estuaries for a period before moving on out to sea (Dorcey, Northcote and Ward, 1978). Miller and Brannon (1982) considered that young chinooks could move to sea at virtually any time during their first few months of life, but their ability to do so may differ between different stocks of the species. Chinooks spend several years at sea, and may return at anything from two to nine years of age, most usually three to five years. The return migration has peaks in both spring and autumn, and though spawning occurs primarily in the autumn, it can occur in most months of the year.

In the sockeye salmon, *O. nerka* (again widely distributed on both sides of the northern Pacific), the young migrate from the spawning streams to a lake, where feeding and early growth takes place for one or two, occasionally three, or even four years, before migrating to sea (Miller and Brannon, 1982). However, Ricker (1972) also referred to stocks of sockeye in the USSR that move to sea soon after emerging from the redd gravels. Sockeye typically migrate to sea at a larger size than other species of *Oncorhynchus*. Thus in the pink and chum salmon, the movement takes place without feeding and soon after emergence, in the coho and chinook there is freshwater life, with feeding and growth lasting several months as the fish move downstream, while in the sockeye there is a definite feeding migration of the young followed by another distinct migration to sea.

The masou, or cherry salmon, *O. masou*, is found in eastern Asia, from Japan, Korea, and Siberia. Although it is essentially migratory like other species of *Oncorhynchus*, Miller and Brannon (1982) indicated that in riverine populations there may be a significant number of fish that never leaves freshwater rivers. The non-migrant, river-resident part of the population is largely, though evidently not exclusively male (Berg, 1962) — possibly 80 per cent of the males and 20 per cent of the females may be non-migratory. Berg found that only males mature in fresh water in the north but that to the south both males and females are involved. They spend only a year at sea, with a sea migration of limited extent. The spawning run of the masou salmon begins in late winter–early spring (February), and continues through the spring and summer. Okada (1960) suggested that masou continue to feed and mature during their freshwater spawning migration (unlike all other species of *Oncorhynchus*), and this was supported by Miller and Brannon (1982). Spawning takes place during late summer through autumn (July – October), and as with other *Oncorhynchus* all the spawners die. The emerging young may spend a year, sometimes two years, in fresh water and they return as spawning adults only a year or so after going to sea.

The structures of the life histories of various anadromous trouts and salmons are summarised in Figure 4.5 Deviations from these basic life cycle patterns are several and varied. In fact Miller and Brannon (1982) suggest that differences between stocks of a given species are nearly as great as among species. Hoar (1976) reported the situation thus:

Salmonids show a spectrum of migratory types ranging from wholly freshwater species that are relatively sedentary in their habits to forms that invariably migrate between lakes and rivers, rivers and oceans, or rivers, lakes and oceans. There are the strictly freshwater species, the obligatory and the facultative anadromous species ....

Sockeye salmon (*O. nerka*) are best known for their deviation from the basic anadromous life cycle, and three life history types can be recognised:

1. The normal anadromous form;
2. Some deviants of anadromous parents that remain in freshwater lakes and grow there to maturity — so-called 'residuals'; and
3. Stocks that have long had a landlocked ancestry and which tend to be morphologically distinct from anadromous and residual fish — known as 'kokanee'.

Foerster (1947) showed that sockeye that have a kokanee parentage can nevertheless go to sea and may return as typical anadromous fish. They have retained the ability to live a marine existence and to find their way back to their natal stream. The occurrence of estuarine spawning in the pink salmon (Hanavan and Skud, 1954) means that at that point, this species becomes only marginally anadromous. However, such spawning seems to be exceptional. It is interesting that this has occurred in a species in which migration to sea occurs almost immediately following emergence of the young from the gravels, i.e. the young are already well adapted to coping with the fresh water/sea water transition very early in life, in contrast with some other species of *Oncorhynchus*. No salmonid has become totally marine, although Rounsefell (1958) said that 'The pink salmon in some localities spawn in the tidal flats off the mouths of streams so that many pass their entire lives without entering strictly fresh water.' In various measures and in differing ways, however, species have modified their basic life history strategies. Some species have established lacustrine populations, either enforced by the impoundment of lake outlets (natural, or man-made), or without such compulsion. Rounsefell (1958) stated that amongst the salmonids 'A few species cannot reproduce, or cannot do so

51

Figure 4.5: Life history strategies in various of the Salmonidae: S spawning location; broken arrow, juvenile migration to sea; thin arrow, juvenile migration in fresh water; broad arrow, adult spawning migration
Source: After Rounsefell 1958

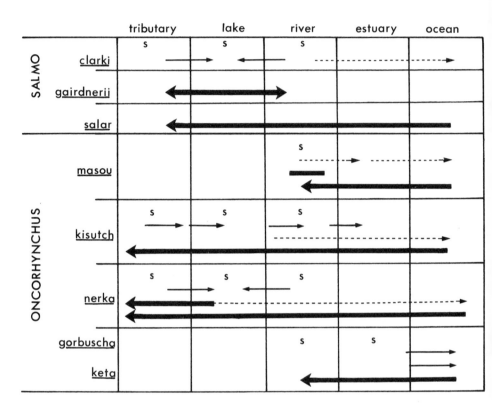

with sufficient success to maintain a natural population without residing for some time in a marine environment'. These he designated 'obligator[il]y anadromous', and he listed pink and chum salmon in this category. Other species have established lacustrine or fluviatile populations, including sockeye salmon, masou salmon, Atlantic salmon, cutthroat, rainbow trout, and other trouts, several chars, various ciscoes and whitefishes, and others. Rounsefell (1958) considered that the coho shows little tendency to develop natural freshwater populations in North America though Ricker (1972) mentioned several populations associated with a lake ... which regularly produce lake maturing 'residual fish'. However, the only self-supporting lake population he knew of is one in Siberia (Dvinin, 1949). It seems that, in general, coho 'residuals', although maturing, do not reproduce successfully. Peck (1970) argued that until coho became established in Lake Superior, there has been 'no documented evidence of reproduction in a natural environment by coho salmon which spent their entire life in fresh water'. Lake populations of coho are obviously rare.

Rounsefell (1958) decided that chinooks should 'perhaps be included' as obligatorily anadromous. However, pink, coho and chinook salmon, have all successfully established lake stocks in the Great Lakes of North America (Peck, 1970; Kwain, 1987), although pinks and chinooks have failed to do this within their natural ranges in western North America. Chinooks have established lake stocks in New Zealand, voluntarily, after anadromous stocks of this species were introduced and became established in the early 1900s (McDowall, 1978a). However, some observers (Stokell, 1962) have considered that the continued viability of these lacustrine populations of chinooks may depend on a continued input of genetic material from anadromous fish. The lake populations, although self-sustaining for 60 years or more, are often not very robust, and this is not too divergent from Rounsefell's (1958) view that there is a serious question concerning the chances of successful maintenance of a natural population. Nevertheless there have long been self-sustaining chinook populations in several New Zealand lakes, some with no access, others with only minimal and intermittent access to anadromous salmon stocks.

While establishment of lacustrine stocks of salmonids is not rare, rather fewer species are facultative as regards loss of anadromy within river systems. Most successful in this regard is the brown trout, which seems to be highly flexible in the life cycle adopted. Jonsson (1982, 1985) identified different stocks of brown trout in Norway. In an evidently anadromous population, one part of the stock was found to smoltify and migrate to sea while another part remained in the streams to mature at stunted size. The non-migrants could be either all males, or include both sexes. Briggs (1953) showed that both anadromous and non-migratory stocks of rainbow trout occur in some streams, while Northcote (1969) described how a population above an impassable fall was able to survive there because of a genetic tendency for some fish in the stock not to migrate to sea. Neave (1944) had earlier shown that the migratory/non-migratory tendencies in this species were inherited, though not invariant in any one stock — a conclusion supported more recently by Ricker (1972) who believed that in Cowichan River rainbow trout, the choice between anadromy and freshwater life was mainly, or perhaps entirely, genetically controlled. Nordeng (1961, 1983) recognised the same phenomenon in Arctic char, while rainbow trout and brook char also successfully establish non-diadromous fluviatile stocks. The cutthroat, though having anadromous stocks, has very numerous non-migratory stocks, and R.J. Behnke (personal communication) has advised that he recognises 14 sub-species of which only one, *S. clarki clarki*, is anadromous. Johnston (1981) discussed sympatric stocks of coastal/resident and anadromous cutthroat, and expressed uncertainty about whether there was gene flow between the populations.

The inconnu is also said to have sympatric riverine and anadromous stocks in some Siberian rivers (Berg, 1962). Ustyugov (1972) identified three distinct forms of the Siberian cisco, *Coregonus albula*, typically anadromous fish, coastal fish, and river fish.

There is not only local variation in the occurrence of anadromy in northern populations, but also anadromy tends to disappear altogether towards the south of species' ranges. The brook char, like the brown trout, is anadromous at the northern extreme of its range, but is less so further south, and anadromy disappears altogether towards the southern limits of its occurrence (Kendall, 1935). Montgomery, McCormick, Naiman, Whoriskey and Black (1983) described the seaward migration of brook char in a Gulf of St Lawrence river as being temporally synchronous within the population, but further south, others observed that migration was more variable in both timing and duration (Mullan, 1958; Smith and Saunders, 1958; McCormick *et al.*, 1985). Arctic char were shown to become non-anadromous at a latitude of about 65°N in Norway (Nordeng, 1961) and are not anadromous in Baltic Sea drainages; nor are they anadromous south of about 49°N in Newfoundland, in eastern Canada (Bigelow, 1963).

In many of the above instances the non-anadromous stocks are riverine (not lacustrine), so that in regard to several salmonids, Baker (1978) is incorrect in claiming that when there is loss of diadromy in fishes, a downstream migration always ends in a lake — this seems to be most true of *Oncorhynchus* in which most non-anadromous species seem to be lacustrine, but the masou is said to have fluviatile, non-anadromous stocks in Japan (Masuda, Amaoka, Araga, Uyeno and Yoshino, 1984).

Interspecific variation within the various salmonid genera follows the same general pattern, there being distinct species in nearly all areas and in nearly all generic groups that have excluded the marine migratory phase from the life cycle, derivative species either becoming lacustrine and migrating into tributaries to spawn, or becoming wholly fluviatile. This is true of *Salmo*, *Salvelinus*, *Hucho*, and *Coregonus* but not *Oncorhynchus*, unless the Japanese species *O. rhodurus* is accepted as being distinct from *O. masou* (not the case with most authorities, e.g. Behnke, Ting Pong Koh and Needham, 1962). It could easily be assumed that the standard pattern of evolution of non-anadromous salmonid species, in genera such as *Salmo*, *Salvelinus*, *Coregonus*, etc., is by the establishment and isolation of landlocked populations. An example of a landlocked anadromous derivative species is *Salmo aquabonita*, said by R.J. Behnke (personal communication) to be allied to and derived from the anadromous *S. gairdneri*. Rounsefell (1958) suggested that species such as *Salvelinus aquassa*, *S. aureolus*, and *S. marstoni* are freshwater counterparts of *S. alpinus* or closely related forms. Although this has undoubtedly occurred in various salmonid lineages, accounts of the relationships of some of the inland species of *Salmo* in western North America (Behnke, 1966, 1972; Miller, 1972a; and others) indicate a significant amount of speciation amongst already non-anadromous species, Miller (1972a) pointing to *S. chrysogaster* as having primitive origins with other species being derived from it. Whether such species as *S. chrysogaster* have anadromous origins cannot now be determined.

5. SMELTS — FAMILY OSMERIDAE

The smelts are smallish (to about 450 mm), elongate, silvery, shoaling fishes, usually of compressed and slender form. The family is widely distributed in the cold to cool Northern Hemisphere, on all continents and in all oceans. The family contains about 12 species in six genera, and is variously anadromous, freshwater or marine; anadromy is a well-recognised phenomenon in the family.

The rainbow smelt, *Osmerus mordax*, is representative of anadromy in this family (McKenzie, 1964). It occurs in eastern North America from the Labrador coast of Canada, south to Delaware Bay in the United States, and also in some Arctic drainages of northern Canada and sub-Arctic drainages of the northern Pacific coast of the United States. Its distribution spreads south to Vancouver Island in British Columbia, Canada, and on the west Pacific coast of Asia south as far as Wonsan in Korea. Many populations are anadromous but some are entirely lacustrine in habit. In New Brunswick (Canada) adult rainbow smelt move inshore during the autumn, but further south, where temperatures are warmer, migration may be earlier, as early as late February. The fish move into brackish bays and river estuaries, and then migrate upstream in spawning shoals in the late winter as the ice begins to melt. Spawning occurs in fresh water during the summer (May–June), usually not very far above tidal reaches, but nevertheless above the influence of the tide. McKenzie (1964) cited distances of up to 13 km as being amongst the longer migrations for this species in the Miramichi River, but McAllister (1984) described it as migrating 1,000 km up rivers. Although there is a communal, shoaling migration, it appears that spawning occurs in pairs or small groups, one female with several males. After spawning the adults return downstream to the river estuaries, where they spend the summer, before moving out to sea again as the rivers freeze. At sea, they inhabit largely coastal and inshore waters, Bigelow and Schroeder (1963) saying that:

they are confined to so narrow a belt that none has ever been reported more than six miles or so out from land and seldom below 2–3 fathoms. Many of them spend their entire growth period in estuarine situations, including the tidal reaches of rivers.

They probably move further out to sea and into deeper waters in the southern parts of the range, seeking cooler waters during the summer. Although there is post-spawning mortality (mostly of males), many of the adults survive spawning; McKenzie (1964) showed that most fish of both sexes reach maturity at age two, practically all at age three, but that some fish reach an age of four and a few five or even six years old. Thus the Miramichi spawning run, which McKenzie studied, comprised 66 per cent, 30 per cent and 4 per cent two-, three- and four-year-olds respectively.

The eggs of the rainbow smelt hatch in about two weeks, and the larvae are carried quickly down into the tidal reaches of the river, where they feed, and presumably disperse to sea.

The very similar and related European smelt, *Osmerus eperlanus*, found in some rivers of the White Sea in Arctic northern Russia, south through the Baltic to the Bay of Biscay, has a similar life history. Wheeler (1969) wrote of this species that it is rarely found far from shore and that some may spend all of their lives in river estuaries. Migration from the sea follows soon after ice melt in northern localities. Spawning is apparently not far from the estuaries, during May, and the young move quickly downstream through the estuaries to the sea (Berg, 1962).

Other smelts are similarly anadromous. The Japanese shishamo, *Spirinchus lanceolatus*, runs in summer into rivers of the east coast of Hokkaido, Japan, while the longfin smelt, *Spirinchus thaleichthys*, is anadromous along the Pacific coast of North America, from Alaska to northern California, although some populations, such as that in Harrison Lake, British Columbia, are non-migratory. This species runs into fresh water during the late autumn and winter, spawning close to the sea; few apparently survive spawning. The eulachon, *Thaleichthys pacificus*, migrates for the first time at age three, and spawns in spring in the rivers of this same area, shortly after ice-melt. Again, inland penetration is slight; Scott and Crossman (1973) say up to about 30 km. Most adults die after spawning, though a few survivors return to the sea, and may recover to spawn a second time (very occasionally a third time — a few five- year-olds have been reported); Most of their lives are spent in the moderate depths of the ocean not far from the shore (Scott and Crossman, 1973). A Japanese smelt *Hypomesus transpacificus* might appear to be less explicitly anadromous in the eastern Pacific where it is known in brackish and fresh waters and may spend nearly all its life in fresh water (McAllister, 1963). However, it is described as anadromous in Japan by Hamada (1961). He described four different 'forms' (referring to it as *H. olidus*); these included:

1. Spring migrants in which all the spawners died;
2. Autumn migrants which did not spawn until the following spring, in which not all spawners died, and which may live for two, rarely three or four years;
3. The progeny of anadromous parents which remain in fresh water — a small percentage of the population more or less equivalent to 'residuals' of the sockeye salmon; and
4. Landlocked populations which have not been to sea for many generations, and in which all the spawners die.

However the status of these various types of population is uncertain, as there is evidence of two sibling species in Japan, that are referred to as *H. transpaci-*

ficus, one of which is more anadromous than the other (D.A. McAllister, personal communication) .

Deviation from the anadromous life history is wide in other osmerids also. Anadromous species are frequently known to be landlocked (*O. mordax, O. eperlanus, H. transpacificus*). Bigelow and Schroeder (1963) reported successful transplants of anadromous *O. mordax* to lakes, with a fishery developing for this fish in Lake Erie, and McAllister (1963) discussed successful transfers of *H. transpacificus* within Japan. Some congeners of anadromous species may be landlocked, e.g. *H. olidus* in Arctic Canada, eastern Asia from Siberia south to Korea, and in northern Japan, and *O. spectrum* in eastern North America (Lenteigne and McAllister, 1983). At least four osmerids, the white-bait smelt, *Allosmerus elongata*, (Washington and California), the surf smelt, *H. pretiosus*, (California to Alaska and Kamchatka to Korea), the capelin, *Mallotus villosus*, (Washington to Alaska, Siberia to Korea in the Pacific, northern Scandinavia in the Atlantic, and around the Arctic of North America, Asia and Europe), and the night smelt, *Spirinchus starksi*, (eastern Pacific from California to Alaska) are all entirely marine in habit and these seem to be the only salmonoid-osmeroid fishes in which an entirely marine life is well-authenticated. Landlocked, freshwater populations of some of these species are known.

Flexibility in the life history is especially marked in *Hypomesus* and *Spirinchus*, in which genera there are entirely marine, anadromous, and entirely freshwater stocks.

6. ICEFISHES — FAMILY SALANGIDAE

The icefishes (or noodlefishes, as Roberts (1984) calls them) are bizarre, aberrant, probably neotenic fishes, of very elongate and slender form (Figure 4.6). They are of small size, reaching about 160 mm in length. This is a very poorly understood group of fishes that is found in the cool-temperate northwestern Pacific/eastern Asia (Siberia, Japan, Korea, China, Vietnam). Nelson (1976) listed salangids as anadromous and fresh water whereas Roberts (1984) thought them anadromous:

> While some species are primarily marine or at least brackish water inhabitants (e.g. *Protosalanx chinensis*) and many spend a part of their lives in the sea, others are restricted to fresh water or have populations which presumably repeat their life cycle without leaving fresh water.

Lindberg and Legeza (1969) believed that the fishes of this family live in salt water, but that most are well acclimatised to fresh water.

Salangichthys microdon, known in Japan, China and the Pacific coast of Siberia, is the best studied species, but even so, is not well-known. Fang

Figure 4.6: *Neosalanx jordani*, male above, female below (family Salangidae)
Source: Wakiya and Takahasi (1937)

(1934) described it as a typical form of the Salangidae, but to what extent this is true remains unknown. Senta (1973a) reported that the eggs were found on sand bars 'exposed at low tide ... 5.2 km upstream from the river mouth' [and that they developed] successfully in spite of the wide range of fluctuations of chlorinity (0.8 [ppt] at low tides to 16.8 [ppt] at high tides)'.

Okada (1960) thought that 'it keeps to coastal waters and enters rivers and estuaries for spawning at the beginning of May ... probably ascending the streams to spawn ... in summer'.

However, Senta (1973a) found eggs present on the spawning grounds in March (early spring) and (Senta, 1973b) reported on the fishery that operated for the species at that time; one may, perhaps, presume that this was during the upstream spawning migration from the sea. Berg (1962) described *S. microdon* as entering rivers and said that it spawns in the 'littoral zone', presumably in fresh water. The adults appear to die after spawning (Fang (1934) described it as an annual species) and the larvae are likely to be carried to sea soon after hatching, but little is reported. *S. ishikawae* seems also to be anadromous in Japan; Okada (1960) said that it likes to live in more salty water than the common species (*S. microdon*), and that it spawns in April and May. He quoted reports that the adults die after spawning. Okada (1960) reported *Salanx ariakensis* from both rivers and the sea, and Lindberg and Legeza (1969) from the Sea of Japan and rivers in northern China, so this species is perhaps anadromous too. Minimal data on the habitats and distributions of *Protosalanx hyalocranium*, *Neosalanx jordani*, *Hemisalanx prognathus*, *Salanx cuvieri*, and *S. acuticeps* all indicate the possibility that these species spend parts of their life cycles in both the sea and fresh water (Lindberg and Legeza, 1969), although anadromy remains undemonstrated. Wakiya and Takahasi (1937) indicated that virtually all the species occur at some time in fresh water, believing that they typically live in brackish water but that most of them become, in places, well acclimatised to fresh water. Spawning times were variously nominated as January–February (*Protosalanx hyalocranium*), February–March (*Neosalanx reganius*), March–May (*Salangichthys microdon, Neosalanx jordani*), April–May (*Salanx ishikawae, Neosalanx andersoni*), and October–November (*Salanx ariakensis*). Fang (1934) made the rather extraordinary comment that in *Protosalanx hyalocranium* 'the mouth is probably used as a buccal incubator'. This has never been confirmed or mentioned for other salangids.

There is some mention of wholly freshwater salangids (Roberts, 1984), as well as landlocked populations of anadromous species, but the data are sparse and the details very limited; even so, it seems that there is considerable flexibility in the life histories of salangids, as species, and for the family as a whole.

The even more aberrant family Sundasalangidae occurs only in the freshwaters of Borneo and is interpreted as a derivative of the Salangidae (Roberts, 1981).It is non-anadromous, and represents a rare invasion of tropical fresh waters by the freshwater and otherwise cool-temperate to sub-polar salmoniforms.

7. SOUTHERN SMELTS— FAMILY RETROPINNIDAE

The southern smelts (Australia and New Zealand, Figure 4.7) closely resemble the osmerids in both appearance and life history strategy. They are small fishes reaching about 150 mm in length. In the New Zealand common smelt, *Retropinna retropinna*, mature/ripe adults migrate from the sea into rivers during the spring and summer, to spawn in upper-tidal or supra-tidal waters (Figure 4.8), usually within a few kilometres of the sea, but sometimes as far as 30–50 km (McDowall, 1978a, 1979). In some populations, immature juveniles return to fresh water during the summer, but in most systems the migrants comprise primarily mature yearlings, a few immature yearlings, and a few two-year-olds. The latter are probably fish that failed to mature at one year. The migrants spend a period of some weeks to months in fresh water prior to spawning, during which period they continue to feed (Eldon and Greager, 1983). In this regard they differ from most other anadromous salmonoids. Post- spawning mortality is probably complete. The eggs are laid over sandy beaches and develop and hatch in fresh water; the tiny, newly hatched larvae (less than 10 mm long) are carried immediately to sea, where virtually the entire life (one or occasionally two years) is spent (McDowall, 1979). *Retropinna tasmanica* in Tasmania has a similar life history, but the mainland Australian species, *R. semoni*, is less explicitly anadromous. It is certainly highly euryhaline, but clearly defined migratory patterns have not been described for this species (McDowall,

Figure 4.7: Distribution of family Retropinnidae

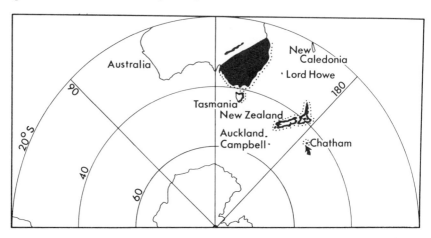

1979). The related Stokell's smelt, *Stokellia anisodon*, from southeastern coasts of New Zealand, has a cycle similar to that of *R. retropinna,* although its inland penetration is less extensive (a few kilometres only), and the spring migration into fresh water consists only of mature to ripe fish that will reproduce over the following summer. Mortality of post-spawning fish is complete (McDowall, 1978a).

At least one species, the New Zealand common smelt is frequently lacustrine, and becomes landlocked very easily. Non-diadromous populations occur in both brackish/coastal and freshwater inland lakes, some at considerable altitudes. Anadromy in *R. retropinna* is highly facultative, and on several occasions anadromous stocks have been transplanted to lakes and have freely and rapidly established landlocked populations (McDowall, 1979). Lake-limited, non-migrant and anadromous populations may be sympatric in some river/lake systems, but it is not known whether or not they interbreed. Another species of *Retropinna* (*R. semoni* — Australia) seems to be highly facultative, euryhaline, and is perhaps not specifically migratory. Landlocked populations are widespread. Stokell's smelt (New Zealand), which has essentially the same life history as the common smelt, has no landlocked populations, and the life history seems to be invariant (McDowall, 1978a, Eldon and Greager, 1983). The Tasmanian smelt, *R. tasmanica*, possibly has one lacustrine population.

8. TASMANIAN WHITEBAIT — FAMILY APLOCHITONIDAE

At the outset it should be recognised that this family is possibly diphyletic, and that the relationships between the two aplochitonid genera are far from certain (McDowall, 1969).

The Tasmanian whitebait, *Lovettia seali*, is an aberrant little fish (to about

Figure 4.8: Life History of common smelt, *Retropinna retropinna*

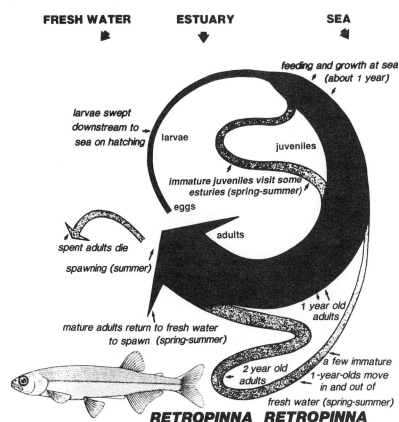

FRESH WATER   **ESTUARY**   **SEA**

feeding and growth at sea
(about 1 year)

larvae swept
downstream to
sea on hatching   larvae

juveniles

immature juveniles visit some
esturies (spring-summer)

eggs

adults

spent adults die

spawning (summer)

1 year old
adults

mature adults return to fresh water
to spawn (spring-summer)

2 year old
adults

a few immature
1-year-olds move
in and out of
fresh water (spring-summer)

RETROPINNA RETROPINNA

75 mm long), that is clearly a neotenic derivative, possibly of some other, un-
determined southern salmoniform. Mature adults of this species migrate from
the sea during the spring, at an age of one year (Figure 4.9); the fish cease
feeding at migration, and move upstream to spawn in low elevation rivers. The
adults all die after spawning. The newly hatched larvae go straight to sea,
where all the feeding and growth apparently takes place (Blackburn, 1950),
though nothing is known about the marine stage of the life cycle.

There are no known deviations in the life history of *L. seali* from that described above, and no landlocked populations.

9. SHADS, HERRINGS, ETC. — FAMILIES CLUPEIDAE AND PRISTIGASTERIDAE

The clupeids are mostly smallish (300–400 mm), silvery, shoaling fishes. The

Figure 4.9: Life history of Tasmanian whitebait, *lovettia seali*

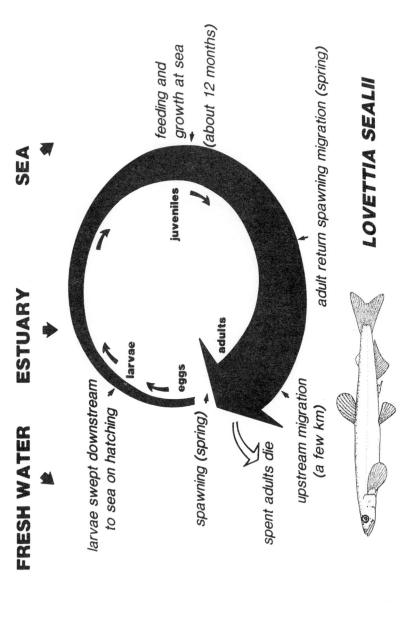

FRESH WATER   ESTUARY   SEA

feeding and
growth at sea
(about 12 months)

juveniles

larvae swept downstream
to sea on hatching

larvae

eggs

adults

spawning (spring)

spent adults die

upstream migration
(a few km)

adult return spawning migration (spring)

LOVETTIA SEALII

family is a large one with about 180 species in 50 genera, which have a very wide distribution in tropical and temperate seas and fresh waters. Anadromy in the family is a well-known phenomenon, particularly as regards species along the Atlantic coast of North America (shads and alewives, Bigelow and Schroeder, 1953) and also some species of *Hilsa* in tropical southeast Asia, but anadromous species are present in many areas. Life histories of tropical clupeids of diverse genera are poorly known (Whitehead, 1985) so that the extent of anadromy is difficult to assess and probably greater than is at present confirmed.

The temperate North American and European clupeids (various species of *Alosa* and *Dorosoma*) are frequently anadromous, migrating from the sea as mature, pre-spawning adults, some just a short distance, but other hundreds of kilometres. The alewife, *Alosa pseudoharengus*, of eastern North America (Labrador coast and Newfoundland, south to South Carolina) has extensively studied and well-reported habits. Adults move into both large rivers and small coastal streams to spawn during the spring and summer, beginning to migrate as early as March in the south, and about June further north; migration may persist until August (Odell, 1934; Kissel, 1974; Tyus, 1974). Migrations are known to be vast: 'They are frequently so numerous on the rapids for some miles above the head of the tide that it is difficult to wade about without treading on them.' (McKenzie, 1959.) The migrants cease feeding at the onset of migration and do not feed again during the period in fresh water (Durbin, Nixon and Oviatt, 1979). Inland penetration is usually modest, although Kissel (1974) wrote of them reaching lakes 40 km inland. M.J. Dadswell (personal communication) advised that they may penetrate 120 km inland in some North American rivers. Age at migration varies with latitude, Tyus (1974) saying mostly two to four years of age with a dominant age class of three in North Carolina, whereas Havey (1961) found most maiden spawners were five, much further north in Maine. The fish spawn in gently flowing reaches in streams and also in ponds and lakes, and the surviving adults return to the sea swimming 'quite rapidly downstream in great schools'. Life in fresh water is said to vary in duration from a few days to nearly three months (Kissel, 1974). Bigelow and Schroeder (1953) described the early spent adults migrating down stream, passing late spawners moving upstream on their way to the spawning grounds. Spawning mortality is very high, Tyus (1974) and Havey (1961) finding around 80 per cent or more for maiden spawners in North Carolina and Maine respectively, but Kissel (1974) around 50 per cent in Connecticut.

The eggs develop and hatch in fresh water. Initially they are adhesive, but this is lost and the eggs may drift downstream where flows are turbulent and swift. The young spend some months feeding and growing in fresh water before moving downstream and to sea over the summer and autumn (June to October in Connecticut waters (Kissel, 1974) and Rhode Island (Richkus, 1975)), or as late as mid-November (Bigelow and Schroeder, 1953). Richkus (1975) showed that size at migration varies between about 40 and 60 mm. Life at sea is little reported. Bigelow and Schroeder (1953) reported them to occur in large shoals:

> the majority remain in the general vicinity of the fresh water influence of the stream mouths and estuaries from which they emerge [although] catches of up to 3,000 to 4,000 pounds per haul were made by otter trawlers some 80 miles off shore.

Hildebrand (1963) considered that their home may be some distance offshore and completed his account of the natural history of this species in the sea with 'However, further study is needed'.

The American shad, *Alosa sapidissima*, occurs from the Labrador coast in northern Canada south to Florida, and has a life cycle comparable with that of the alewife. They move widely at sea, Leggett (1973) noting some fish that had travelled nearly 4,000 km at sea. The spawning shad move into rivers all along this coastline, and spawning takes place over a wide area, from areas just above the tide, to headwater tributaries (Bigelow and Schroeder,1953); inland penetration exceeding 600 km is known. For the period in fresh water, they cease feeding, resuming only when (if) they return to the sea. Spawning occurs in spring, with migration inshore beginning as early as November or December in the south, and becomes progressively later with increasing latitude (Talbot and Sykes, 1958). The fish may still be spawning in June in the colder, northern latitudes of the species' range in Canada (Rulifson, Huish and Thoeson, 1982b). The spent adults may survive spawning and return to sea soon after spawning and will spawn again in the following years. Hildebrand (1963) believed that they may spawn repeatedly, year after year, reaching a considerable size (600 mm and 6 kg); however in the southern part of the range (North Carolina (Sykes, 1956; Warburg; 1956, 1957)), almost all the shad die after spawning, although not in Canada (Vladykov, 1950) where they may live for up to 13 years and spawn up to seven times (M.J. Dadswell, personal communication). Leggett (1973) considered spawning mortality to be complete in rivers south of Chesapeake Bay, with an increase in survival north of that. It appears that the proportion that spawns more than once increases with latitude, but at the same time fecundity declines (Leggett and Carscadden, 1978). The eggs are demersal and non-adhesive, and are dispersed downstream by the river flows, as are the larvae, which hatch in fresh water but drift on downstream and eventually into the sea (Rulifson Huish and Thoeson, 1982a). The young remain to feed and grow in fresh water for some time, moving to sea during the autumn following hatching. They take several (three to six) years to attain sexual maturity (Leggett, 1973).

Other eastern North American species of *Alosa* are similarly anadromous, although less well known. The blueback herring, *Alosa aestivalis* (distributed from Nova Scotia to northern Florida), penetrates into fresh water far less than either the shad or the alewife but may move offshore at sea as far as 200 km

(Bigelow and Schroeder, 1953). They migrate into fresh water to spawn during the spring and summer (April – September), and do so upstream of brackish water influence, sometimes well upstream (sometimes more than 200 km, (Loesch and Lund; 1977, Loesch, 1987)). Spawning occurs in swiftly flowing waters, during spring. The swiftness of the water chosen tends to be less where this species is not sympatric with the alewife - competition evidently forces the blueback into the swifter flows (Loesch, 1987). Timing of spawning is increasingly later with rising latitude and cooling temperatures, beginning in January in Florida, but persisting until mid-July in northern areas (Rulifson *et al.*, 1982b). The eggs are initially demersal but lose their adhesiveness after several hours and then are swept downstream. The spent fish return to the sea shortly after spawning, where they feed and recover condition. Adult blueback may spawn two or three times. The young also move downstream to sea after a period of some months feeding and growing in fresh water, reaching the sea at the end of the summer (September–October).

The gulf shad, *A. alabamae*, is very closely related to the American shad (rivers of the coast of the Gulf of Mexico, from the Suwannee in Georgia, west to the Mississippi). It once migrated great distances up the rivers of the southern United States (it is now restricted by dams, weirs, etc.), and is similarly anadromous. The skipjack herring, *A.chrysochloris*, is less certainly anadromous in rivers of the Gulf of Mexico (Rulifson et al., 1982a,b) whereas the hickory shad, *A. mediocris* (Bay of Fundy to Florida), although little known, is regarded as anadromous by Lee, Gilbert, Hocutt, Jenkins, McAllister and Stauffer (1980). Rulifson *et al.* (1982b) said that it has adhesive, semi-demersal eggs that become bouyant in turbulent waters, and thus are presumably carried downstream. Larvae are found in fresh and brackish river waters, as are the juveniles, perhaps spreading into estuaries and inshore waters. The oceanic distribution of adults is not understood, but they return to fresh water at age two or older, to spawn during the spring and summer.

Alosa is found also in the northeastern Atlantic. The allis shad, *Alosa_alosa* (central-northern Norway – North Sea – United Kingdom, and south to the Bay of Biscay, plus the Mediterranean), enters rivers of these areas during the spring (May), and spawns at night in swiftly flowing fresh waters, sometimes migrating far upstream to do so (Wheeler, 1969, Whitehead, 1985). The adults return to the sea, while the young may spend from 12 to 24 months in fresh water before they go to sea. The twaite shad, *A. fallax* (southern Iceland, the Faroes, Norway and the Baltic, south through the North Sea to the Bay of Biscay, and also the Mediterranean), is believed to migrate into rivers, but spawns in May-June just above the tidal reaches (Berg, 1962; Wheeler, 1969). The fry move downstream to the sea during early growth.

The habits of other temperate Eurasian clupeids seem sparsely described, but Berg (1962) and Svetovidov (1963) mention some that are possibly anadromous including several Caspian/Black Sea species. Whitehead (1985) concurred. The blackspined shad, *Alosa kessleri*, winters in the sea and moves into rivers during the spring. It is described as having long upstream spawning migrations during which feeding ceases. At migration from the sea the degree of maturity of fish varies. Some are very fat and not fully mature sexually, and their gonads develop at the expense of the fat deposits as the fish migrate long distances upstream. Others enter the rivers already ripe and spawn soon afterwards close to the sea, some in the river deltas. The larger fish migrate upstream further than the smaller ones (Nikolskii, 1961). Spawning occurs in the summer, and spawning mortalities are high, though some post-spawners survive and may spawn up to three times altogether. The bentho-pelagic (semi- bouyant) eggs evidently develop as they drift downstream again, hatching in a day or two, and the larvae/fry feed and grow in fresh water as they move further downstream and into the sea. The Pontic shad, *A. pontica*, also inhabits the Black and Azov Seas and their tributary rivers; it behaves in a manner similar to *A. caspia*; there is a spring migration from the sea to spawn up to 500 km upstream from the sea. The fry return quickly to the estuaries where they remain until winter (Whitehead, 1985). The Caspian shad, *A. caspia*, seems less explicitly anadromous, as it evidently spawns in the sea in some stocks while others spawn in fresh or brackish water. However, stocks in both the Black and Azov Seas enter rivers to spawn. Whitehead (1985) regarded it as euryhaline and migratory, chiefly in brackish water, but entering fresh water to spawn. There are not massive post-spawning mortalities like those reported for many herrings.

Berg (1962) also mentioned the tyulka herring, *Clupeonella cultiventris* (as *C. delicatula*), which, he says, enters rivers and lakes, and spawns widely from estuarine to fresh waters in the Volga during the spring. It is perhaps marginally anadromous, but it is more probable that it is facultatively euryhaline and wanders in and out of river estuaries. Whitehead (1985) classified it as semi-anadromous, euryhaline, and essentially a brackish water species; purely freshwater forms occur in rivers and lakes

The gizzard shads (genus *Dorosoma* in North America) may also be anadromous. Miller (1960) described the life history of the North American gizzard shad, *D. cepedianum* (eastern North America from New York south to the Rio Panuco in Mexico). This species evidently prefers brackish water around river mouths to the open sea, and migrates into rivers to spawn during the spring. The young live in fresh water for several years before venturing towards the sea. They reach maturity in two or three years, but may live for 10 years, so they are clearly repeat spawners. The related threadfin shad, *D. petenense* (Florida to Guatemala and Belize in Central America), is even less anadromous, living mostly in estuaries and migrating upstream to spawn in both spring and autumn, as for the gizzard shad.

A Western Australian gizzard shad or Perth herring, *Nematalosa vlaminghi* (western Australia between Broome and Perth, Allen (1982)), has been labelled anadromous by Miller (1960), but Chubb and Potter (1984) believed it to be 'semi-anadromous', and Whitehead (1985) found no evidence of its occurrence in fully fresh waters.   Shoals of this species enter the estuaries of Western Australian rivers during the spring, penetrating into the upper estuaries to spawn, during the summer (December–January).   Chubb and Potter (1984) found that the young gradually disperse downstream through the late summer and on into the early autumn, and presumably they, along with any surviving spent adults, move out into the sea again.   However, Chubb and Potter (1984) did not know whether the spent adults recovered to spawn again in later years. *N. vlaminghi* is scarcely anadromous   Other species of *Nematalosa* may also be more or less anadromous, although Whitehead (1985) did not explicitly classify any species in this way.   Al-Hassan (1986) referred to the fact that *N. nasus* ascends the rivers of the Arabian Gulf during the spring.   However, he did not indicate whether or not this was a spawning migration.   Whitehead (1985) also mentioned entry into fresh water and noted an instance of breeding there.   However, details are, again, meagre.   The Galathea shad, *N. galatheae* (India to Malaysia), is primarily marine in habit, but there is limited evidence that it penetrates into and breeds in fresh water.

Tropical clupeids of the genus *Tenualosa* are anadromous in some areas. Best known of these is the Indian hilsa, *T. ilisha* (Red Sea east through India to Burma), which lives in the sea for most of its life, but migrates at least 1,200 km up some river systems in India and 700 km in Burma (Pillay and Rosa, 1963).   Distances of 50–100 km are more typical (Whitehead, 1985).   Migration takes place in India during the monsoon rains, when the rivers are full, but also in some areas when the rivers are swollen from snow thaw.   The fish cease feeding for the duration of the spawning migration.   The eggs are deposited in fresh waters, and the larvae and juveniles make their way downstream to the sea during a period of several weeks or months, feeding and growing on the way.   They go to sea and spend most of their lives there, mostly in coastal seas, less than 25 km from the shore.   Spawning occurs at age one or two, but they may live to an age of six or more years.   Adults that survive spawning return to the sea.   Some of the far upstream populations may be non-anadromous (Whitehead, 1985) while some of the main stocks of this species are purely estuarine in distribution and habit (Pillay, 1967).

The longtail shad, *Tenualosa macrura* (Indonesia, Malaysia), has a similar life history, as does the Reeves shad, *T. reevesi* (China and Thailand).   The toli shad, *T. toli* (India to Indonesia), may also be anadromous, though Whitehead (1985) seemed less certain;   he considered it 'perhaps anadromous', but various other workers do not agree.   The genus *Herkotslichthys* contains several Indo-Pacific herrings, most of which are coastal/marine pelagic fishes, but Goto's herring, *H. gotoi*, which inhabits waters draining into the Timor Sea north of Australia, may enter rivers there;   Koningberger's herring, *H. koningsbergeri*, may do the same.   Both may eventually prove to be anadromous.

West African clupeids of the genus *Pellonula* are possibly also anadromous. They occur in the rivers of that area, and perhaps coastal seas, possibly entering rivers to spawn, e.g. *P. leonensis* and *P. vorax*, but again, details of these life cycles are sparsely understood. FAO (1984) also listed *Clupanodon_thrissa* (northwestern Pacific from China to Thailand) and *Anodontostoma chacunda* Red Sea east to the Philippines, northern Australia and New Caledonia) as diadromous clupeid species, but as with so many others life history details are sparse. Whitehead (1985) provided little support for diadromy.

There is some evidence also for anadromy in herrings of the family Pristigasteridae (they were until recently included in the Clupeidae). They occur in tropical and subtropical seas, mostly, although not exclusively in the Indo-West Pacific. Most species of *Ilisha* are marine but a few are fresh water. *I. novalcula* penetrates 650 km up the Irrawaddy River, in Burma, but may be entirely riverine. Several marine species enter estuaries or fresh water (*I. africana*), but *I. megaloptera* (India to Indonesia) ascends and spawns in rivers (Whitehead, 1985). *Pellona* includes both marine and freshwater species, but some, like *P. ditcheli*, the Indian pellona (Africa to the Philippines, Australia and Papua New Guinea), penetrate rivers. FAO (1984) listed it as diadromous, presumably anadromous.

Deviations from the basic life histories of these clupeid species include the following. Several species of *Alosa* become landlocked; they have done so naturally in North America, Europe, and the inland Caspian/Black/Azov Seas, and also have done so following transplantation by man in North America. Eiras (1983) described a landlocked population of *A. alosa* in Portugal that developed in a river system when a dam was constructed, forming a lake and preventing movement to and from the sea. Some *Alosa* species, especially those in the inland seas of Russia, seem to be wholly marine in habit, although these seas are, of course, of reduced salinities. No species of *Alosa* is described as being restricted to fresh water. The genus *Dorosoma* has landlocked, wholly freshwater populations but none that is wholly marine. In contrast *Nematalosa* is marginally anadromous in one perhaps two species (*N. vlaminghi, N. nasus*), but otherwise may be either entirely marine or entirely fresh water (Nelson and Rothman, 1973; Whitehead, 1985)). The same is true of *Tenualosa* which is widely marine with some anadromous species having wholly freshwater populations.

10. ANCHOVIES — FAMILY ENGRAULIDAE

Anchovies are small, silvery, shoaling fishes of herring-like appearance and affinities. They are mostly marine fishes, a few are euryhaline/estuarine, some are fresh water, and one, at least, seems to be anadromous. The family is worldwide/tropical in distribution with about 110 species in 20 genera.

The Brazilian anchovy or manjuba, *Anchoviella lepidentostole*, while little-studied, is known to migrate into rivers along the coast of Brazil during the spring and autumn, to spawn (Giamas, Santos and Vermulem, 1983). The mature adults occur in the rivers for two or three months, and then disappear

(Nomura, 1962; N. Menezes, personal communication). Little more seems to be known, but the life cycle appears to be an anadromous one. Bonetto (1986) described *Lycengraulis olidus* as anadromous in the Parana River system of Brazil, and Di Persia and Nieff (1986) said this of both *L. olidus* and *L. simulator* in the Uruguay River, but further details of the migrations of these species are elusive.

11. FORKTAILED CATFISHES — FAMILY ARIIDAE

The ariid catfishes comprise a very widespread subtropical and tropical family and are amongst the very few ostariophysan fishes that are frequently present in marine waters. According to Nelson (1976) there are numerous species in numerous genera.

Ariid catfishes are well known for the diverse salinities which they occupy. However, as in many groups, the life histories are not well understood or described. Rimmer (1985) said that anadromous movements associated with breeding have been reported for several estuarine and marine species, and Rimmer and Merrick (1983) listed *Arius felis* (from North America), *A. heudeloti* (from west Africa) and *Osteogenieosus militaris* (from India and southeast Asia). However, the classification of these species as anadromous seems to be dubious. Lee (1937) discussed reproduction of *A. felis* and found that it spawns in bays with brackish water. And Pantulu (1963) said that *O. militaris* migrates into estuaries to spawn. Neither species seems to enter fresh water to spawn and neither can therefore be described as anadromous. A Yanez-Arancibia (personal communication) investigated ariid catfish life histories in Terminos Lagoon, on the Mexican Gulf coast of Mexico. Again, in these species, although there are movements from the sea into estuarine/brackish lagoons, it seems that life cycles are, at best, only marginally Anadromous. *Arius melanopsis* was found to occur around the less saline margins of the lagoon and never occurred in the sea; the juveniles were more frequent in the fresher water. *A. felis* was found to spawn near the lagoon outlet, and the young move into the estuary to feed and grow before returning to the sea. *Bagre marinus* was wholly marine during the adult stage but the juveniles entered the brackish estuarine waters, to feed. *Arius latiscutatus* was found 275 km up the Gambia River in West Africa, but was classified as a species of marine origin by Lesack (1986). Rimmer (1985) examined the life history of the Australian-Papua New Guinean species *Arius graeffei*, and summarised evidence that indicates a seasonal change in abundance in fresh water. Numbers were low during the late autumn through early spring, and Rimmer believed that higher abundance during the summer represents anadromous breeding migrations. The implication is that adult *A. graeffei* leave the sea and move to fresh water to spawn over the spring and summer, although such migrations have not been observed.

Other ariid catfishes may prove to have similarly anadromous life histories, though at present little is known. *Arius madagascariensis* from Madagascar may be an example (Moreau, 1986); Moreau thought a period at sea necessary for breeding, but that the adults enter fresh water to breed during the spring; it is a mouth-brooder, like other species of *Arius*.

12. STICKLEBACKS — FAMILY GASTEROSTEIDAE

The sticklebacks are tiny fishes, sometimes reaching over 100 mm but often only half that size. They are distributed widely on all continents in the cool temperate Northern Hemisphere. There are only about eight species of which two species are described as anadromous — the threespined stickleback, *Gasterosteus aculeatus* and the ninespined stickleback, *Pungitius pungitius*. The threespined stickleback is found very widely, from Baja California in northern Mexico, north to Bristol Bay in Alaska, also on the Arctic coast of Alaska, and in eastern Asia/western Pacific from Siberia south to Korea; it also occurs from Hudson Bay and Baffin Island in the Arctic Atlantic, throughout Atlantic North America south to Chesapeake Bay, and in most of western Europe south to Spain, the Mediterranean, and northern Africa. Its latitudinal range is from about $35°$ to $70°N$, in all of these areas (Wootton, 1976).

The life cycle (and morphology) of the threespined stickleback are enormously variable, so variable that they have caused considerable variation with the taxonomy of the species. It occurs in three intergrading forms, referred to as:

1. '*trachurus*' - which is often anadromous, although not always (it may be fully marine), and which has strong development of lateral skutes;
2. '*leiurus*' - which is not anadromous and which has weakly developed lateral skutes; and
3. '*semi-armatus*' which is also not anadromous, and which has moderate development of lateral skutes (Wootton, 1976).

The anadromous '*trachurus*' form is confined to northern latitudes within the range of the species whereas the other forms, although predominantly more southern, also occur much more widely, including the far north. '*Leirus*' occurs, for instance, in Greenland. In California '*trachurus*' reaches south as far as $37°N$ but '*leiurus*' reaches $32°N$.

Threespined sticklebacks that are migratory seem to be quite clearly anadromous, although migration to sea is not crucial to sexual development (Baggerman, 1957). Adults that overwinter in the sea migrate into fresh waters during the spring; the migration may be as early as February in the Netherlands (Mullem and Vlugt, 1964), April–May in England (Craig-Bennett, 1931), or as late as May–June in British Columbia, Canada (Hagen, 1967). At this stage most of the population are one year old, although some may live for up to three years

(Mullem and Vlugt, 1964). Spawning in migratory populations usually takes place in freshwater streams, seldom far upstream, especially if there are swift flows or difficult barriers hindering migration. After hatching, the young feed and grow in fresh water for some time over the summer and, along with any adults that might have survived spawning (spawning mortality is very high), they return to the sea in the late summer and autumn (August–September).

Initially the young stay close inshore, but they may move offshore in substantial numbers and great distances (up to 800 km) and Morrow (1980) questioned whether these fish that go so far from shore ever find their way back. Regan (1911) claimed that 'towards the south [it] is more generally a freshwater fish and there is reason to doubt that it ever goes to sea at all in the Mediterranean.' Ekman (1935) also regarded the species as more strictly fluviatile with decreasing latitude. The three-spined stickleback is known to have wholly freshwater stocks, and these are not always in lakes (cf. Baker, 1978). In spite of evident wide variability in the life history strategies of the various forms of threespined stickleback, and even though non-migrant, freshwater populations of the anadromous '*trachurus*' form are well-known, Wootton (1976) showed that:

> *trachurus* stocks that normally migrate to sea in autumn show very poor survival if kept in water of low ionic conditions during autumn and early winter, but in spring they tolerate these conditions perfectly well. In late summer and autumn, *trachurus* fish undergo a physiological change which makes them unable to survive long periods in fresh water while a change in spring takes place that enables the fish to survive and reproduce in water of low salinity. In autumn ... the fish loses its ability to osmo-regulate in fresh water ... In spring [hormonal changes] induce a preference for fresh water.

The interesting aspect of this is that in what seems to be a highly facultative and euryhaline diadromous species, individual fish demonstrate much less euryhalinity than the species seems to show — they are what Fontaine (1975) described as amphihaline rather than euryhaline fishes at the individual level.

Several studies have shown that in some populations there is reproduction of the threespined stickleback in fully marine waters (Black and Wootton, 1982, Vancouver Island; Weeks, 1985, Isle of Shoals, Massachusetts), and such populations are non-anadromous, just as are those that are wholly fresh water in habit. These saltwater spawning populations include the morph that has reduced numbers of lateral skutes, which is characteristic of purely freshwater populations (Bell, 1979, a population in a saline lagoon, near Santa Cruz in California). Kedney, Bopule and Fitzgerald, (1987) described a marine population in the Gulf of St Lawrence in which the adults migrated away from their marine feeding grounds, but in which some of the fish entered freshwater streams to spawn while others migrated into marine-salinity tide pools to spawn. They found no evidence that the different migratory stocks came from different gene pools.

The ninespined stickleback, *Pungitius pungitius*, is very widely distributed throughout the northern cool temperate of both Eurasia and North America. It spreads south into temperate waters along the Pacific coast of Asia, into the Baltic Sea in the north-eastern Atlantic, and along the Atlantic coast of North America. The life history does not seem as well elucidated as that of the threespined stickleback, *Gasterosteus aculeatus*, but it is known that there are both marine and freshwater forms. Scott and Crossman (1973) reported that it spawns in fresh water during the summer, which implies that there is an anadromous movement from the sea into streams prior to that. However, these authors also say that the ninespined stickleback is less tolerant of high salinities than the threespined stickleback. Neither Scott and Crossman (1973) nor Lindbergh and Legeza (1969) described this species as anadromous, but it may be marginally so.

Other species in this family may be entirely fresh water, or marine in habit, and some are non-diadromous but highly euryhaline (Wheeler, 1969; McPhail and Lindsey, 1970; Scott and Crossman, 1973).

12. PIPEFISHES AND SEAHORSES — FAMILY SYNGNATHIDAE

The pipefishes and seahorses are smallish fishes that are widely distributed in all oceans, especially the tropics and warm temperate, and are most often found in shallow coastal waters. They are primarily marine, but invade estuaries and occasionally fresh water. There are about 175 species in 36 genera.

One species of pipefish, *Microphis brachyurus*, appears to be *possibly* anadromous. It occurs widely in eastern North America and the Caribbean south to Brazil, and equally widely in the Indo-Pacific, from Africa to India, north to Japan, and east to Papua New Guinea and the islands of the Pacific. This species apparently demonstrates a late winter movement of maturing young adults into fresh water in preparation for a summer spawning. Gilbert and Kelso (1971) found the brooding adults in fresh or low salinity waters. According to Hildebrand (1939), males bearing eggs are found in Gatun Lake, in Panama. Gilmore (1977) reported that spawning occurs in fresh water, and that when the young are released from the male's brood pouch, they are swept down stream and apparently undergo early growth in marine salinities. On the basis of this migratory pattern, *M. brachyurus* might be described as anadromous, although C.E. Dawson (personal communication) expressed doubts and considered that there are no instances of diadromy in the pipefishes.

13. NEEDLEFISHES — FAMILY BELONIDAE

The needlefishes are distinctive, elongated fishes that are widely distributed in tropical and warm temperate sea and fresh waters. In the Australian region, the long-tom, *Strongylura kreffti,* was discussed by Merrick and Schmida (1984) as follows:

> This species has frequently been considered a marine visitor to fluviatile environments; however the occurrence of small juveniles of this distinctive animal long distances inland, upstream of effective impoundment barriers, constitutes indirect evidence that *S. kreffti* breeds in some fresh waters ... Nothing is known of the reproductive biology of *S. kreffti* ... and no long migrations associated with breeding are known.

Allen (1982) and Munro (1967) both describe the long-tom as fresh water, and Roberts (1978) found it 717 km up the Fly River. It could be anadromous, but there is insufficient information available yet to be any more certain.

14. CODS — FAMILY GADIDAE

The cods are mostly rather stocky to elongated and tubular fishes that are mostly bottom-living species and almost entirely marine in habit. The family comprises about 55 species in 21 genera. Only one species is fresh water, while another is anadromous.

The tomcod, *Microgadus tomcod,* is described as anadromous along the eastern seaboard of North America, from Newfoundland south to Virginia. Peterson, Johansen and Metcalfe (1980) outlined the life cycle of the tomcod. The adults are marine, being largely coastal/inshore in habit. The mature adults migrate into fresh waters to spawn, during the winter, often under ice (December–January in New Brunswick, Canada). Inland penetration by the spawning adults is regarded as not distant, in some instances being just beyond or even at the fringes of tidal influence. The eggs are largely demersal, although some are carried downstream by stream flows. The fate of the latter, should they be carried to sea, is doubtful — Peterson *et al.* (1980) thought that they would not develop normally in sea water. However, the larvae require salt water for normal growth and development. The eggs hatch in fresh water during periods of high river flows occasioned by the thaw in the spring. The larvae are positively phototropic and are thought to be carried rapidly downstream to the sea. Tomcod are repeat spawners, the peak downstream migrations of the spent adults taking place in late January (McKenzie, 1959).

Wholly freshwater/landlocked populations of tomcod are known, in spite of the fact that the larvae are believed to require salt water for survival.

A related species, *M. proximus,* is entirely marine along the Pacific coast of North America. Apart from the tomcod, the only gadid that lives in fresh water is the burbot, *Lota lota.* Although this species was listed as 'diadromous' by Morin *et al.* (1980) Jager *et al.* (1981) described it as a well established freshwater species which avoids brackish water completely, even though they also found that its eggs have a 'high tolerance limit to brackish water'.

Figure 4.10: Life history of striped bass, *Morone saxatilis* (Haedrich)

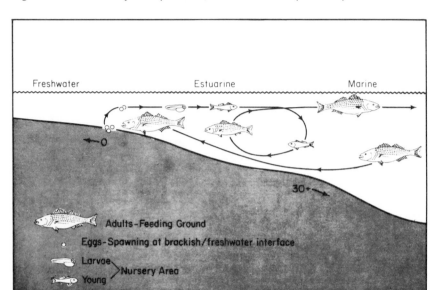

15. TEMPERATE BASSES — FAMILY PERCICHTHYIDAE

The percichthyids are large, predatory, heavy-scaled fishes — they are typical, if primitive, perch-like fishes. The family is widely distributed with representatives in many temperate lands as far apart as eastern North America, Chile and Australia. The monophyly of the family is open to question, and it may not constitute a natural unit; as at present recognised the family consists of about 40 species in 17 genera.

The American striped bass, *Morone saxatilis*, occurs naturally on the east coast of North America from the Gulf of St Lawrence in Canada south to northern Florida (St Johns River), and also along the coast of the Gulf of Mexico from western Florida to Louisiana. The males of the species mature early, at one to two years, and the females later, at three to five years. The adults live in the sea (Figure 4.10), usually inshore not many kilometres from the coast, and often in large estuaries and bays. They leave the sea and migrate into rivers during late winter (February in the south in Florida) to early summer (June– July further north in the St Lawrence). The adults survive spawning and may repeat spawning many times. The eggs are shed freely into the water in locations where the current is swift enough to keep the eggs suspended. Talbot (1966) and Rulifson *et al.* (1982a) indicated that the eggs of striped bass must be kept suspended if they are to hatch. They are semi-bouyant, so that where river currents are sufficiently swift, the eggs may be kept in suspension and carried downstream. They take only two or three days to hatch (Englert,

Lawler, Aydin and Vachtsevanos, 1976). In some instances they may not hatch before reaching brackish-water estuaries (Talbot, 1966). (The downstream transport of bouyant or semi-bouyant eggs is a rare phenomenon in diadromous fishes.) The larvae may initially be fluviatile but float near the surface for a day or two, so often are carried quite rapidly out to sea. They are found in shoals at sea, mostly close inshore through the first summer, but may disperse more widely as they grow (Setzler, Boynton, Wood, Zion, Lubbers, Mountford, Frere, Tucker and Mihurskey, 1980). The adults return to sea after spawning, but excessive sea temperatures in summer may result in striped bass not going to sea at all over the summer (Rulifson, et al., 1982a); Dudley, Mullis and Terrell, (1977) found that the incidence of anadromy decreases in southern (warmer) latitudes. The species is non-anadromous in rivers entering the Gulf of Mexico (Rulifson et al., 1982a). Numerous landlocked populations of the striped bass are known, many of them having been introduced by man (Setzler et al., 1980), although Ommanney (1964) reported that there were no land-locked populations until some were trapped in rivers and a formative lake, following the construction in 1941 of the Santee-Cooper Dam on the Suwannee River in the southeastern United States.

The related white perch, M. americanus, has a similar, if less definite, anadromous migration. It was described by Mansueti (1964) and Lee et al. (1980) as a predominantly estuarine/brackish water species, but Scott and Crossman (1973) wrote of anadromous populations which spawn in fresh water during the late spring and early summer. They reported sea-run populations, especially in Chesapeake Bay in the eastern United States, but thought them less common further north in southern Canada, and virtually absent in the Bay of Fundy on account of the cold temperatures. Mansueti (1964) described it as 'semi-anadromous', and he observed that there are spawning runs into lakes and ponds during the spring (April–May). However, it will spawn in tidal, fresh, and slightly brackish water and landlocked populations are known. Beyond this, little has been reported on anadromy in M. americanus. Other species of Morone are entirely fresh water in habit in the United States but marine in Europe.

European/Mediterranean sea basses, Dicentrarchus labrax and D. punctatus (in the closely related family Serranidae), are essentially marine/coastal species that some studies show to be at least estuarine/euryhaline, but it does not seem that that they are in any sense diadromous (Chervinski, 1975; Dando and Demir, 1985).

16. GOBIES — FAMILY GOBIIDAE.

Although mostly regarded as amphidromous (see Chapter 5), some goby species seem to be anadromous, although very little is known. Welcomme (1979) described Batanga lebretonis (from western tropical Africa) as

'diadromous', and reported it as entering rivers to spawn on the flooded vegetated fringes of the lower reaches during the flood season (Welcomme, 1985).
The young presumably return downstream at some stage after they hatch. As
thus described, *B. lebretonis* is anadromous, but details are sparse. He said the
same of *Dormitator latifrons*, but others have described it as catadromous.
The Mexican fat sleeper, *Dormitator maculatus,* is littleknown although Nordlie (1981) pointed to reports that it is anadromous (McLane, 1955). Darnell
(1962) was of that view also. Details of its life history are few, although it is
known to penetrate inland in fresh waters to an altitude of 80 m in Puerto Rico
(Erdman, 1984). *D. latifrons*, however, is elsewhere described as amphidromous.

The Japanese ice goby, *Leucopsarion petersi,* is a tiny species (45–50 mm)
that is widely distributed in central and southern Japan, and in Korea. Mature
adults migrate from the sea, beginning in early spring (February), and
continuing until April. Movement takes place at night on an ebbing tide.
Final maturation of the fish takes place over several weeks, after migration and
in fresh water, during which time the adult fish do not feed. The eggs are laid
in the typical, primitive goby nest, and the eggs take 10–18 days to develop.
On hatching the larvae are positively phototactic, and so move towards the surface. River flows carry them to sea where all feeding and growth occurs. The
ice goby probably spends most of its life close inshore and not far from river
mouths (Tamura and Honma, 1969; Matsui, 1986).

Darnell (1962) also classified *Eleotris pisonis*, occurring in the rivers of
Central America's Caribbean coast, as anadromous, but did not elaborate on
this. Nordlie (1981) regarded it as more probably amphidromous.

5

Detailed Analysis of Diadromy: Catadromy

1. FRESHWATER EELS — FAMILY ANGUILLIDAE

The freshwater eels are well known as very elongated, sinuous and flexible-bodied fishes. There are about 15 species in the single genus *Anguilla,* spread widely in the western north Pacific and the entire Indo-West Pacific, from Africa east to Tahiti (Figure 5.1). The Indo-Pacific distribution extends north into cool temperate waters in Japan and similarly into southern cool temperate waters in Tasmania and New Zealand; two rather more isolated species occur in the north Atlantic. (Because Atlantic eels are so much more intensively studied than those of the Indo- Pacific, the North Atlantic is often regarded as the home of anguillid eels; there are far more Indo-Pacific species.)

The anguillid life history involves a migration of mature adults from freshwater habitats to usually far distant marine spawning grounds. Spawning takes place at moderate depths in the ocean (500–600 m), usually over very deep oceanic waters far from land. Both species of Atlantic eel are known to spawn in the Caribbean, whereas Japanese eels are thought to spawn in the northwestern Pacific. Those from New Zealand, Australia and the islands of the Pacific spawn in the western-central Pacific, and those from eastern Africa in the western Indian Ocean, east of northern Madagascar (Tesch, 1977). Where the other, numerous, Indo-Pacific tropical eels spawn has not been resolved. It seems certain that the adult eels all die after spawning. Little is known and most views are conjectural. The eggs hatch as they rise through the water column towards the surface, and the distinctive leptocephalus larvae develop. This stage grows and develops in the sea as it makes its way back towards land over a period of several years. It has long been believed that the leptocephali are carried helplessly in ocean currents; however, in recent years analyses of likely breeding locations for various eel species and knowledge of the ocean current patterns in these areas has led some workers to hypothesise that the movement back to the freshwater rivers is at least in part an active migration by the leptocephali (Jellyman, 1987; McCleave, Kleckner and Castonguay, 1987). As they approach land, the leptocphalus larvae metamorphose

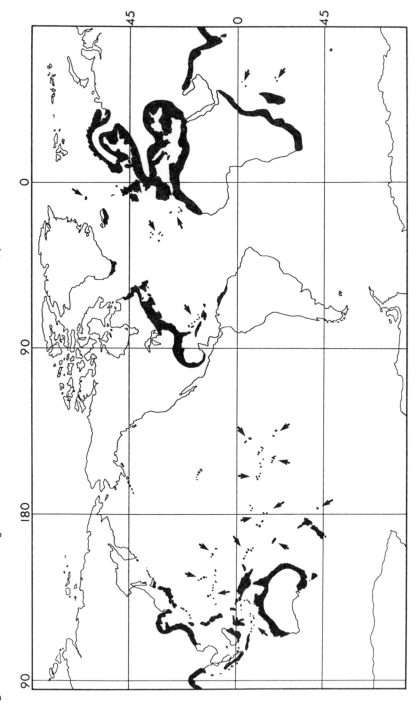

Figure 5.1: World distribution of anguillid eels — note concentration in the Indo-West Pacific, plus the North Atlantic

producing the transparent, slender glass eel, which migrates back into fresh water. Sinha and Jones (1975), reporting the studies of Fontaine and Callemand (1941) and Callemand (1943), said that at the completion of metamorphosis the elvers are dehydrated and spend a period of time in brackish water to rehydrate and thus adjust to life in fresh water — it appears that the elvers prefer to stay in diluted sea water for some time before moving into truly fresh water. Glass eels may penetrate inland enormous distances, overcoming difficult obstacles, and the eels grow to maturity over a long period in fresh water (up to at least 60 years, (Todd, 1980)). When mature, the eels undergo another morphological change, both form and colour changing distinctly; they cease feeding, make their way downstream to the sea, and head off to their distant oceanic spawning grounds.

The basic structure of the life history of anguillid eels is invariant, although it might be argued that the eels of some other anguilliform families (e.g. Congridae) could represent wholy marine variants of the diadromous life history strategy of *Anguilla*. There are no wholly freshwater anguillids (or even anguilliforms).

The moray eels (family Muraenidae) are essentially marine eels (100 species in 12 genera), but Jehangeer (1986) described the occurrence of the Mauritius species *Muraena manubeena* in the rivers of that small Indian Ocean island. Little seems to be known of its life history, or the extent to which the freshwater phase is obligatory. If *M. manubeena* is diadromous at all, it is likely to be a marine spawner, and therefore catadromous.

2. TARPONS — FAMILY MEGALOPIDAE

The tarpons are large, silvery, superficially herring-like fishes; they are streamlined fast-swimming fishes, mostly of the open oceans of the tropics and subtropics in both the Indo-Pacific and Atlantic Oceans. There are just two species in two genera.

The life cycle of the oxeye herring, *Megalops cyprinoides*, is not well understood. Roberts (1978) recorded the species over 900 km up the Fly River in Papua New Guinea, and Coates (n.d.) reported it in the Sepik River, also in Papua New Guinea. Roberts found that substantial populations of sub-adults were always present in the upper reaches of the Fly River, while there were regular collections of juveniles in the lower and middle reaches. Individuals collected in fresh water in the Sepik were all either juveniles or sexually inactive adults (Coates, n.d.). Roberts recorded cyclic appearances of larvae and small juveniles in coastal shallows, estuaries and the lower reaches of rivers and thought that the oxeye herring moved downstream to estuarine and shallow seas for spawning in the summer months. After spawning, the adults did not return to fresh water, but remained at sea, feeding there, and recovering condition. Coates (n.d.) concluded that these fish are either temporary visitors into fresh water or they spend their early life there and return to the estuary and sea to mature. Subpopulations of oxeye herring occur in the fresh waters of southern Africa in a similar way. Bruton and Kok (1980) discussed *M. cyprinoides* in African lakes that had been cut off from sea access for two years, which demonstrates this species' ability to tolerate long periods in fresh water, but there are no suggestions of reproduction or landlocking there. These observations appear to make *Megalops cyprinoides* 'sometimes catadromous', at least in Papua New Guinea and South Africa (and perhaps elsewhere), but too little is known to be more than tentative about this. Whether the life cycle is really catadromous, rather than a facultative entry to fresh water by some members of the population, remains uncertain. If there is a return migration of juveniles to the sea, as seems to be suggested, then this species is marine amphidromous rather than catadromous.

3. SOUTHERN WHITEBAITS AND GALAXIIDS — FAMILY GALAXIIDAE

The galaxiids are a smallish family of cool-temperate southern salmoniforms, with less than 40 species widely spread in all the main southern lands (southern Australia, New Zealand, Patagonian South America, and southern South Africa), as well as some of the small islands associated with these lands (McDowall, 1970, 1971b; McDowall and Frankenberg, 1981). They are smallish, scaleless fishes commonly reaching about 250 mm, occasionally over 500 mm.

This family has widely been cited as being catadromous, but this has been on the basis of limited understanding of one atypical species, the inanga, *Galaxias maculatus* (Figure 5.2). This species migrates downstream from low elevation, swampy or riverine freshwater habitats, and it usually spawns in tidal estuaries. Some spawning in fully freshwater habitats is recorded, but it does not spawn 'in the surf' as Ommanney (1964) suggested. Spawning salinities may vary from 'fresh' (<1 ppt) to 'sea' (>30 ppt), and the eggs are able to develop in this wide range of salinities (McDowall, 1968b, 1978a). Most, if not all the adults die after spawning. A distinctive feature of the spawning of this fish is that it takes place among terrestrial vegetation on low elevation tidal estuarine margins when this vegetation is inundated by water at the high spring tides. There is a definite lunar periodicity in the migration and spawning behaviour of the adult fish (Figure 5.3). After the tide has ebbed, the eggs are left out of water and develop in the humid atmosphere of the vegetation amongst which they are deposited (Figure 5.4). They are normally not again covered with water until the subsequent set of spring tides, some 14 days (or a multiple of 14 days) later. If temperatures are low development is extended, and the eggs are able to survive for at least six weeks without re-immersion. The eggs hatch when they are again covered with water, and the larvae (less than 10 mm long) go immediately to sea in the river/tidal flows. The larvae spend about six months at sea, returning to fresh water as 50 mm-long

Figure 5.2: Life history of the inanga, *Galaxias maculatus*

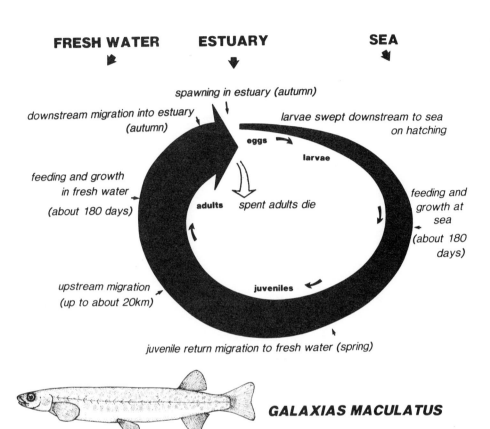

GALAXIAS MACULATUS

juveniles. A further six months (sometimes 18 months, rarely 30 months, Burnet, 1965) are spent feeding and growing in fresh water, following which the mature adults migrate downstream again, to spawn in tidal estuaries. The essential cue for spawning seems to be tidal flux rather than salinity as there are reports of spawning in fully fresh water. As the inanga does not actually go to sea to spawn, it can be described as 'marginally catadromous'. In this regard it differs from all other diadromous galaxiids, *none of which are catadromous* in spite of the common generalisation that galaxiids are catadromous on the basis of knowledge of *G. maculatus*.

Deviant populations include freshwater landlocked stocks (Pollard,1971; McDowall,1972; Campos, 1974). These include one Australian population in which there has been a reversal in the direction of the migration — an upstream

Figure 5.3: Lunar periodicity (both new and full moons) in the spawning migrations of the inanga, *Galaxias maculatus*
Source: After Burnet 1965

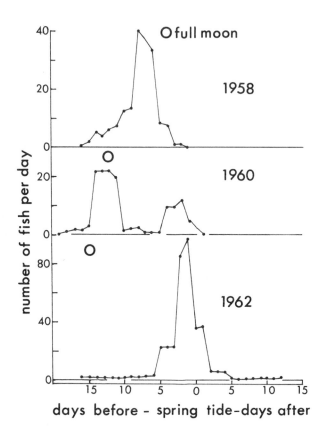

days before - spring tide-days after

migration of spawning adults in a landlocked population of *G. maculatus* (Pollard, 1971). Migration and spawning take place during floods in lake tributaries so that the eggs are deposited on stream-side vegetation that is submerged only when the spawning stream is in flood. Development takes place out of water and hatching occurs at the next suitable flood that covers the eggs. There is also at least one landlocked *Galaxias* species derived from *G. maculatus* (McDowall, 1972).

4. SHADS, HERRINGS, ETC. — FAMILY CLUPEIDAE

Although known primarily for their anadromy (see p.61), there is some evidence to suggest that at least one clupeid is catadromous. The Australian

Figure 5.4: Eggs of the inanga, *Galaxias maculatus*, amongst the marginal estuarine vegetation

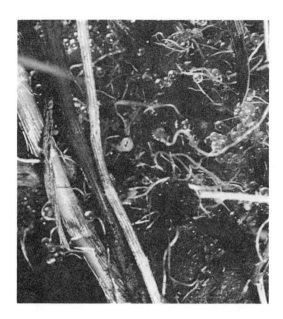

freshwater herring, *Potamalosa richmondia,* possibly migrates from upstream, freshwater habitats to estuaries to spawn, the young presumably migrating back upstream again to adult habitats; however, very little is known (Merrick and Schmida, 1984).

Ethmalosa fimbriata is a west African clupeid, discussed briefly by Whitehead (1981). He said that one-or two-year-old fish migrate 200 km up the Gambia River (in March and January respectively). This would necessitate a downstream spawning migration at some stage (if spawning is marine), and indicates the possibility that this fish is catadromous.

5. ANCHOVIES — FAMILY ENGRAULIDAE

The anchovies are smallish, herring-like fishes. The family is largely marine found in nearly all tropical to temperate oceans, there being about 110 species in 20 genera.

The large Australian and Papua New Guinean anchovy, *Thryssa scratchleyi,* is suggested as being catadromous by Merrick and Schmida (1984), on the basis of Roberts' (1978) study of fishes in the Fly River in Papua New Guinea. This freshwater species is said to migrate from the lower Fly River into the sea in the Gulf of Papua to spawn, and the small juveniles are found in the estuarine deltas, presumably moving back upstream as they grow. Again, far too little seems to be known to be categorical, but perhaps *T. scratchleyi* is catadromous.

6. SILVERSIDES AND HARDYHEADS — FAMILY ATHERINIDAE

The atherinids are mostly small, silvery fishes that most often occur in tropical and temperate waters and which are known for their occurrence locally in great abundance.   They occupy diverse habitats but prefer marine inshore waters, sheltered bays and estuaries. Many species are marine, some are euryhaline/ brackish-water fishes, and others occur in fresh water.   The family is a large one with about 156 species in 29 genera.

Atherinids are well known for their euryhalinity and it is surprising that various forms of diadromy are not well reported for the family. This does not seem to have happened.   Prince and Potter (1983) discussed the life histories of several western Australian atherinids and concluded that *Atherinosoma presbyterioides* and *Pranesus ogilbyi* are predominantly marine species that use river estuaries as a nursery area. In this regard such species can perhaps be regarded as marginally catadromous, returning from estuaries (rather than from fresh water) to the sea to spawn.   As strictly defined, diadromy does not seem to occur in any atherinid.

7. SNOOKS — FAMILY CENTROPOMIDAE

The centropomids are, like the percichthyids, basal level perciforms; they are largish, spiny-finned predators and are mostly marine in temperate to tropical areas of all the major oceans.   There are about 30 species in nine genera.

In this family the Asian/Australian sea bass or barramundi, *Lates calcarifer,* lives in fresh water until adulthood, and migrates downstream to tidal estuaries and coastal seas, to spawn (Figure 5.5). The young grow in these areas for several months to a year or more, before moving upstream into riverine habitats. The upstream migration, over a period of years, may take the fish very long distances up rivers, Roberts (1978) recording the species 800 km up the Fly River in Papua New Guinea.   Growth to maturity is in these freshwater habitats, and is followed by a return, downstream migration of the pre-spawning adults (Moore, 1982; Moore and Reynolds, 1982).   Spawning occurs in spring and summer (October to February in Australian waters).   It takes place close inshore in shallow water (2–3 m), the adults often moving only a short distance at sea (Griffin, 1987). Davies (1986) found no evidence of a movement of spent adults back into fresh water.   The larvae, growing from their tiny size at hatching (about 1.5 mm), move initially into brackish to freshwater coastal swamps, at a length of about 5 mm.   Later, they enter rivers, at the age of about

Figure 5.5: Life history of the barramundi, *Lates calcarifer*
Source: Morrissey (1985)

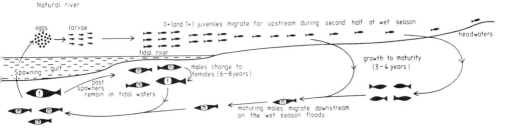

a year and disperse up river systems (Roberts, 1978).

A distinctive feature of the life cycle in *L. calcarifer* is that the 'females' never live in fresh water — they are six- to eight-year-old, sex-reversed males, that remain in tidal waters after spawning. The implication of this is that the post-spawning adults do not return to fresh water habitats, but this is not explicit from available information. Possibly the males can be described as catadromous, but the females strictly non-migratory. If this is a proper classification of migratory strategies in the barramundi, it appears to be unique. The adults presumably survive spawning, to recover condition and spawn again, either as males, or later in life, as females. Landlocked populations of *L. calcarifer* do not seem to be reported. A related species the Nile perch, *L. nilotica*, is an entirely freshwater species found in northern Africa. Other centropomids are apparently non-diadromous, some occurring in fresh water, others being marine.

8. TEMPERATE BASSES — FAMILY PERCICHTHYIDAE

As noted earlier (see p. 74), this family may be an assemblage of unrelated fishes; it is probably something of a catchall for primitive, unspecialised perciforms. Some are anadromous (see p. 173), but others are catadromous. The temperate eastern Australian percichthyid, the Australian bass, *Macquaria novemaculeata*, migrates downstream to tidal estuaries to spawn, and the progeny gradually return upstream as juveniles. Wal (1985) has shown that the eggs require saline waters to develop successfully. As such, this species is 'marginally' catadromous. This life cycle seems obligatory, as there are no landlocked populations, and when stocks are cut off from the sea by impoundments, they rapidly disappear; nor are there truly marine populations (Harris, 1984). A related species, the estuarine perch, *M. colonorum*, also in temperate

85

eastern Australia, is largely estuarine in habit, while other related Australian species are exclusively fresh water (Llewellyn and MacDonald, 1980).

The Japanese sea bass, *Lateolabrax japonicus*, is described as catadromous by Chan (1968). It occurs in the lands bordering the Sea of Japan, and southwards into Korea and China. Spawning is marine, and yearlings migrate from the sea into fresh water during the early spring (March–April, Matsumiya, Mitani and Tanaka, 1982). Later they migrate back to sea over the winter, returning to fresh waters again the following spring. There are several movements of this type to and from fresh water during growth to maturity, but the mature adults return to the sea for an autumn spawning. This species grows to a large size (more than 1,200 mm), so it clearly spawns repeatedly over several to many years.

9. MULLETS — FAMILY MUGILIDAE

The mullets are distinctive, largely marine and estuarine fishes — Thomson (1966) described them as 'typically estuarine'. They are small to moderate-sized, silvery fishes with largish scales, and are found in large shoals, mostly in coastal seas of tropical and warm temperate lands throughout the world. There are about 70 species in 13 genera.

The mullets are widely described as catadromous. As far as is known all mullets must breed in the sea, so that to the extent which species enter fresh water, they must also return to the sea to spawn, and so are more or less catadromous. Torricelli, Tongiorgi and Almansi (1982) and De Silva (1980) classified all of the grey mullets (Mugilidae) as catadromous, with few exceptions. However, this does not seem to be quite correct, and Blaber (1987) considered that only about 10 of the 70 mullet species move into fully fresh waters. As with numerous tropical groups, details of the life cycles of many species are elusive, but some mullets may be better described as highly euryhaline rather than diadromous; they often move facultatively and erratically from the sea into river estuaries (some being marginally catadromous). Rather fewer probably move regularly and more or less obligatorily into fresh waters as a consistent phase of their life cycles.

In spite of being described (Thomson, 1966) as the 'best studied species' of mullet, even the grey or striped mullet, *Mugil cephalus*, seems still to have a poorly understood life cycle. Although several authors agree with Miller (1959) that *Mugil cephalus* is diadromous, there is no evidence that the freshwater phase is obligatory (Thomson, 1966). *M. cephalus* is very widespread in tropical and warm-temperate seas (from about $42^\circ N$ to $42^\circ S$, in all oceans). It enters fresh waters very freely; Johnson and McLendron (1980) found it 190 km up the Colorado River in the United States, and Gunter (1938) reported it 180 km up the Rio Lempa in El Salvador. Merrick and Schmida (1984) said that in Australia it penetrates to the uppermost reaches of many streams in all drainages except those of the internal division. It seems to be catadromous, but to what extent is not clear, in spite of all the studies (reviewed by Thomson, 1966). Behaviour of *M. cephalus* is very variable; sometimes it is present in fresh water only in the juvenile stages (Bok, 1979) — which would make it marine amphidromous rather than catadromous, but in other populations the fish grow to adult size in fresh water before going to sea to mature and spawn (Shireman, 1975 in the USA; Thomson, 1966 in Australia) — in which case they are catadromous. Shireman (1975) found that both sexes of *M. cephalus* will mature in the fresh waters of Louisiana, but that if the fish are prevented from going to sea, they begin to resorb their gonads. Blaber (1987) discussed large aggregations of ripe adults in river estuaries. Chubb, Potter, Grant, Lenanton and Wallace (1981) found that in western Australia the fish found in fresh waters were primarily 0+ and 1+ fish. The 1+ fish migrated downstream into the estuaries during the spring and summer, and probably lived there and in the sea, where they feed and range widely (Merrick and Schmida, 1984). Grey mullet spawn at sea, probably during the autumn and on through the winter at an age of 3+ or older, but where they spawn is contested; some believe that they spawn in coastal shallows, others over deep ocean (Anderson, 1958). The young are first encountered in fresh water at a length of 20–30 mm, and during the early winter (May onwards in Australia Thomson, 1966) suggesting that they may take several months to move upstream through the estuaries to reach fresh waters). Merrick and Schmida (1984) believed that the spent adults return to river systems for the greater part of the year; here feeding and growth (which occur chiefly in summer) resume. However, entry to fresh water may be erratic, it may occur on a daily-tidal basis in some parts of the population and may not be necessary. If the above interpretation of the life cycle of *M. cephalus* is correct, the species can be described as perhaps catadromous. Thomson (1966) reported that when a population became landlocked in an estuary it did not reproduce, and stocks in some New Zealand lakes have not reproduced, as far as is known (McDowall, 1978a). Other *Mugil* species may also be catadromous. Moreau (1986) considered that in Madagascar, *M. robustus* lives in fresh coastal waters throughout the year and that there are possibly migrations between interior waters and the coastal zone.

The north and central American mountain mullet, *Agonostomus monticola,* is also poorly known, but seems to migrate from rivers into the sea to spawn. Inland penetration extends well beyond tidal reaches of rivers, Erdman (1984) reporting that it reaches altitudes exceeding 80 m. Early development and growth take place at sea, the smallest examples found in fresh water being 22 mm long (Anderson, 1957). Gilbert and Kelso (1971) question marine spawning, saying that it is possible that the species spawns in fresh water and that the eggs are carried into the ocean where they hatch. No other mullet has been described as spawning in fresh water. Loftus, Kushlan and Voorhees (1984) argued this question at some length, noting that adult mountain mullet have not been collected from marine waters, and reporting 'probable' but unconfirmed spawning behaviour in streams of Puerto Rico. Ripe adults have apparently always been taken in fresh water (Corujo-Flores, 1980) Thus, Loftus *et al.* (1984) reached a conclusion that spawning occurs in the freshwater streams, followed by passive transport of the eggs or larvae to sea. If this life history is found to be correct, then *A. monticola* is amphidromous, not catadromous. Little is known about a second species of *Agonostomus, A. hancocki*, described from fresh waters of the Galapagos Islands by Seale (1932); this species is possibly catadromous also. African/western Indian Ocean species of *Agonostomus* are also poorly known, but Thomson and Luther (1984) said that *A. catalai* and *A. telfairi* inhabit fresh water but may occur in estuaries. They seem to be freshwater/euryhaline species, in contrast with other mullets that are mostly marine/euryhaline. Presumably they nevertheless eventually go to sea, to spawn. Another Central American species, the hognose mullet, *Joturus pichardi,* occurs in the rivers from Mexico to Panama (Miller, 1966). It ascends the rivers as juveniles (Darnell, 1962), and is found in large numbers at obstructions to upstream migration, like waterfalls (Jordan, 1907). It descends to river mouths to spawn, the spent adults returning upstream in November and the juveniles in December (Erdman, 1974).

The Australian 'freshwater mullet', *Myxus petardi*, seems to have an essentially similar life history (Merrick and Schmida, 1984), although this species also remains poorly known. The South African species, *Myxus capensis,* which is distributed in southern and eastern Africa (Bok, 1979) is reported to live in fresh water as a juvenile and sub-adult, penetrating as far as 120 km upstream. The young, 10–40 mm long, enter fresh water soon after hatching mainly during the late winter and through the spring, and they stay there until they reach adult size and are nearly mature. However, full gonad development does not occur in fresh water but takes place after a migration to the estuaries. It depends on extensive fat reserves as feeding ceases. Spawning takes place in all months of the year and is said to be adapted to the unstable rainfalls and river flows of southern Africa. The fish reach maturity at age two to five years, and there is no post spawning survival. Bok (1979) believed that the adults leave fresh water and probably spawn in inshore/coastal areas. He thought that the freshwater stage in the life cycle of *Myxus capensis* is important and possibly obligatory.

Torricelli *et al.* (1982) reviewed the upstream migrations of the juveniles of several species of mullet in the Arno River in Italy, including *Liza ramada, L. aurata, L. saliens* and *Chelon labrosus*. To what extent these migrations are an obligatory or even general feature of the life cycles of these species is unclear, nor is it obvious how far they penetrate into fresh water, or to what extent they remain and grow there before returning to the sea to spawn (presumably). Lasserre and Gallis (1975) found that in culture situations *C. labrosus* had only a limited ability to live for long periods in fresh water before suffering osmoregulatory problems. Sarojini (1957, 1958) described moderate penetration of Indian rivers by *Mugil parsia* and *M. cunnesius*, but again, life history details are sparse and the status of these species as diadromous, or not, is unclear.   Payne (1976) found both *Liza falcipinnis* and *L. dumerilii* in the fresh waters of Sierra Leone, with both species showing a greater tendency to penetrate areas influenced by fresh water than *M. cephalus*.   Thomson and Luther (1984) considered *L. abu* (Tigris, Euphrates and Indus Rivers) as occurring mainly in fresh water.   Hickling (1970) listed *Crenimugil labrosus, Liza ramada*, and *L. aurata* as penetrating wholly freshwater habitats.   He cited reports of *L. ramada* 200–300 km up rivers in Morocco, and considered that it lives for months long distances up rivers in France.   It ascends rivers and remains in fresh water to feed, often for several months.   *C. labrosus* also enters fresh water, especially when young.   These species were observed to migrate to sea, to spawn.

Other species of mullet are known to enter rivers — Liza diadema in Papua New Guinea (Merrick and Schmida, 1984), *Valamugil speigleri* in India, *V. seheli* in Africa (Thomson and Luther, 1984), *Aldrichetta forsteri* in New Zealand (McDowall, 1978a), etc, but these have not been shown to be catadromous. Certainly, the yelloweyed mullet, *A. forsteri,* is a euryhaline wanderer in Australia and New Zealand, and is in no sense catadromous.   Clearly, many mullets do enter fresh water and some of them seem to be catadromous;   others of the more poorly known mullets may also prove to be when their life cycles are elucidated.

10. FLAGFISHES OR AHOLEHOLES — FAMILY KUHLIIDAE

The kuhliids are another small family of lower-level percoid fishes that are widespread in the marine and fresh waters of the tropical Indo-Pacific Oceans. There are only about 20 species in three genera.   Like those of so many tropical perciform fishes, life histories are not well elucidated but there are some suggestions that in some species the adults live in fresh water and migrate downstream into estuaries to spawn.   This was reported for the jungle perch, *Kuhlia rupestris*.   This species is very widespread from the eastern coast of Africa, through southeast Asia, and east into the Pacific as far as Tahiti. Recent studies in Fiji and Australia (Lewis and Hogan, 1987) have shown that a catadromous life cycle occurs.   A freshwater phase is regarded as being obligatory prior to maturation, followed by a downstream movement of the adults into the estuaries and coastal waters for spawning.   Other species of *Kuhlia*, such as *K. sandvicensis* in Hawaii may be similarly catadromous;   this remains uncertain and study is required to clarify this.

11. THERAPONS — FAMILY THERAPONIDAE

Therapons are small to moderate-sized perciform fishes that are widely marine/coastal and freshwater in the temperate to tropical Indo-Pacific; they are especially well represented in the fresh waters of Australia and Papua New Guinea. The family is a small one with about 20 species in three genera.

Recent studies of the Fijian endemic *Mesopristes kneri* have shown that its life history is comparable with that of the Australian percichthyid, *Macquaria novemaculeata*. *M. kneri* lives most of its life in the rivers of Fiji, but migrates downstream into estuaries and coastal seas to spawn; the juveniles move back into fresh water to grow to maturity. Life histories of other theraponids are in need of study.

12. SNAPPERS — FAMILY LUTJANIDAE

Lutjanids are generally larger, heavy-bodied, spiny-rayed perciforms found throughout the warmer oceans of the world, with about 23 genera and 230 species. Although primarily a marine family, various species are described as entering rivers, especially the mangrove jack, *Lutjanus argentimaculatus,* in the Indo-Pacific. Life history details suggest that this species is best regarded as a facultative, euryhaline wanderer. However, a Papua New Guinean endemic, *Lutjanus goldei*, is believed to spend virtually all of its life in fresh water, apart from reproduction, which is marine (A.D. Lewis, personal communication). If studies confirm this life cycle, then *L. goldei* is catadromous.

13. SCORPIONFISHES — FAMILY SCORPAENIDAE

Scorpionfishes are distinguished, in part, by the presence on the head region of numerous sharp, and often venomous spines; they are otherwise rather typical bottom-living, perch-like fishes. The family is widely distributed in all tropical and temperate seas with over 300 species in about 60 genera. Scorpaenids are very largely marine.

The Australian bullrout, *Notesthes robusta,* was listed by Harris (1984) as being catadromous, but I am unaware of any explicit documentation of this. Parker (1980) and Merrick and Schmida (1984) said that it can be found in a wide range of environments from brackish estuaries upstream in rivers for long distances, and also that juveniles have been found above impoundments that would, in their view, constitute barriers to upstream migration by the bullrout. This observation may be interpreted as indicating that a life cycle confined to fresh water is possible in the species, but it does not exclude catadromy. J.H. Harris (personal communication) has observed that upstream distribution of the bullrout in some Australian river systems is limited by the construction of impoundments, and that there may be accumulations of fish below such dams, implying an upstream migration. Harris has regularly encountered small juveniles in brackish estuaries during the spring. Lake (1971) regarded it as essentially marine in habit. With this diversity of views, it is obvious that much remains to be learned. Catadromy in the bullrout remains possible but uncertain.

14. SOUTHERN ROCK CODS — FAMILY BOVICHTHYIDAE

Bovichthyids are bottom-living, often stocky and rather spiny percoid fishes with only a few species - about six in four genera. They have a sub-Antarctic and cool southern-temperate distribution and are almost exclusively marine.

The only exception to this is the tupong or congolli, *Pseudaphritis urvilii* (Figure 5.6), of southeastern Australia and Tasmania. It is little known, but Sloane (1984b) described the young as entering fresh water from the estuary or sea and discussed the possibility of progressive upstream movement over a number of years. W. Fulton (personal communication) considers that the tupong is diadromous. He has found it to be very common in estuaries and that it moves upstream considerable distances. Juveniles have been collected among the upstream migrations of the Tasmanian whitebait, *Lovettia sealii*. Fulton advised that the spawning site is undiscovered but is regarded as probably estuarine. On the basis of this limited knowledge of the tupong, it appears to be marginally catadromous.

Figure 5.6: Tupong, *Pseudaphritis urvilii* (familiy Bovichthyidae)

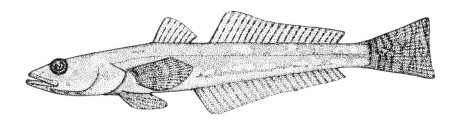

15. SCULPINS — FAMILY COTTIDAE

The sculpins are stockily built, largely bottom-living fishes. The distribution of the family is primarily marine, northern, cool-temperate and sub-Arctic, but there are some southern-temperate marine species and some that occur in fresh water. The family contains about 300 species in 67 genera.

Diadromy in the cottids is not clearly described, but it seems that some regular catadromous movements between fresh water and the sea do take place. Morrow (1980) distinguished two forms of the prickly sculpin, *Cottus asper*, which occurs along the Pacific coast of North America from California to Alaska. The inland form, he said, does not migrate to estuaries to spawn while the coastal form spawns in brackish water. It seems that in the latter form, the males move down into the estuaries during spring and summer, followed by the females. McAllister and Lindsey (1959) listed several observations of downstream spawning migrations of prickly sculpins into estuaries. They considered that a downstream spawning migration is probably a common feature of coastal populations. After spawning, the females return upstream while the males guard the nests where the eggs were laid. The larvae are planktonic for a period of 30–35 days, and Morrow (1980) thought that they are 'probably swept up and down the estuary by the tidal currents, but [that] any carried out to sea doubtless perish as none have been taken in the ocean'. Whether or not this is correct remains unknown. After completion of spawning, the adults make their way back upstream, during the autumn. The upstream and downstream movements are evidently quite short. McAllister and Lindsey (1959) considered that the coastal form penetrates only about 16 km upstream. Although Morrow (1980) contrasted these estuarine spawners with 'non-anadromous forms in lakes', the life cycle as presently understood is probably catadromous, but possibly amphidromous. Krejsa (1967) described the prickly sculpin as a 'catadromous, brackish-water spawner'.

The coastrange sculpin, *Cottus aleuticus,* also present from California up the west coast of North America to Alaska (including the Aleutian Islands), is described by Morrow (1980) and McAllister and Lindsey (1959) as having a life cycle similar to that of the prickly sculpin, though it seems less clearly understood. Morrow mentioned that spawning is in the lower reaches of the streams and in the estuaries, but may occur well upstream as well. He did not draw a distinction between two types, as he did for the prickly sculpin. He regarded the young of the coastrange sculpin as being planktonic and occurring in the intertidal region of estuaries. The Japanese river sculpin, *C. kazika* is similarly 'more or less' catadromous. The adults move down to river estuaries in winter, to spawn in brackish water, and the larvae go to sea. The juveniles return at a length of 15–30 mm, during the spring (Okada, 1960).

The fourhorn sculpin, *Myoxocephalus quadricornis,* is circumpolar in Arctic seas and drainages and has both marine and freshwater forms. The marine form is described as spawning in estuaries (Morin *et al.,* 1980), but migratory patterns are little described and apparently not understood. Diadromy in any form is unconfirmed.

16. GOBIES — FAMILY GOBIIDAE

The gobies are small, stocky, scaled, intertidal and subtidal fishes that are found very widely in tropical and warm temperate waters throughout the world. Various gobiid fishes have been described as catadromous, but I have been able to verify few of these or obtain explicit details of their life histories. Merrick and Schmida (1984) said that *Tasmanogobius lordi* inhabits 'brackish estuaries or freshwater reaches where there are mud bottoms. ... Apparently this goby migrates to lower marine estuaries to spawn in spring'. Possibly it is catadromous, but far too little is known to be definite.

Nordlie (1981) studied goby life histories in the Tortugueros Estuary of the Mexican Gulf coast of Costa Rica. Five gobiid fishes occur in the estuary and its freshwater tributaries, and their juveniles are taken in the 'tismiche' - the massive migrations of tiny fishes and crustaceans that enter this estuary (Gilbert and Kelso, 1971). It is clear that at least some of these are amphidromous, but in the best-known species, *Gobiomorus dormitor*, the cycle is possibly catadromous. Nordlie found the juveniles (all bigger than 21 mm) to occur in the lower estuary and Darnell (1962) showed that the adults return from the rivers to spawn in the estuarine lagoons, a finding with which Nordlie concurred. He regarded it as an opportunistic species that is found primarily in estuaries and their associated rivers and which is capable of completing its life cycle when restricted to fresh water. Erdman (1984) described it as reaching an elevation of at least 80 m in Puerto Rico.

Although details are sparse, *Mugilogobius abei* and *Redigobius bikolanus* are regarded as catadromous in Japanese waters (N. Mizuno, personal communication).

17. FLOUNDERS AND SOLES — FAMILIES PLEURONECTIDAE AND SOLEIDAE

The flounders and soles are typical flatfishes, with the body laterally compressed, and both of the eyes on one side of the body. They are very widely distributed throughout the world with about 99 species in 42 genera in the Pleuronectidae, and 117 species in 31 genera in the Soleidae.

Various flounders and soles are present in fresh waters, and if their life cycles are typical of the group, with a pelagic, upright-swimming larval stage, there is a high likelihood that this stage is marine — there are no reports of lacustrine larval flounders; it is possible that some species undergo their larval life up to metamorphosis in the estuaries of large river systems, but this is not reported either. However, Gery (1969) listed a species of *Achirus* 3,200 km up the Amazon, and in such instances it is possible that development as far as metamorphosis has been completed before the little fish are swept out to sea. It seems to me highly likely, however, that many freshwater flounders and soles complete part of their life cycles in the sea, and that therefore they are possibly catadromous to some extent.

This is certainly true of the black flounder, *Rhombosolea retiaria*, found in New Zealand fresh waters. The newly metamorphosed young enter river mouths during the spring and may move 50–100 km upstream, where they occupy varied habitats (moderately swift-flowing gravelly runs to still or gently flowing sandy pools), and there they feed and grow to maturity. The maturing adults migrate downstream and are believed to go to sea to spawn, and they probably spend the remainder of their lives there, recovering condition to spawn repeatedly. Not much is known (McDowall, 1978a).

Merrick and Schmida (1984) listed three freshwater soles in Australia — Asseragodes klunzingeri, Brachirus salinarum and *B. selheimi* - and said that *A. klunzingeri* has the typical flatfish life history, so presumably the adults go to sea to spawn. There seems to be very little information on whether these fish normally live in fresh waters, or are just facultative wanderers. McHugh (1967) referred to the starry flounder, *Platichthys stellatus,* as having young which frequently move into rivers, and Hart (1973) described it as noteworthy among flatfishes for its tolerance of low salinities. It occurs in both the Fraser River in North America, the Amur River in Siberia, and no doubt elsewhere in the intervening areas. It moves in and out of fresh water, and it has been suggested that the movements may be seasonal; catadromy is possible but uncertain.

Berg (1962) listed several flatfishes as occurring in the fresh waters of Russia, but again, details are very limited. *Platessa platessa* and *Liopsetta pinnifasciatus* were described as entering rivers, while *Pleuronectes flesus* was listed as the 'river flounder'. It occurs widely in western Europe from Scandinavia to Spain. The young of this species stay in brackish to fresh water but the adults migrate to sea to spawn (Nikolskii, 1961; Berg, 1962). Beaumont and Mann (1984) found that the numbers present depended on the immigration of O-group fish and the emigration of older ones; the young arrived at their study area in summer and increased in abundance through the autumn and winter. Post-larval *P, flesus* acquire increased tolerance to fresh water as they develop, and after metamorphosis they actively swim towards fresh rather than sea water. Dando (1984) found that in the Tamar River in England, the young may spend less than two weeks at sea. The adults usually spawn within 25 km of the mouths of estuaries and return to their home range in the river/estuary. Hartley (1940) described a spring movement of estuarine flounders into areas of reduced salinities, and described *P. flesus* as a 'typical catadromous fish. ... They grow and feed in fresh and brackish water until the onset of sexual maturity, when they go down to the sea to breed.' Ripe fish leave the estuaries in winter (January–February).

The western Atlantic species, *Pseudopleuronectes americanus* is even less clearly catadromous. Pearcy and Richards (1962) said that it migrates to spawn in estuaries in late winter and early spring, but it was also described as an estuarine resident. It is not clear where the migrants came from. Guelpen and Davis (1979) described how it lives in shallow waters, bays, and estuaries during the winter, moving offshore to cooler waters in the summer. Catadromy seems unlikely.

The hogchoker, *Trinectes maculatus* (Soleidae), is widespread in eastern North America from Connecticut to northern coasts of South America. It is

Figure 5.7: Migratory movements of the hogchoker, *Trinectes maculatus*, in the Patuxent River estuary, with special reference to salinity gradients
Source: Dovel *et al.* (1969)

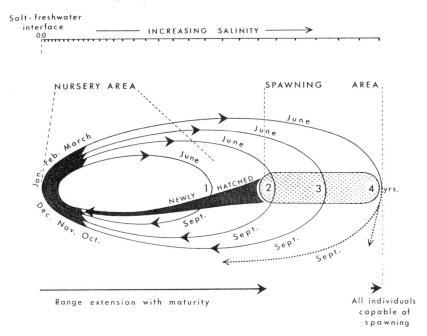

said to ascend rivers into fresh water (Figure 5.7). There is a spring downstream migration to the sea to spawn each year, and in the autumn, the adults make their way back upstream (Dovel, Mihursky and McErlean, 1969; Koski, 1978). The eggs develop in less than 36 hours and this probably limits the extent to which they disperse offshore (Dando, 1984). The larvae probably ascend rivers soon after metamorphosis.

6

Detailed Analysis of Diadromy — Amphidromy

1. AYU — FAMILY PLECOGLOSSIDAE

This family contains only the ayu, *Plecoglossus altivelis*, an osmerid-like fish found in the streams and coastal seas of Japan, southern China, northern Taiwan, and southern Korea. The ayu is a small fish growing to about 250 mm. It is distinctive and specialised amongst the freshwater salmoniforms in being herbivorous (see also Prototroctidae p. 100). Ayu have very specialised jaw dentition, adapted to scraping algae from rocks, and a much elongated alimentary canal to increase the efficiency of digestion (Okada, 1960; Kawanabe, 1969). Spawning takes place during the late autumn in fresh water over gravelly reaches in rivers after a downstream migration by the adults to lower elevation areas. The eggs are demersal and become fastened to the gravel by adhesive membranes. The larvae develop in two to three weeks, hatch in fresh water, and are carried immediately to sea by the river flows at a length of only about 6 mm. The young fish live for several months over the winter in the sea, mostly in coastal areas. During the spring there is an upstream return migration of juveniles 50–70 mm long into the rivers, where they feed and grow over the spring and summer; they mature the following autumn, and spawn. There is complete post-spawning mortality, so the life cycle is annual and semelparous. Landlocked populations are known widely in lakes, at least in Japan (Okada,1960; Kawanabe, 1969; Azume, 1981).

2. SOUTHERN WHITEBAITS AND GALAXIIDS — FAMILY GALAXIIDAE

The galaxiids have wide distribution in the cool temperate Southern Hemisphere (Figure 6.1). In addition to the single marginally catadromous species (*Galaxias maculatus* see p. 80), at least seven species of *Galaxias* are amphidromous (McDowall, 1970, 1971b; McDowall and Frankenberg, 1981; Fulton, 1986). The amphidromous species, which are known from Australia (*G. truttaceus, G. brevipinnis, G. cleaveri*), New Zealand (*G. brevipinnis, G.*

Figure 6.1: World distribution of family Galaxiidae

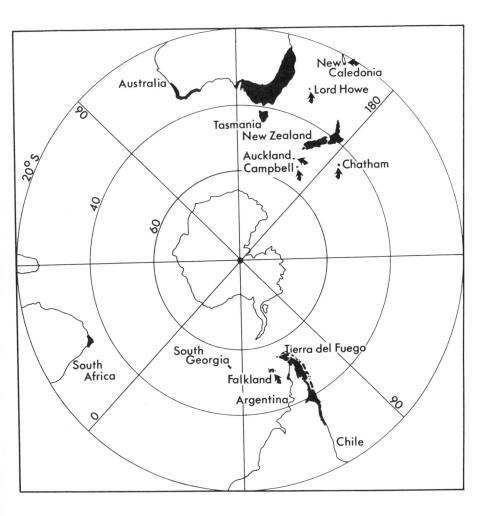

argenteus, G. fasciatus, and *G. postvectis*) and possibly South America (*G. platei?*), tend to be rather bulky, secretive species, and they do not shoal as adults like *G. maculatus* does. Not much is known but it seems likely that they spawn in or near typical adult freshwater habitats, often a considerable distance upstream from the sea (some more than 200 km). Mitchell and Penlington (1982) have described spawning in the New Zealand banded kokopu, *G. fasciatus.* The adults move onto forest litter that accumulates along the margins of the streams, when this is inundated by small floods, and the eggs are laid amongst this litter. When the flood abates, the eggs remain amongst the litter

97

Figure 6.2: Life history of the banded kokopu, *Galaxias fasciatus*

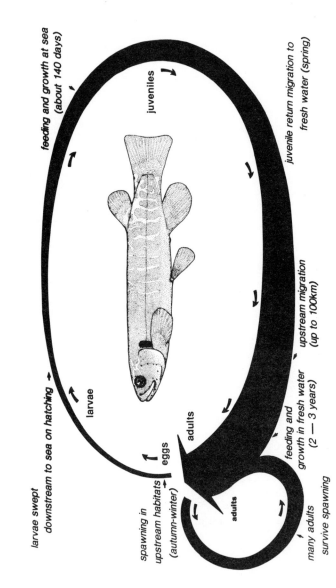

FRESH WATER   ESTUARY   SEA

feeding and growth at sea (about 140 days)

juvenile return migration to fresh water (spring)

juveniles

larvae swept downstream to sea on hatching

larvae

upstream migration (up to 100km)

spawning in upstream habitats (autumn–winter)

eggs

adults

feeding and growth in fresh water (2 — 3 years)

adults

many adults survive spawning

GALAXIAS FASCIATUS

and develop in humid air; they hatch when re-immersed by a later flood. On hatching, the larvae of the banded kokopu are washed downstream during floods (Ots and Eldon, 1975), and presumably into the sea as no larval or early juvenile stages are known in fresh water (Figure 6.2).

Observations on the other species give little indication of spawning migrations of any significance (McDowall, 1978a; McDowall and Frankenberg, 1981), and it is likely that their early life history strategies are similar to that described for *G. fasciatus*. Growth of the larvae and early juveniles of these species takes place in the sea, lasting for several months, and is followed by mass migrations of small juveniles (about 50 mm long) into fresh waters in mixed species shoals. Penetration of fresh waters may be extensive, and is often aggressive, large barriers to migration, including falls tens of metres high being passed by the migrating fish. In freshwater habitats there is further feeding and growth for 18 months or more (Hopkins, 1979a) before the fish mature. The cycle is completed by spawning in fresh water, and (unlike *G. maculatus*), repeat spawning over several years is normal (Hopkins, 1979a,b; McDowall, 1978a).

Deviation from this life cycle involves the frequent establishment of landlocked stocks in several of the amphidromous species. In general the landlocked stocks live as adults in tributaries to high country lakes, spawn there, and the larvae move down to spend the early months of their lives in the lake pelagic zone. There is a return migration to the lake tributaries of juveniles, but some may mature and remain in the lakes until adulthood. This modified life cycle is particularly characteristic of *G. truttaceus* in Tasmania, and *G. brevipinnis* in both Tasmania and New Zealand. In both of these species landlocked populations have also been formed in recent times by man-constructed impoundments on rivers.

There are no wholly marine galaxiids (even though galaxiids are not infrequently, but quite erroneously, said to have marine origins), but there are substantial numbers of species in the family that are exclusively freshwater in habit.

3. PELADILLOS — FAMILY APLOCHITONIDAE

The South American peladillos, *Aplochiton* spp. (southern Chile, Argentina, Tierra del Fuego and the Falklands, McDowall, 1971a) are poorly known, although Campos (1969) has shown that *A. taeniatus* (Figure 6.3) spawns in fresh water. Eigenmann (1928) thought that *A. marinus* spawns in the sea, but he did not document evidence for this view. I regard *A. marinus* as a junior synonym of *A. taeniatus* (McDowall, 1971a), so clearly there is some confusion here that needs to be resolved. Recent collections of larval *Aplochiton* (species uncertain) from the sea (McDowall, 1984a), indicate that some movements to sea do take place. If only the young are at sea, then *Aplochiton* is amphidromous.

Figure 6.3: Peladillo, *Aplochiton taeniatus* (family Aplochitonidae)

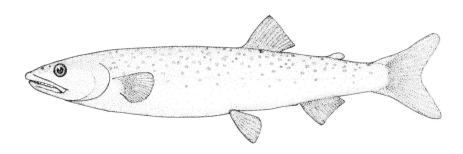

Zama and Cardenas (1984) discuss *Aplochiton* in southern Chile close to the sea, but not in fully marine waters. If these should be adults returning to fresh water to spawn, then the fish would be anadromous. However, virtually nothing is known of migratory behaviour, spawning or post-spawning behaviour. Campos (1969) properly argued that the discovery that reproduction takes place in fresh water (in contrast with the view of Eigenmann, 1928), excludes the possibility that *Aplochiton* is catadromous. The population that Campos studied was in a large lake not far from the sea, and the possibility remains that *Aplochiton* in his population was a deviant, landlocked one. Landlocked populations of at least *A. zebra* do occur in Chile, and possibly also of *A. taeniatus* (Campos, 1969; McDowall, 1971a)

4. SOUTHERN GRAYLINGS — FAMILY PROTOTROCTIDAE

The two species of *Prototroctes* (one each in Australia and New Zealand, the latter extinct) are smelt-like fishes reaching a typical maximum size of about 300 mm. Examples much larger than this are reported, though not reliably (McDowall, 1976, 1978a). Little is known explicitly of this small family; what is known suggests that the life cycle is comparable with those of *Galaxias* and *Plecoglossus* (see above). Spawning seems to be in fresh water, and is probably during the autumn (Berra, 1982). Males in the Australian species may mature at age one, but females not until they are two. Spawning mortalities are little understood, but repeat spawning may occur, and more than one age class seems to be present in the spawning population. This might imply spawning survival, but might also imply varied ages at maturity. Berra found fish up to five years old in the Tambo River in Victoria, Australia. The larvae probably go to sea to feed and grow for some months (Berra, 1982; McDowall, 1984a). A return upstream migration of juveniles seems likely to occur during the spring, this being reported for both New Zealand (*Prototroctes oxyrhynchus*, Clarke, 1899) and

100

Australia (*P. maraena*) where W. Fulton (personal communication) has taken juveniles returning from the sea amongst mixed-species shoals of adult *Lovettia sealii* and juvenile *Galaxias* species. These fish probably move upstream into forest-covered streams where they graze on encrusting algae and the invertebrate fauna associated with it. Like *Plecoglossus* it seems that *Prototroctes* is at least partially herbivorous in fresh water (McDowall, 1976, 1978a; Bishop and Bell, 1978); it has highly specialised jaw dentition (although quite unlike that of *Plecoglossus*), and has an extended alimentary canal. Two full loops occur in the intestine, a feature very unusual in salmoniform fishes and presumably related to the occurrence of herbivory (McDowall, 1976).

Landlocked populations of *Prototroctes* have not been described, and the life cycle is, as far as is known, invariant.

5. HERRINGS — FAMILY CLUPEIDAE

This family has already been discussed at some length under anadromy (p. 61) and catadromy (p. 83). McHugh (1967) described the menhaden, *Brevoortia patronus* (along the east coast of North America) as behaving as follows in Chesapeake Bay:

> They spawn offshore in late fall or winter and the young move into the inshore estuaries soon after hatching, often penetrating fresh waters in the rivers as much as 25 to 35 miles. In fresh or brackish water they spend the winter and as they grow, the young move slowly down the tributaries to the Bay. ... Most of these young, now about one year old, leave the Bay with the adults in the fall.

This life cycle could be classified as marginally amphidromous, although it is essentially marine.

There are two white sardines in the genus *Escualosa*. One that is widely distributed from western India to the Philippines and northeastern Australia (*E. thorocata*) has juveniles that enter rivers, but which later return to the sea (Whitehead, 1985). It is not stated whether this is a typical habit of *E. thorocata*; if it is, then it can be classified as marine amphidromous.

6. SANDPERCHES — FAMILY MUGILOIDIDAE

This family consists of moderate-sized, stoutly built, hard-scaled perciform fishes that are largely bottom-living in the coastal seas of the Indo-Pacific; there is a single freshwater species in New Zealand, the torrentfish, *Cheimarrichthys fosteri* (Figure 6.4), which is amphidromous, although its life cycle is not well known (McDowall, 1973, 1978a). The adults segregate by

Figure 6.4: Torrentfish, *Cheimarrichthys fosteri* (family Mugiloididae)

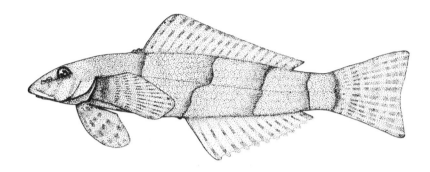

sex in riverine habitats, the females in the headwaters and the males at low elevations, and sometimes spread over distances of up to 100–150 km. The sexes evidently 'get together' by a downstream migration of females to the lower reaches of the river systems (this is not certain, but something like this must happen!). Wheeler (1985) suggested that the adults migrate downstream in autumn and that spawning is probably in the sea, but neither of these has been shown to be true. (If they were true, *C. fosteri* would be catadromous.) Spawning appears to take place in fresh water, and from observations of the eggs after artificial stripping (they are demersal), it seems likely that egg development is in fresh water and that a movement of larvae to sea takes place soon after hatching — larvae have not been found in freshwater habitats. After several months, presumably at sea, the small juveniles (about 25 mm long) are found making their way into river mouths; they move inland occupying typical adult habitats where they feed and grow for a few years, before maturing and spawning. They may penetrate substantial distances upstream (more than 100 km), and the size distribution of the populations becomes increasingly large with distance inland (McDowall, 1973). Repeat spawning, probably over several successive years, is highly likely. There are no landlocked populations, and the life cycle seems invariant. All other mugiloidid species are marine.

7. SCULPINS — FAMILY COTTIDAE

Although sculpins are primarily catadromous (if diadromous at all) it appears that the Pacific staghorn sculpin, *Leptocottus armatus*, may be amphidromous; it occurs in the eastern Pacific from Baja California north to the Bering Sea. Morrow (1980) described it as occurring in the 'sub-tidal zone and tide pools'. He found that the eggs will develop in varying salinities between 10 and 34 ppt, and in varying localities, from estuaries and bays to the open ocean. Unlike patterns reported for other sculpins, the young evidently move from these saline/brackish areas into fresh water for up to six weeks, although Morrow considered this movement to be non-obligatory. To the extent that this cycle can even be described as diadromous, it would be classified as amphidromous, and would appear to be one of very few instances of marine amphidromy, in which spawning is in the sea and early juvenile life is spent in fresh water as a result of a brief and temporary invasion by the newly hatched young.

The Japanese river sculpin, *Cottus hongiongensis*, has recently been shown to be amphidromous (Goto, 1981, 1987). Adults reach maturity at about two to three years old, with the males living further upstream in the rivers than females, and tending to reach maturity later in these upstream habitats. The males must migrate downstream to join the females on the spawning grounds. Spawning is in spring and the pelagic larvae are carried to sea soon after hatching, returning to fresh water after about a month. Another Japanese *Cottus* is also evidently amphidromous, although details are very sparse. Mizuno (1963a) described *C. japonicus* as being amphidromous with tiny larvae which hatch from small eggs and are washed downstream into the sea. This would constitute freshwater amphidromy.

8. SLEEPERS AND GOBIES, ETC. — FAMILIES ELEOTRIDAE, GOBIIDAE AND RHYACICHTHYIDAE

These families contain numerous rather similar species. Rhyacichthyids are regarded as primitive gobioids (Miller, 1973b). Life cycles in many fishes in these two families are reported to be amphidromous. Mizuno (1980), for instance, commented that:

Almost all of [the freshwater gobies in Japan] are originally not so-called freshwater fish but migratory ones showing amphidromous migrations between rivers and seas; although their planktonic larval stages can be spent in lakes and ponds instead of seas, they may be considered to be pure freshwater fishes secondarily.

He listed diverse gobies that appear, on the basis of rather limited knowledge, to be amphidromous, but many details await confirmation or clarification. Even so, about 20 Japanese gobies seem to be amphidromous, a figure substantially larger than those species indicated to be amphidromous by Akihito, Hayashi and Yoshino (1984). Information on Philippines' species is based on limited detailed information, but a variety of species seems to be amphidromous and the actual number may turn out to be far larger than is indicated by present information. Knowledge of goby life histories in Indonesia, Papua New Guinea, Melanesia, Polynesia and the western Indo-Pacific seems even more sparse than in the Philippines and Japan, apart from isolated cases, and there is a high likelihood that many additional gobies will eventually be discovered to be amphidromous. The present list should be seen as minimal.

In some of the species of the New Zealand and Australian genus *Gobiomor-phus* (Eleotridae), e.g. the redfinned bully, *Gobiomorphus huttoni*, most of the life is spent in fresh water, where the fish reach adulthood and spawn, mostly during the spring (August onwards). Many of the adults survive spawning, re-cover condition, and may spawn over several subsequent years. Egg develop-ment occurs in fresh water (taking several weeks), and the tiny larvae (about 5 mm long) are swept to sea in river currents soon after hatching. They do not feed in fresh water. Little is known about life in the sea, but the small juveniles, about 15–20 mm long, are encountered making their way back into fresh water during the later spring and early summer, sometimes in huge aggre-gations. They move upstream into adult habitats, where they resume feeding and grow to maturity over the next 18 months, or thereabouts. Over a period of years the growing, maturing fish gradually move upstream, so that size/frequency distributions show increasing average size with distance upstream. A life history strategy of this type is known for four species of *Gobiomorphus* in New Zealand (Figure 6.5) (McDowall, 1975, 1978a), and possibly also some in Australia. Hoese, Larson and Llewellyn. (1980) reported that juveniles of *G. australis* are found in estuaries or the lower reaches of rivers, and thought that the young were carried downstream, migrating back up-stream again later in life. This may also be true of the Australian Cox's gud-geon, *G. coxii*. Whether the larvae actually go to sea in Australian populations as they do in New Zealand, is not known, so amphidromy remains uncertain in Australian *Gobiomorphus*.

There is a series of related, tropical, gobioid fishes in the genera *Sicyopterus*, *Sicydium*, and some other, similar genera; they are very widely distributed from the Caribbean through the tropical mid-Pacific to the western Indian Ocean, and they are also widely reported to be amphidromous. J.A. Maciolek (personal communication) considers that all the sicydiine genera are diadromous, including *Lentipes*, *Sicydium*, *Sicyopterus*, *Sicyopus*, and *Stiphodon* Miller (1973a) implied the same. The identity and distribution of many of these fishes are poorly understood though some are reasonably well

Figure 6.5: Life history pattern in the New Zealand redfinned bully, *Gobiomorphus huttoni*

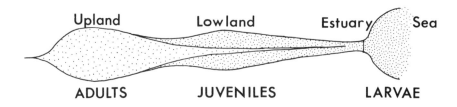

Upland   Lowland   Estuary   Sea

ADULTS   JUVENILES   LARVAE

known. Their life cycles resemble those of the Australasian *Gobiomorphus* discussed above.

Sicyopterus extraneus (Gobiidae) was thoroughly studied by Manacop (1953), who showed that the fish was amphidromous in Philippines waters (and not catadromous as earlier workers had asserted, e.g. Herre, 1927). It forms the basis of the 'ipon' fishery there. Manacop found that spawning is in fresh water and that the eggs develop rapidly, in a day or two. The tiny larvae, little more than a millimetre long, are swept to sea where they live for a month or so before they return to fresh water and move upstream, by this time being about 25 mm long. Manacop considered that *S. lachrymosus* and *S. fuliag* had the same life history strategy in Philippines rivers (although Herre, 1927, had argued that in these and other gobies the adults migrated to sea to spawn). The young of these fishes also form part of the 'ipon' fishery.

Erdman (1961) was less explicit, but reported a similar life cycle in *Sicydium plumieri* in the Caribbean. Clark (1905) believed that the adults migrated downstream towards the sea to spawn (making *S. plumieri* virtually catadromous), but this was not corroborated by Erdman (1961). The adults of *S. plumieri* live upstream in bouldery rivers, where Erdman (1986) believed that they spawn, with the larvae being swept downstream to sea. A return upstream migration of post-larvae/juveniles takes place at a length of about 25 – 27 mm during August to December.

Dotu and Mito (1955) suggested similar habits in *S. japonicum* in Japan. They thought that it spent the whole winter in the sea, following an autumn spawning, and that the young returned to freshwater in large numbers, at a length of about 32 mm. N. Mizuno (personal communication) advised that although not explicitly confirmed in Japan, amphidromy is also considered probable in *Sicyopterus macrostetholepis*, *Sicyopus leprurus*, *S. zosterophorum*, *Stiphodon elegans*, and *S. stevensoni*.

Schultz (1943) discussed *Sicydium taeniurus* from Tahiti, and concluded that 'these small fishes appear in enormous numbers during the first half of November and early December of each year and go into the mouths of the rivers'. Dr G. Marquet (personal communication) confirmed the occurrence of similar migrations in present day Tahiti, advising that occasionally the young of *Eleotris fusca* occur with them. The young fish are about 30 mm long at migration. Tomihama (1972) examined the life history of the Hawaiian species *Sicydium stimpsoni*, finding that it, too, is amphidromous. The adults spawn over the winter; the larvae hatch in only a day or so and are immediately swept to sea. They return after about five months at sea when about 25 mm long, and spend the remainder of their lives in fresh water. They live to a considerable age (perhaps 12 years). A similar fishery on Reunion Island in the eastern Indian Ocean (near Madagascar) is based on *S. lagocephalus* (Aboussouan, 1969), a species also present on the Comoro Islands (Teugels, Janssens, Bogaert and Dumalin, 1985). Reports of 'whitebait' from Samoan rivers (McDowall, 1984c) and of large numbers of gobioid fry in New Guinea (Lynch, 1965) may also refer to this or related species of goby. P. Ryan (personal communication) advised that several gobies migrate into the rivers of Fiji as juveniles — including species of *Stiphodon* and *Sicyopterus* (*micrurus*?) and possibly others. He found that the large females contain tens if not hundreds of thousands of minute eggs and he hypothesised that either the eggs or larvae are washed to the sea.

Koumans (1953) discussed *Sicyopterus gymnauchen*, which occurs widely in tropical southeast Asia and the Pacific islands. It, too, is the basis for a fishery for juveniles swimming up the smaller rivers at certain times of the year. Deraniyagala (1937) described it as entering rivers in Sri Lanka (Ceylon), mostly in May. He considered it to be catadromous but, in view of the habits of other species of *Sicyopterus*, amphidromy seems more probable. Erdman (1984) alluded to this study, and also to those by Herre (1927), Annandale and Hora (1925), Blanc, Cadenat and Stauch, (1968) and Aboussouan (1969), noting that these authors all assumed that the species they studied were catadromous. Perhaps by association with his study of *S. plumieri*, he implied that these species are amphidromous. The same seems to be true of *Chonophorus guamensis*, *Sicyopterus macrostetholepis*, and *Microsicydium elegans* in Guam (Kami, 1986). Study is needed.

The Hawaiian goby, *Lentipes concolor*, lives well inland in steep and swiftly flowing streams. Spawning is in fresh water (possibly throughout the year), and development takes place there. The larvae are carried to sea soon after hatching and they live there for a month or so before they return as post-larvae, up to about 20 mm long (Maciolek, 1977). The Japanese species *L. armatus*, inhabits fast-flowing streams like *L. concolor*, but its life history does not seem to have been described. It, too, may be amphidromous (N. Mizuno, personal communication).

Other groups of gobies are also evidently amphidromous. *Acanthogobius flavimanus* is a goby found in the Orient, from Japan, China and Korea, where Akihito *et al.* (1984) described it as occurring mostly in brackish water and along the shores and bays of estuaries, some reaching the freshwater parts of rivers. Okada (1960) said that it spawns in January to March, with the larvae pelagic in the sea, then becoming benthic at about 15–20 mm length. Lee *et al.* (1980) described this species as ascending into streams and lakes in North America (where the fish has become acclimatised, possibly accidentally as a result of ballast water discharged from ships). C. Swift (personal communication) advises that acclimatised populations in California thrive in estuaries without access to fresh water, and that this species only rarely penetrates above tidal fresh water. *A. flavimanus* is possibly amphidromous like other gobies discussed above. Akihito *et al.* (1984) listed *Rhinogobius brunneus* as amphidromous in Japan. The newly hatched larvae move to sea, and return several months later. Dr N. Mizuno (personal communication) advises that what has in the past been regarded as a polytypic species, *R. brunneus*, is more likely to constitute a species group with up to 11 species, seven of which might be amphidromous. *R. giurinus*, which occurs in central Japan, Korea, and south to Taiwan, is also amphidromous. Spawning is in winter and spring (July to October), in fresh water, and the larvae go to sea, and return in summer and autumn (September to November) (Akihito *et al.*, 1984). Other Japanese gobies that may be amphidromous include *Tridentiger obscurum*, *T. brevispinis*, *Chaenogobius castaneus*, two undescribed species of *Chaenogobius*, *Luciogobius pallidus*, and *L. guttatus* (N. Mizuno, personal communication).

Manacop (1953) listed goby species as possibly being similarly amphidromous in the Philippines, including *Chonophorus occellaris*, *Glossogobius giurus*, *G. celebius*, *Ophiocara aporos* and *Eleotris melanosoma*. Smith (1937) described *G. giurus* as being equally at home in the sea and fresh water and said that it breeds freely in captivity, perhaps implying fresh water. Bruton and Kok (1980) found *G. giurus* to be common in the fresh waters of southern Africa, and thought that it bred there.

R.E. Watson (personal communication) added species of *Stenogobius* to the list of amphidromous gobies, and considered that *Awaous* may also be. *A. stamineus* in Hawaii spawns from August to December; the adults migrate to the lower reaches to spawn and the eggs hatch in 24–48 hours. The larvae are swept to sea and return about four months later, when 15–20 mm long. They mature at age one. Inland penetration is only moderate as regards distance but they reach altitudes of about 500 m (Ego, 1954). Yerger (1978) described the life history of the American river goby, *Awaous tajasica* (Florida to Brazil), as far as it was then understood. He regarded the young and adults as occurring in fresh water, the adults moving to sea to spawn, with the larvae being marine and pelagic in habit. As such, this is a catadromous cycle, but if the adults do not migrate to sea to spawn, as Yerger assumed, the cycle is amphidromous like so many other gobies.

R.E. Watson (personal communication) observed that the fry of *Eleotris sandwichensis* (which occurs in the fresh waters of Hawaii) would perish if they were unable to escape into purely marine conditions. This implies some form of diadromy. Some Japanese *Eleotris* species are possibly amphidromous, e.g. *E. melanosoma*, *E. oxycephala*, (N. Mizuno, personal communication), while Nordlie (1981) concluded that the life cycles of Mexican species, viz. *E. pisonis* and *E. amblyopsis*, were amphidromous in the Tortugueros Estuary. The smallest examples of these two species were taken in the tismiche of the lower lagoon and estuary — the tiny post-larvae were evidently on their way back into fresh water from the sea (Gilbert and Kelso, 1971). The larger juveniles and the adults are found in tributaries of the lagoon. Lee *et al.* (1980) said that *Eleotris pisonis* prefers the low salinity upper estuaries, but may invade fresh water.

Many other gobioid fishes are known to occupy habitats of diverse salinities. Koumans (1953), in his large revision of the taxonomy of Indo-Australian gobies, listed many genera in addition to those already mentioned above — *Pseudoapocryptes*, *Butis*, *Boleophthalmus*, *Brachyamblyopus*, *Bostrichthys*,

Acentrogobius, Stigmatagobius, Bathygobius, Brachygobius, Hemigobius — all of which contain species that occur in both the sea and rivers. However, the taxonomies of these genera, let alone their life histories, are poorly understood. While most cannot yet be designated diadromous, some of them probably are. Clearly, amphidromy is widely present in the gobioid families, but much remains to be learned of just how widely, and about the details of these life cycles. Yerger (1978) considered that no species of true goby can complete its life cycle entirely in fresh water, and if this is correct, a variety of these and other river-living gobiid fishes may be diadromous.

The monotypic gobioid family Rhyacichthyidae contains only *Rhyacichthys aspro*, which is very widely distributed in the western Pacific (Japan, China, and south to New Guinea). Akihito *et al.* (1984) described this fish as amphidromous, but provided no further details.

Variation in the life cycles of gobies is widely present. Landlocked populations are common in some species of the eleotrid genus *Gobiomorphus* in New Zealand, e.g. *G. cotidianus*, but not in others, e.g. *G. huttoni, G. hubbsi,* etc. which are entirely amphidromous (McDowall, 1975, 1978a). Species congeneric with amphidromous species, e.g. additional species of *Gobiomorphus* in New Zealand, are known to be exclusively fresh water. Landlocked or at least freshwater-limited stocks of many other gobies seem to be recognised, especially in Japan, although the details of relationships between amphidromous and non-migratory stocks does not seem well worked out. Species in other amphidromous gobioid genera may be exclusively marine, and other gobioid genera and species are probably similarly variable.

7

Diadromy and Geography

The occurrence of diadromy in fishes is not geographically uniform; the over-all occurrence of diadromy varies geographically, and varies within species or within groups of related species. This variation can be examined in several ways:

1. One way is to examine how presence/absence of diadromy within species changes geographically;
2. It is also possible that the characteristics of diadromy itself may vary latitudinally within species, and this question needs examination;
3. Another issue that needs study is the frequency of occurrence of diadromy in regional fish faunas, with change in latitude; this may involve diadromy as a whole, and also the different forms of diadromy;
4. A further question involves the relative proportions of diadromous species in regional fish faunas with change in latitude.

And there are no doubt further interesting questions that might be addressed and which might provide some answers that would assist in elucidating the nature and significance of the phenomenon.

WITHIN-SPECIES VARIATION

Some diadromous fishes are clearly facultative in the occurrence of diadromy — they are diadromous only over a part of their geographical range, or diadromous and non-migratory forms may be widely sympatric, or have overlapping distributions.

Geographical variation in the presence/absence of diadromy within species has been discussed in several groups. Rounsefell (1958) summarised the situation for salmonids, finding that different species inhabit different ranges of latitude, but within each species there seems to be a greater degree of anadromy towards the north. Kendall (1935) reported that on each coast of both the

Figure 7.1: Varying life history strategies in differing populations of brook char, *Salvelinus fontinalis*, showing falling longevity and decreasing movement to sea with decreasing latitude
Source: Power 1980

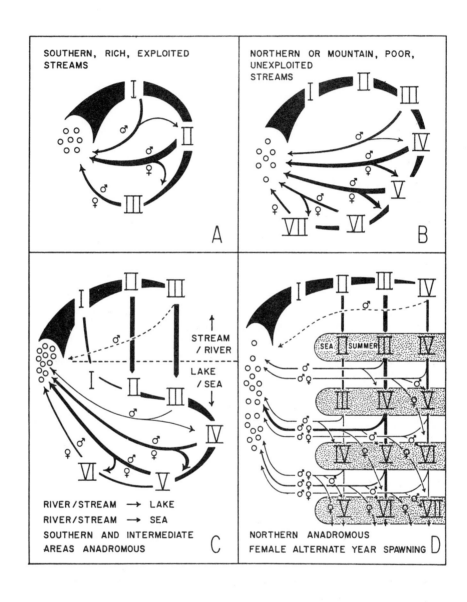

Pacific and Atlantic Oceans there are certain extents of southwards projections of the anadromous- *Salvelinus* zones. But these marine forms gradually disappear, becoming almost or quite exclusively freshwater inhabitants at the southern terminus of each range. As an example, the brook char, *Salvelinus fontinalis*, in eastern North America is anadromous in Canada and the northern New England states of the USA, but becomes restricted to cool mountain streams in Pennsylvania, Virginia and Georgia (Figure 7.1). The arctic char, *S. alpinus*, also tends to be more exclusively freshwater in habit in the southern parts of its range (Scott and Crossman, 1973); Vladykov (1963) suggested that in this species there is anadromy across the Arctic north of Canada, with the species becoming wholly freshwater in New England and the north coast of the Gulf of St Lawrence. Nordeng (1961) discussed the loss of anadromy in southern populations of alpine char in Scandinavia, and suggested that it disappears in areas that are warmed by the North Atlantic drift — at about 65°N along the coast of Norway. Loss of anadromy in southern populations of dolly varden, *S. malma*, is also likely (Scott and Crossman, 1973).

Similarly in Europe the brown trout, *Salmo trutta*, is widely although facultatively anadromous in the northern parts of its range, e.g. in Scandinavia and the British Isles. But at the southern end of its range, in the Mediterranean, no anadromous stocks of brown trout are reported. Outside its native range, in Australia and New Zealand at least, brown trout are confined more sharply to fresh water in northern (warmer) regions and become increasingly 'sea-run' in the south, where 'sea-trout' constitute important angler fisheries (McDowall, 1984b). Brown trout also certainly run to sea in southern Chile (Zama and Cardenas, 1984) and the Falkland Islands (Stewart, 1980), but I am unsure of its status further north in South America. The masou, *Oncorhynchus masou*, of Japan is known for the fact that a significant proportion of the young fail to go to sea, and mature instead in fresh water. Berg (1962) reported that only males behave in this way at the northern extremities of the range, but that both males and females do so further south.

Ekman (1953) reported that the threespined stickleback, *Gasterosteus aculeatus*, becomes more strictly fluviatile (i.e. non-anadromous) as latitude decreases; McKeown (1984) described it as almost entirely marine and only entering fresh water to breed in northern parts of its range, whereas in the south it is almost entirely fresh water in habit. Bell and Baumgartner (1984) found that in eastern North America it occurs in marine waters south as far as Chesapeake Bay, but generally is absent from fresh waters south of Maine. The striped bass, *Morone saxatilis*, is similarly more anadromous in the northern part of its range (Dudley *et al.*, 1977). Populations in rivers draining into the Mexican Gulf are all non-anadromous (Figure 7.2, Rulifson *et al.*, 1982a,b).

There are also some instances of loss of diadromy at northern (cold) extremes of species' ranges. This is reported, for instance, of the American bass, *Morone americanus*, which is diadromous around Chesapeake Bay, but not in the colder waters of the Canadian maritime provinces (Scott and Crossman,

Figure 7.2: Distribution of striped bass, *Morone saxatilis*, showing native range with anadromy in the northwest and non-anadromy in Gulf of Mexico populations, and with acclimatised Californian populations also anadromous
Source: After Setzler *et al* 1980

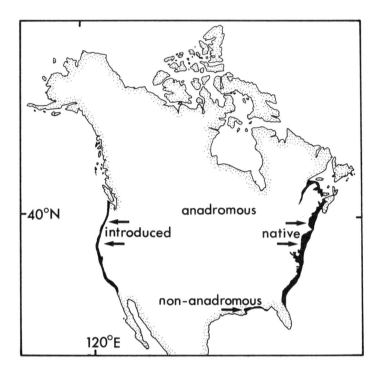

1973); loss of anadromy was attributed to the cold temperatures. Dadswell (1979) considered that anadromy was less well developed in northern populations of the shortnosed sturgeon, *Acipenser brevirostrum*, along the Atlantic coast of North America. Power *et al.* (1987) ascribed the failure of Atlantic salmon to move out to sea beyond estuaries in rivers flowing into Ungava Bay (Arctic Canada) to the very low water temperatures there. Dando (1984) showed that in the United Kingdom, the flounder, *Pleuronectes flesus*, is entirely estuarine except for its marine spawning and larval development in southern regions (Tamar River), but that in eastern Scotland it spends the summer at sea, while in the west of Scotland it is entirely marine in habit.

The life history strategy adopted by species varies geographically in other ways. Species from high latitudes tend to spawn first at greater age and have longer intervals between spawnings — often spawning is not annual. This is true, for instance, of Arctic char (Johnson, 1980). Dempson and Green (1985) found that Arctic char are generally alternate year spawners, although consecutive spawning may occur in southern latitudes. In the rainbow trout comparison of populations at northern (colder) and southern (warmer) localities in western

Figure 7.3: Variation with latitude in various life history features of rainbow trout, *Salmo gairdneri*
Source: After Rounsefell

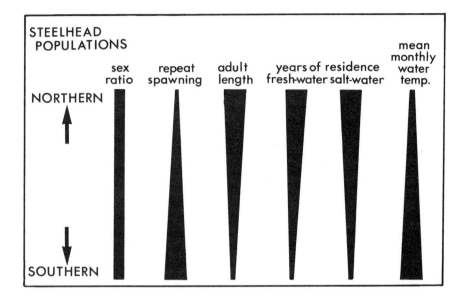

North America has shown the following variation: the number of repeat spawnings by individuals increases from north to south, adult size attained increases as does the length of time spent during both the marine and freshwater phases (Figure 7.3). Thus a longer time is taken in reaching maturity in cold, northern localities, resulting in larger adult size, presumably higher annual fecundity, but a higher level of commitment to one or few successful spawnings (Withler, 1966). The same may be true of the beluga, *Huso huso* (Berg, 1962). Leggett and Carscadden (1978) described how the strategy of the American shad, *Alosa sapidissima*, changes with latitude, from semelparous in southern, warm locations to iteroparous in northern cooler latitudes. There is also clinal variation in the timing of shad migrations with spawning being earlier in the south than in the north. This is probably widely true of diadromous species with broad latitudinal ranges. Thus, the runs of *Galaxias* whitebait in New Zealand are later in the south than in the north (McDowall, 1984c).

Baker (1978) has argued that 'where landlocked demes of otherwise anadromous species occur they are invariably found to terminate their downstream migration in a lake'. If Baker uses the expression 'landlocked' in a restricted sense to mean obligatorily confined to a lake system by the presence of an impoundment that prevents downstream sea access, then the statement is virtually a self-evident truth. In such circumstances the populations are prevented from going to sea by a dam which forms a lake at the downstream end of the species' range. If, as I think likely, he is using it more generally, to refer to normally diadromous species that have ceased going to sea, then the statement quoted above is only partly true. There are quite a few instances where species have voluntarily forsaken diadromy but in which there is not a downstream migration terminating in a lake. Well-known examples include fluviatile populations of rainbow trout/steelhead, *Salmo gairdneri*, in western North America, as well as in many lands where this species is acclimatised. Similarly this is true of arctic char, brook char, brown trout, striped bass and ninespined stickleback, amongst well-known northern hemisphere groups. It is true also of the southern inanga, *Galaxias maculatus*, (in Western Australia), and the Australian smelt, *Retropinna semoni*, (in eastern Australia — if this species is properly described as diadromous). Baker's generalisation is quite widely true, but is not, as he would seem to claim, 'invariably' true.

BETWEEN SPECIES VARIATION

It is widely recognised that there is latitudinal variation in the occurrence of diadromy. Several, workers, e.g. Nikolsky, 1963, have suggested that anadromy is primarily a phenomenon of high latitudes. Northcote (1979) concluded that reproductive migrations of fish from marine feeding areas to freshwater spawning habitats (anadromy) do not seem to be nearly so common in tropical waters as in temperate and arctic regions.

Others, however, have gone rather further than this in their claims. Gunter (1956), for instance, believed that all anadromous species are found in the northern temperate zone and further northwards towards the Arctic. Day, Blaber and Wallace (1981) claimed that:

> anadromous fish are virtually absent in the tropics and as far as we are aware, there are no endemic anadromous fish in the southern hemisphere, although the lamprey, *Geotria australis*, is known from Australia, New Zealand, and South America.

Berg (1959), in contrast, thought that anadromous fishes are peculiar to the temperate and to some extent the polar latitudes of both hemispheres and that the tropics lack them. Norman and Greenwood (1963) referred to a lack of diadromy of any sort in the Mediterranean, and attributed this to water temperatures. Haedrich (1983) on the other hand, recognising that catadromy is a 'rare phenomenon', and having observed that four of New Zealand's 27 freshwater fishes are catadromous, concluded that 'it may be that catadromy is really a feature of cold temperate waters in the Southern Hemisphere. In the Northern Hemisphere only the eels and mullets are catadromous. A comparison might prove interesting but the data are lacking'. Baker (1978) claimed that anadromy is maximal in polar-temperate environments and catadromy in the tropics, and suggested that only eels are catadromous in the tropics. McKeown (1984) accepted Baker's (1978) assertion and agreed that although there is an overlap, the majority of anadromous migrations occurs in cold-temperate and subpolar regions whereas catadromous migrations are more evident in tropical and warm temperate regions.

What data these views were based on is not clear in the above accounts, but some of the assertions are manifestly untrue. Berg (1959) and Day *et al.* (1981) are both wrong in saying that there is no anadromy in the tropics, Baker (1978) is wrong in claiming that only eels are catadromous in the tropics, and Day *et al.* (1981) are similarly wrong in arguing that there is none in the Southern Hemisphere.

The validity of assertions about the distribution and relative abundance of anadromy and catadromy in relation to latitude depends on what the assertions actually mean. Arguments that anadromy is more abundant in the cooler and catadromy in the warmer regions could refer to changes in the number of species that are anadromous with changes in latitude, or to the relative abundance of anadromy and catadromy (and amphidromy!) with change in latitude. The answers to these two questions may be quite different. Alternatively the question may relate to the relative proportions of the faunas that are anadromous, catadromous, or amphidromous, and to how these proportions vary with latitude. Each of these questions needs to be examined separately, and it should be noted that none of the above authors has given any consideration to the abundance and whereabouts of amphidromy.

The most straightforward of the above questions concerns, simply, where do anadromy, catadromy, and amphidromy occur, in relation to latitude, and how frequently? This matter can be examined by charting the latitudinal distribution of fish species in each of the categories of diadromy, and enumerating the number of species that are diadromous at each latitude. The relevant species are listed in the Appendix, together with their described latitudinal ranges, as far as I have been able to elucidate these. The numbers of species that are anadromous, catadromous, and amphidromous at each latitude have been plotted along the 90°N – 90°S gradient at 5° intervals. The data very clearly show that both by the proportion of diadromous species in the three categories (Figure 7.4) and by species number (Figures 7.5 – 7.9):

1. Anadromy is very strongly dominant in the northern sub-polar/cool temperate (Petromyzontidae, Acipenseridae, Salmonidae, Osmeridae, Cottidae etc.), declines through the northern warm temperate, the tropics and the southern warm temperate (primarily Clupeidae), and occurs sparsely in the southern cool temperate (Geotriidae, Retropinnidae, Aplochitonidae) (Figures 7.5 and 7.6).

2. By comparison, catadromy is present though sparse in the northern cool temperate (primarily Anguillidae and Mugilidae), rises through the warm temperate and tropics to a peak in the southern subtropics/warm temperate

(primarily Anguillidae, plus more Mugilidae, Centropomidae), and declines rapidly in the southern cool temperate where it becomes rare (some Anguillidae, Galaxiidae) (Figures 7.7 and 7.8).

3. Amphidromy is sparsely described from the northern cool temperate (Plecoglossidae, Cottidae, Gobiidae), but appears strongly in the northern subtropics and the tropics (primarily Gobiidae, Eleotridae), declines in the southern warm and recurs strongly in the southern cool temperate (Galaxiidae, Aplochitonidae, Prototroctidae, Mugiloididae, Eleotridae) (Figure 7.9).

These patterns are based on rather limited existing data of questionable reliability and which are likely to change considerably as knowledge of fish life history patterns grows; thus the data discussed here differ substantially from those I published earlier (McDowall, 1987a), which comprised a previous analysis of this question.

The data offer some support for Baker's (1978) view that anadromy is mostly cool temperate and catadromy mostly tropical, though the occurrence of anadromy in the tropics and catadromy in the cool temperate is stronger than he and most other authors have realised. If the data are broken down by group, a somewhat altered perspective is obtained, although it does not negate the above generalisations. This breakdown shows that lampreys, sturgeons, and

Figure 7.4: Geographical variation in the occurrence of diadromy, as expressed in anadromy, catadromy and amphidromy; the proportion of diadromous fish at any latitude which belongs in each category of diadromy is shown

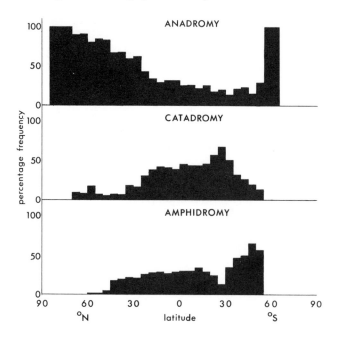

Figure 7.5: Frequency distribution of anadromy, with latitude

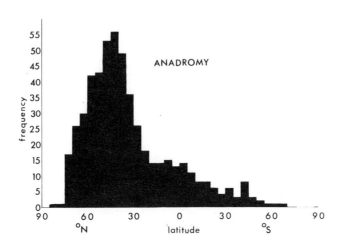

Figure 7.6: Breakdown of frequency distribution of anadromy with latitude by major taxonomic group

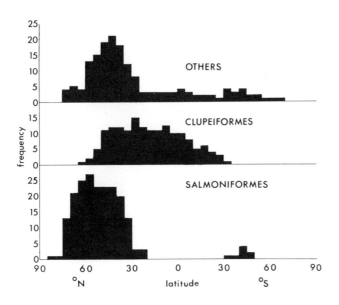

Figure 7.7: Frequency distribution of catadromy with latitude

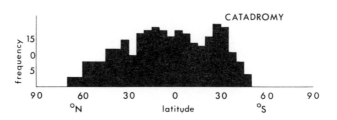

Figure 7.8: Breakdown of frequency distribution of catadromy with latitude by major taxonomic group

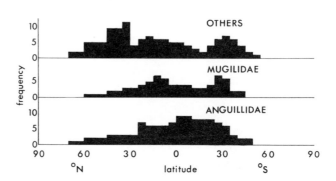

Figure 7.9: Frequency distribution of amphidromy with latitude

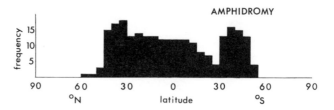

several salmoniform families are cool temperate anadromous fishes (Figure 7.6) but these families are cool temperate fishes anyway. Clupeiformes are shown to be widely anadromous from the northern cool temperate through the tropics to the southern warm temperate, and possibly this, too, is representative of the distribution of the group. Analysis of catadromy (Figure 7.8) shows that a large majority of catadromous fishes, in temperate and tropical latitudes, comprises anguillid eels and grey mullets, and both these groups are temperate to tropical in range. Thus, in arguing that anadromy is characteristic of colder areas and catadromy of the tropics, we are saying little more than that certain groups (which happen to be cool temperate) are anadromous, while others (which happen to be catadromous) are essentially warm temperate to tropical in range. Whether there is a causal connection in these relationships, is a much more complex question that is not easily examined rigorously. Analysis of aquatic biological productivity shows that it varies with latitude, and that it varies clinally. Marine habitats are more productive at high latitudes and less so towards the tropics, whereas freshwater habitats have greatest productivity towards the tropics and less towards the poles (M. Gross, personal communication). This is consistent with the direction of migrations in diadromous fishes — they are anadromous at high latitudes spending most of their lives feeding at sea, and are catadromous at low latitudes, living for most of their feeding lives in fresh water.

PROPORTIONS OF FAUNAS THAT ARE DIADROMOUS

The question of the proportions of faunas that are anadromous, catadromous and amphidromous, and how these proportions vary geographically, is also complex, and data to clarify the question are sparse. The freshwater/diadromous fish faunas of most areas of the world are suprisingly poorly documented in a way that makes calculation of the proportions that are diadromous possible. What is needed to collate the information is a fauna of which the distribution is well documented, together with details of which species are diadromous and over what extent of their distributional range. These faunas need to be continental faunas, because I believe that the proportion of the fauna that is diadromous becomes distorted on islands. Useful and interesting information on this matter can be obtained for the Atlantic (east) and Pacific (west) coasts of North America from Lee *et al.* (1980).

Using Lee *et al.* (1980) as a basic resource, I have enumerated the number of species occurring at each degree of latitude, separately for the east and west coasts of North America. Species listed include only those freshwater species that occupy coastal drainages between Point Barrow and Cape San Lucas (west coast) and between Cape Chidley and the tip of the Florida Peninsula (east coast). Species were subdivided according to whether they were diadromous or not, and diadromous species were separated into the three subcategories

Figure 7.10: Distribution of frequency of anadromy (shaded) and total fish diversity along the Pacific coast of North America

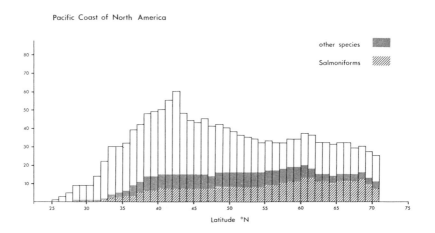

anadromous, catadromous, and amphidromous.

For the west coast of North America, the following points are evident (Figure 7.10). Faunal diversity varies widely but is quite low, with a maximum of about 60 species at about latitude 42°N, and a broad peak between latitudes 35 and 50°N. The absolute abundance of anadromy along this coast is relatively stable at about 10–15 species from about latitude 38°N northwards to the limits of analysis at 70°N. Amongst these anadromous species the proportion of salmoniforms (Salmonidae, Osmeridae) rises, though slightly, to the north and of other groups to the south; thus most northern anadromous species are salmoniforms. Catadromy is sparsely represented and shows slight reduction in frequency with increasing latitude. Amphidromy is represented by only the occasional species and no significant trend is evident.

On the east coast of North America (Figure 7.11), overall diversity is much greater, reaching a peak of about 125 species at latitude 35°N, with a concomitantly much greater decline to the north and south where diversity is much lower. Diversity at high latitudes is comparable with that of the west coast, with 20 – 30 species, north of about latitude 50°N. Anadromous fishes contribute relatively sparsely to the total fauna, and reach a peak of about 20 species at latitudes 40–45°N, this being distinctly north of the peak diversity for all species (35°N). Major contributors to anadromy are clupeids and salmoniforms (Salmonidae and Osmeridae), and these have overlapping but complementary distributions with clupeids being more southern and salmoniforms tending to replace them to the north. Catadromy in eastern North America is

Figure 7.11: Distribution of frequency of anadromy (shaded) and total fish diversity along the Atlantic coast of North America

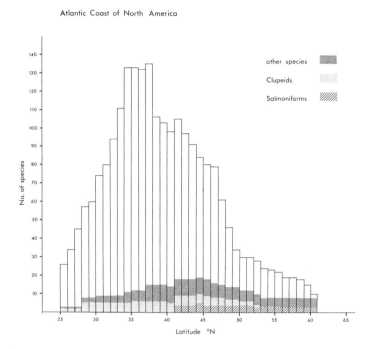

Atlantic Coast of North America

sparsely represented, as on the west coast, and declines with latitude; amphi-dromy is even rarer.

Taken together these data suggest that through warm temperate to subpolar environments, anadromy is present in a small and relatively constant number of fish species but the relative contribution of anadromy to the biotas varies inversely with species diversity and is greatest towards the polar extremes of the region. Most of this diadromy is anadromy, and catadromy is so sparsely represented that it can scarcely be regarded as a significant phenomenon amongst North American freshwater fishes. Amphidromy can largely be discounted on the basis of present knowledge, but this view might change significantly as the faunas become better known.

A comparable analysis of South American fish faunas would be very interesting, but data are insufficient to make it possible. However, there is enough information to show that in far southern Patagonia, all the species present may be diadromous (Geotriidae, Galaxiidae, Aplochitonidae which are anadromous or amphidromous). However, the fauna itself is very sparse; Tierra del Fuego has no more than about five species, four of them certainly, perhaps all of them, diadromous. These southern groups extend northwards in range for varying distances, gradually disappearing, and at the same time the fauna becomes augmented by additional, non-diadromous groups. These additions include

121

only two non-diadromous representatives of the southern diadromous families (the galaxiids *Galaxias globiceps* and *Brachygalaxias bullocki*), and the other species in the fauna are primarily southern outliers of the vast ostariophysan faunas of the tropical-temperate Amazon basin and other large eastern South American river systems. In addition, there are a few groups of less obvious origins, particularly a few basal percoid fishes, family Percichthyidae, of uncertain relationships (Campos, 1977). Detailed analysis of these faunas and of the occurrence of diadromy is not possible owing to the meagre knowledge of the ecology of the fish faunas of northern Chile and Argentina, and areas to the north; however, diadromy in any form seems to be sparsely represented, and not nearly as abundant as in either polar-temperate northern lands, or Australasia. What can be said is that both absolute abundance and relative abundance of diadromous fishes declines with decreasing latitude from far southern South America. Were similar data available for sub-Arctic western Europe and eastern Asia, I have little doubt that a similar pattern would emerge.

Haedrich (1983) concluded that with a change in latitude there is a shift in the relative phylogenetic position of the dominant families that are found in estuaries — some of which are diadromous. He suggested that this was in keeping with the general distribution of the world's fish faunas. The more primitive groups are found in estuaries at high latitudes and the more advanced groups nearer the tropics. A pattern of this sort cannot be recognised in the diadromous fishes, and is upset primarily by the fact that the great majority of diadromous fishes has primitive origins. The lampreys, sturgeons, the various salmoniform families, the herrings/shads, and the freshwater eels can all be aptly described as groups that have ancient and primitive origins in the phylogeny of fishes. More advanced groups have occasional diadromous representatives, but they are variously subpolar to tropical in occurrence — groups like cottids and sticklebacks are strongly represented in the colder, high latitudes, while mullets, gobies and others are more tropical in occurrence.

BIOGEOGRAPHICAL PATTERNS IN DIADROMOUS FISHES

Reference was made above to the distortion of the proportions of faunas that are diadromous caused by the faunas being on islands. Although the subject is a contentious one, I believe that diadromy has played an important role in the development of the biogeographical patterns of freshwater fishes, especially in the widely separated and remote lands of the temperate Southern Hemisphere (McDowall, 1978b); it is probably generally true. 'The biological characteristics of an organism which govern its possible means of dispersal are an important consideration when discussing the organism's zoogeography' (Dadswell, 1974). One of these important characteristics is probably diadromy. Dymond and Vladykov (1934) drew attention to the fact that the northern temperate salmonids, which are essentially freshwater in their habits, tend to be strictly limited in distribution to one or other of the northern continents. This is true of *Hucho* and *Brachymystax* in Asia and Europe. Referring to the genus *Coregonus* they noted that the anadromous species tend to be identical on both sides of the northern Pacific, but that the freshwater groups are represented by distinct species in the two areas.

Relatively recent land connections are available and may explain the ranges of the Pacific salmons (*Oncorhynchus* spp.) on each side of the far northern Pacific Ocean. Even so, the very wide-ranging oceanic habits of these species may equally explain the fact that diadromous salmonids and also osmerids tend to occur very widely along both eastern and western margins of the northern Pacific while non-diadromous species tend, rather, to be more locally distributed. Similarly, Atlantic salmon, *Salmo salar* (although not brown trout, *S. trutta*) very widely present in lands bordering the North Atlantic on both the eastern and western sides. Other diadromous species like sea lamprey, *Petromyzon marinus*, arctic char, *Salvelinus alpinus*, threespined stickleback, *Gasterosteus aculeatus*, and ninespined stickleback, *Pungitius pungitius*, also have similarly very broad ranges.

Widely distributed Southern Hemisphere species have much more disjunct distributions. The southern pouched lamprey, *Geotria australis*, is present in Australia, Tasmania, New Zealand, Chile and Argentina (Figure 7.12); the inanga, *Galaxias maculatus*, occurs in all these areas plus Lord Howe Island, the Chatham Islands, and the Falklands. Somewhat less widely distributed is the koaro, *Galaxias brevipinnis*, which is known from southeastern Australia, Tasmania, New Zealand, the Chatham, Auckland and Campbell Islands (McDowall, 1978a). Any land connections relevant to the generation of such broad distributions must date back to the occurrence of Gondwanaland in late Mesozoic and early Tertiary, some 60–70 million years ago. It seems to me that the relationship between the occurrence of diadromy in these fishes, (and also some other southern salmoniforms), and their remarkably wide distributions, is not coincidental, but rather that they have attained these distributions because they are able to live in and disperse through the sea. All of these fish species routinely, some obligatorily, spend a period of several months to several years living and feeding in the sea. Some are known to spread widely in the sea and disperse substantial distances from the source land area (McDowall *et al.*, 1975; Potter *et al.* 1979). In New Zealand, the non-diadromous species of freshwater fish, particularly the galaxiids, tend to have much more restricted distributions than the diadromous ones. These distributions can often be related to known hydrological, geomorphological and tectonic events (McDowall, 1970, 1987), and this suggests that the lack of diadromy has restricted their dispersal, limiting gene flow between isolated populations, and has thereby contributed to localised speciation. By contrast, of 17 diadromous species in the New Zealand fauna, all but one are found around the entire coastline. Several, and only diadromous species, are found on many of the small

Figure 7.12: Broad southern cool-temperate distribution of the pouched lamprey, *Geotria australis*

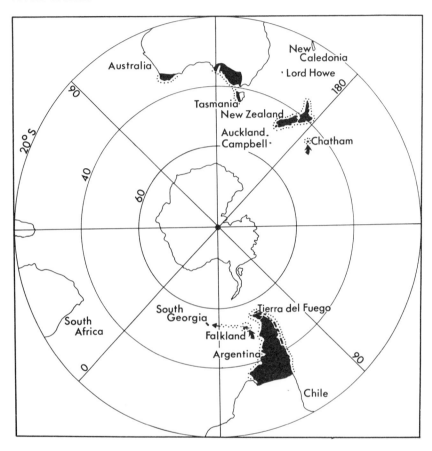

islands associated with the New Zealand continental shelf. It seems to me that the existence of diadromy has facilitated both widespread dispersal and continued gene flow between geographically separated populations, thereby retarding the speciation process in geographically disjunct populations of diadromous species. Thus, although Campos (1973) has suggested that the presence of inanga, *Galaxias maculatus*, on the Chatham Islands, some 500 km east of New Zealand, is due to a vicariance event (and thus a former land connection that is widely accepted as having occurred) examples of this fish have been taken at sea at a distance from New Zealand equal to that of the Chatham Islands (Figure 7.13, McDowall *et al.*, 1975). Thus dispersal of the fish this distance is clearly possible and would satisfactorily explain the presence of the fish on the Chathams. Furthermore, the fact that all the 10 species of freshwater fish known from the Chathams are diadromous is suggestive of the role of dispersal in establishing and/or maintaining that fauna. If this is so, it is dif

Figure 7.13: Sea distribution of inanga, *Galaxias maculatus, around the seas of New Zealand*

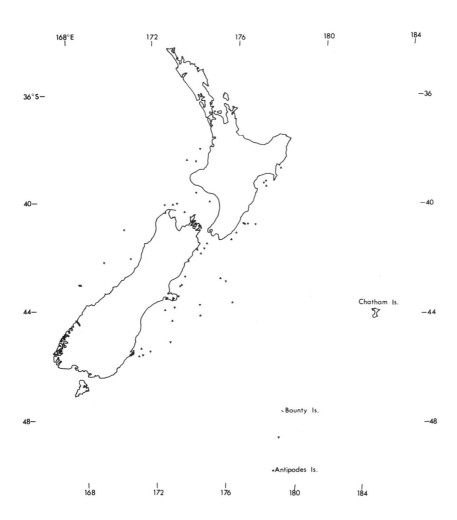

ficult to know with any certainty whether these wide distributions and the presence of diadromous fishes on distant, isolated islands, is a result of an original, long-past dispersal event with the possibility of continuing dispersal/gene flow maintaining the taxonomic similarity of the isolates, or whether the dispersal event has been a relatively recent one. I suspect the latter.

There appear to be some instances of speciation in isolated stocks of diadromous species. The Australian and New Zealand species of southern grayling,

125

genus *Prototroctes*, are distinct, as are the species of *Retropinna* and *Gobiomorphus* in the two regions. Three large, stout-bodied New Zealand diadromous galaxiids seem to have close affinities with the Australian diadromous *Galaxias truttaceus*, and if this relationship is real, it implies genetic isolation and speciation. By contrast, other species discussed above are common to eastern Australia, Tasmania and New Zealand, perhaps implying recent dispersal and/or continuing gene flow.

If, as I argue, the presence of diadromy has played a role in the relatively recent dispersal of extant fishes around the southern temperate (and presumably elsewhere too), and has had a role in the development of existing geographical distribution patterns, then it is equally likely to be true that this has occurred in the past — both recent and ancient — and that diadromy has played a formative role in the establishment and development of the fish faunas of both the main southern land masses of Australia, South America and Africa, as well as the more isolated, smaller, land areas like New Zealand, New Caledonia, the Auckland and Campbell Islands, Chathams, Falklands, etc. This is a hotly disputed question (Croizat, Nelson and Rosen, 1974; Rosen, 1974; Craw, 1978, 1979; McDowall, 1978b). Some observers have drawn comparisons between the distributions of such southern fishes as galaxiids and lampreys, and those of other 'classical southern temperate' groups, like the southern beeches, genus *Nothofagus*, the podocarps, chironomid midges, and ratite birds. Such similarities are certainly real (Figure 7.14). These and other groups are interpreted as having attained their distributions as a result of their ancient occurrence on Gondwanaland (Figure 7.15), and its subsequent dispersal around the southern cool temperate, as Gondwanaland fragmented during the Tertiary. The hypothesis is that the dispersing land masses carried their biotas with them. Whether this is true or not remains hypothetical (though likely), but Sibley and Ahlquist (1981) have suggested that the date of disjunction of stocks of ratite birds is sufficiently ancient to have occurred at the time that South Africa and South America broke away from Gondwanaland. Whether the southern pouched lamprey, and galaxiids such as the inanga, attained their very broad ranges by the same means and at about the same time, is clearly arguable. It would require a remarkable level of phenotypic stability of widely separated populations over a period of 60 million or more years (Horton, 1984), which seems to be an unlikely prospect, especially in galaxiids which now appear to be speciating actively in some areas of the family's range, where stocks become geographically isolated; this is especially true in Tasmania (McDowall and Frankenberg, 1981). The fact that these groups have marine life history stages suggests a plausible alternative — that these fish attained their present distributions much more recently, by transoceanic dispersal of marine stages.

I am hopeful that this question will be clarified in the not too distant future by studies of DNA/DNA hybridisation in the Galaxiidae and related families (Sibley and Ahlquist, 1981) This technique should enable us to establish relative dates at which there was last gene flow between species with disjunct

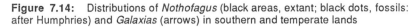

Figure 7.14: Distributions of *Nothofagus* (black areas, extant; black dots, fossils: after Humphries) and *Galaxias* (arrows) in southern and temperate lands

populations or species groups with a common ancestry that are separated by barriers. Consistent relative and great ages of separation in various stocks in areas like Australia, New Zealand and South America will tend to suggest a common, identifiable cause for that separation, such as the fragmentation of Gondwanaland. But the finding of highly varied and/or much more recent dates of separation will point to much more varied causes and we would need to look at both geological causes and ecologically related causes like dispersal. In the meantime the wide ranges of species covering vast ocean gaps, and the fact that these widespread species are, without exception, diadromous, appear to point to a pivotal role for diadromy in creating distribution patterns.

The role of dispersal of diadromous fishes in establishing fish faunas is possibly also indicated by the contribution such species make to the faunas of islands. It is unlikely to be merely a coincidence that all the freshwater fishes in Ireland, Iceland (Wheeler, 1969), Greenland, the Shetlands, Orkneys, and

Figure 7.15: Present-day distribution of galaxiid fishes as represented on a reconstruction of Gondwanaland, about 70 million years ago

Hebrides (Balon, 1968), Hawaii (Sakuda, 1986), Guam (Kami, 1986), Mauritius (Jehangeer, 1986), the Comoros (Teugels *et al.*, 1985), Lord Howe, the Chathams, Aucklands and Campbell (McDowall, 1978a), Tierra del Fuego, and the Falklands are all diadromous or of marine derivation (Figure 7.16). The faunas of some of these islands were probably extirpated during the Pleistocene glaciations, so that the present faunas comprise species that have been able to reach the islands across oceanic gaps since the glaciations retreated.

Caldwell (1966) reported that all the fish in Jamaican fresh waters are either 'secondary' types or are marine species which enter fresh water periodically or at a point of their regular life histories. A detailed analysis of the fish faunas of additional islands in many parts of the world would probably reveal the same sort of pattern — faunas made up largely of species of marine derivation or diadromous species.

Possibly, too, occupation of small islands by fish species is often temporary owing to the instability of aquatic habitats and/or the small sizes of the populations that the streams can sustain. Possibly occupation of small islands is further hazardous because the returning sea-migrants do not always locate the land areas to which they are returning if their navigation during migration is not

Figure 7.16: Islands on which all known freshwater fish species are diadromous (there are probably many others in addition to those indicated)

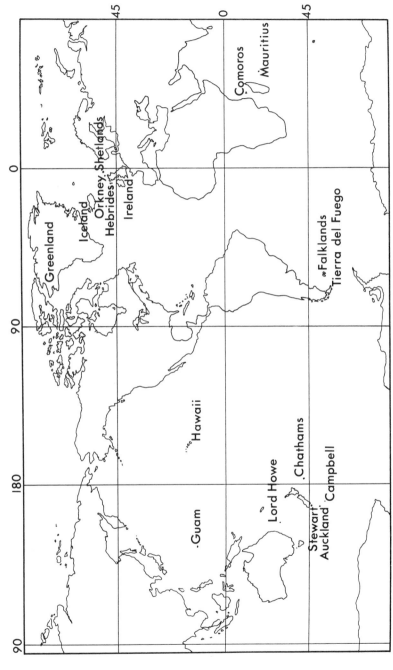

sufficiently accurate. In any of the above scenarios, dispersal can be seen as having an important role in establishing and maintaining the faunas of isolated islands by dispersal from adjacent land.

Although this may seem a strange question to raise, diadromy may also have had a role in species extinction. On essentially theoretical grounds this could be true of the southern grayling, *Prototroctes oxyrhynchus,* in New Zealand. Decline and extinction of this species is likely to have resulted from modification and deterioration in its habitat that followed extensive deforestation, the establishment of pastoral farming, and the release of exotic salmonid fishes. The larvae and juveniles of this fish may have dispersed widely in the sea during the marine phase. Assuming the absence of homing by the fish returning to fresh water (for which there is no evidence), then the progeny, produced in waters where good populations persisted owing to the retention of suitable habitats, are likely to have spread widely into nearby and perhaps distant river systems that were no longer suitable for the species. Return migrations to the home water are therefore likely to have been seriously depleted, this contributing to and accelerating any decline intitiated by general habitat deterioration or other causes. Population decline of this sort would be obviated either by loss of diadromy or the occurrence of homing. Reasons for the extinction of *P. oxyrhynchus* are poorly understood (McDowall, 1978a), but it occurred very rapidly and early in the settlement of New Zealand. This fish was described as depleted in some areas of New Zealand by the 1870s, only about 25 years after European settlement really began; it was in general decline by the 1900s and extinct by the mid-1930s. Some mechanism, related to diadromy as discussed here, may have been critical to the pace of extinction and possibly a contributor to its apparent finality.

8

The Origin and Evolution of Diadromy

Having reviewed the systematic and geographical occurrence of diadromy, it is time to draw some threads together, to determine what the occurrence of diadromy tells us about the phenomenon, and derive some generalisations.

PHYLOGENETIC PATTERNS IN DIADROMOUS FISHES

In the detailed analysis earlier of the occurrence of three forms of diadromy in the various families of fishes, a number of generalisations becomes evident:

1. Certain groups have diadromy as a strongly represented, if not dominant life history strategy. This is true of the three families of lampreys (Petromyzontidae, Geotriidae, Mordaciidae), eels (Anguillidae), sturgeons (Acipenseridae), and several salmoniform families (Salmonidae, Osmeridae, Salangidae, Retropinnidae, Prototroctidae, Plecoglossidae, etc.).

2. Certain other groups have diadromy represented as a rarity — an occasional species or species group is diadromous. This is true of diverse perciform families.

3. In addition, in some groups diadromy is a quite widespread, if not strongly represented phenomenon, such as the herrings and shads (Clupeidae), galaxiids (Galaxiidae) and mullets (Mugilidae).

A further generalisation is that groups in which diadromy has a strong, even majority representation, tend to be phylogenetically primitive. Table 8.1 lists the families in which diadromy has been reported reliably, and indicates the number of species in each family (from Nelson, 1976), with the number of species that is reported to be diadromous, and the percentage that is diadromous in each. These data illustrate the high levels of diadromy in a range of primitive fish families. In most of such families, the number that is diadromous is 50 per cent or more; in the remainder the proportion is usually much lower, frequently less than 10 per cent, and where more than 10 per cent this is

Table 8.1: Proportions of families that are diadromous

Family		Diadromous species	Total species	Percentage diadromous
Petromyzontidae	*	6	33	18.1
Geotriidae	*	1	1	100.0
Mordaciidae	*	2	3	66.7
Acipenseridae	*	11	27	41.4
Anguillidae	+	15	15	100.0
Salmonidae	*	27	68	39.7
Osmeridae	*	6	12	50.0
Salangidae	*	11?	14	78.6
Plecoglossidae	#	1	1	100.0
Retropinnidae	*	3	4	75.0
Prototroctidae	#	2	2	100.0
Aplochitonidae	*#	3	3	100.0
Galaxiidae	#	8	36	22.2
Clupeidae	*+#	32?	180	17.8
Pristigasteridae	*	2?	33	6.4
Engraulidae	*	1?	110	0.9
Ariidae	*	1?	120	0.8
Syngnathidae	*?	1?	175	0.6
Gasterosteidae	*	2	8	25.0
Gadidae	*	1	55	1.8
Scorpaenidae	+	1	300	0.3
Percichthyidae	*+	4	40	10.0
Centropomidae	+	1	30	3.3
Kuhliidae	+	3?	12	25.0
Mugilidae	+	14?	70	20.0
Cottidae	*#	6	300	2.0
Mugiloididae	#	1	26	3.8
Bovichthyidae	+	1	6	18.6
Gobiidae	*+#	45?	800	5.6
Eleotridae	#	9?	150	6.0
Rhyacichthyidae	#	1	1	100.0
Pleuronectidae	+	3?	99	3.0
Soleidae	+?	1?	117	0.8

Notes: * anadromous; + catadromous; # amphidromous; +# galaxiidae;
? highly doubtful.

likely to be due to the presence of one or two diadromous species in very small families. The families in which diadromy predominates include the agnathans (the lampreys) which have likely origins in the Silurian, more than 400 million years ago, and the chondrosteans (the sturgeons, which may date back to the Permian, about 250 million years ago) (Jarvik, 1968). Otherwise strongly diadromous taxa tend to be primitive teleosts — the anguilliform eels, the clupeids, and the various freshwater salmoniform families, all of which probably have their origins long ago in the earliest radiations of the teleostean fishes. This origin is likely to be well back into the Mesozoic, some 100 million or more years ago, e.g. Tchernavin (1939) pointed to an origin of the salmonids during the Cretaceous. Nevertheless there are plenty of ancient and primitive groups in which there is no diadromy, as in the entire Chondrichthyes, the several classes of 'lungfishes', and the ostariophysans, all of which also, probably, have ancient Mesozoic origins.

In other major and more advanced fish groups diadromy is either largely lacking (paracanthopterygians), or is isolated and intermittent in occurrence (acanthopterygians). In no family in these major taxa can diadromy be described as a 'general feature' of the group. Rather, diadromy appears haphazardly amongst the families that are otherwise variously marine, marine/euryhaline, or marine/euryhaline/freshwater. One of the most striking examples of the erratic appearance of diadromy in a perciform family is the New Zealand torrentfish, *Cheimarrichthys fosteri*, a member of the otherwise wholly marine family Mugiloididae, that shows no other tendency to euryhalinity. *C. fosteri* has successfully taken spawning and development into fresh water. Similarly the largely marine families Bovichthyidae and Gadidae have successfully invaded fresh waters, the Bovichthyidae having a single catadromous species while the Gadidae is represented by one anadromous species and another that is freshwater in habit. Associated generalisations about these groups are that where diadromy is a dominating phenomenon in the group, reproduction in that group is often very strongly (though not absolutely) locked into occurring either in fresh water or the sea. Thus, lampreys, sturgeons, and the various salmoniforms seem almost rigidly tied to reproduction in fresh water, whereas freshwater eels are rigidly locked into marine reproduction. It is probably no coincidence that eels as a whole (Anguilliformes), which are, apart from anguillids, almost entirely marine, all breed in marine habitats. And where diadromy is lost in those groups that are strongly diadromous, as in landlocked populations of diadromous species, or in wholly freshwater or marine species derived from diadromous congeners, the entire life cycle normally takes place in the habitat where reproduction of the diadromous species occurs. However, this is not entirely true in the freshwater smelts (Osmeridae) in which there are several marine spawning species, nor in either the herrings and shads (Clupeidae), the mullets (Mugilidae), or the gobies (Gobiidae, Eleotridae), where diadromy is quite widespread and variable in character (anadromy and/or catadromy may be represented). In these families deviation from a diadromous life cycle is equally variable and may result in reproduction in either marine or freshwater habitats.

Families that are diadromous are usually consistently either anadromous or catadromous or amphidromous, although there are exceptions. For example, diadromous galaxiids are largely amphidromous, although one species (*Galaxias maculatus*) has 'taken' its spawning habitat from upstream riverine habitats downstream into estuaries to become 'marginally catadromous'. Most diadromous clupeids are anadromous but it seems that the Australian *Potamalosa* is catadromous and the African *Escualosa thoracata* amphidromous; most diadromous gobiids are amphidromous though some, e.g. *Leucopsarion petersi* and *Batanga lebretonis* may be anadromous and others like *Gobiomorus dormitor* catadromous. Where diadromy is a much more isolated phenomenon in a family, there are no such 'rules' about where reproduction takes place and confamilial or even congeneric species may have diverse reproductive habitats. This is true, for instance, of percichthyids and cottids. Although their diadromy is not well authenticated, there seem to be both catadromous and amphidromous cottids. Of the diadromous percichthyids, at least one is anadromous (*Morone saxatilis*), and one is catadromous (*Macquaria novemaculeata*), but too much should not be made of this as the family Percichthyidae is generally regarded as a 'catch-all', basal-level perciform family that may be polyphyletic. Exceptions to this generalisation are thus quite few.

THE ORIGINS OF DIADROMOUS GROUPS

There has been much and prolonged discussion about where various fish groups evolved, and this is a question that has been both contentious and unresolved with any certainty. Whether or not specific fish groups evolved in fresh water, as has been widely suggested for the salmonids, or in the sea, as seems to be believed for the perciforms, the existence of diadromy, in which there is an alternation between freshwater and marine habitats, implies the secondary evolution of tolerances to both habitat types and to a wide range of salinities — what Baker (1978) described as habitat switching. It seems unlikely to me that diadromy is a primary condition in fishes as a whole.

Several workers have explored the question of the evolution of diadromy with regard to a very limited array of fish groups, primarily the anguillid eels and the salmonids. Only Baker (1978) has attempted to review the evolution of diadromy in a general sense. He discussed at some length what he called the 'anadromy/catadromy paradox', finding it surprising that:

the existence in the rivers of some parts of the world of two groups of fish, one of which breeds in the sea and migrates to the sea to feed, and the other of which breeds in the sea and migrates to freshwater to feed, has received little detailed consideration.

McKeown (1984) echoed this view, with no elaboration. This paradox, if indeed there is one, is further complicated by the existence of amphidromy, which represents a further life history/migratory strategy (or tactic? Gross, 1987) adopted by fish which also coexists with anadromy and catadromy in many river systems or geographical areas. However, to describe the coexistence of anadromy and catadromy simply as a paradox, is grossly to oversimplify the great diversity of strategies within either of these categories, and to seek for a general model that explains either anadromy, catadromy or amphidromy is, I think, a hopeless task. Baker's very restricted understanding of the diversity of behavioural patterns of various diadromous fishes renders his attempts to analyse the selective values of these two general strategies, and to explain their evolution, of limited worth. His analysis is restricted to lampreys, sturgeons, salmonids, eels and sticklebacks, it fails to explore the diverse forms of anadromy and catadromy, even within these families, and does not address the diadromy that is found in a wide range of other groups. Having reached conclusions, on the basis of his rather limited analysis, that anadromy is primarily cold and cool temperate and catadromy warm temperate in distribution, Baker sought to analyse how these habits evolved. He concluded, without obvious evidence, that the latitudinal effect of selection occurs at the 'river/sea boundary', suggesting that:

> Latitudinal variation implies ... that in tropical, subtropical and warm temperate regions the selection for fish to cross from sea to rivers is reinforced and/or selection for fish to cross from rivers to sea is reduced. In cold temperate and subpolar regions, however, the selection for fish to cross from sea to rivers is reduced and/or the selection for fish to cross from river to sea is reinforced. In temperate regions ... the selection may continue to act in both directions.

Baker apparently has not appreciated that all diadromous fishes, whether cold temperate or tropical and whether anadromous or catadromous, undertake a migration in both directions, and, as such, it is hard to see how even his very general suggestions have much significance to the evolution of diadromy. Furthermore, although anadromy is more common in the cold temperate and catadromy is more common in the tropics, both phenomena are present at almost all latitudes to the extent that generalisations with regard to the evolution of anadromy and catadromy in relation to latitudinal variation seem unlikely to be easily developed.

McKeown (1984) suggested that catadromy is derived from oceanodromous migrations, i.e. migrations entirely in the sea, while anadromy is derived from potamodromous migrations, i.e. those that occur entirely in rivers, the implication being that in each case, the migration already existing is extended across osmoregulatory barriers. This is not inconsistent with the observation that some anadromous species seem to have had freshwater origins and catadromous species marine origins, but there is no explicit evidence to suggest that either form of diadromy was preceded by migrations of other sorts. Nor does it cope with the fact that some anadromous or freshwater amphidromous species have marine origins, or that some confamilial species are variously anadromous and/or catadromous and/or amphidromous. Gross (1987) argues that anadromy evolved from, initially, a freshwater wanderer that moved into the sea, became amphidromous as movements became regular, and ultimately anadromous. Catadromy, according to Gross's model, evolved from a marine wanderer that moved into fresh water, the movement becoming regular leading to amphidromy, and ultimately catadromy. Although this model may have quite broad application (both anadromy and catadromy have undoubtedly evolved many times in independent lineages of fishes), there are some instances where the model seems inapplicable, e.g. the inanga *Galaxias maculatus*, in which catadromy seems to have evolved from amphidromous species that spawned in fresh water. Gross's model might be taken to provide an indication of the spawning habitat of the type from which diadromous species originated, but this, too, is not always true, as indicated by freshwater spawning in numerous amphidromous gobiids and eleotrids, and in the amphidromous torrentfish, *Cheimarrichthys fosteri* (McDowall, 1973), all of which can be regarded as having had a marine ancestry.

Baker's general thesis is not helped by an 'escape clause' in which he suggested that the selective pressures involved in specific cases may be sufficiently great to reduce or even reverse the general trends. Baker concluded by stating that:

The hypothesis of residual selection throughout the world for fish to cross the river/sea boundary in both directions to which is added further unidirectional selection, the sign of the gradient which shows latitudinal variation, seems sufficient to account for a large proportion of the characteristics of the anadromy/catadromy paradox.

I suggest that, in fact, it accounts for very much less than this, and I am not sure what!

When seeking causes for his hypothesised changes, Baker is more cautious admitting that 'suggestions are scarce though temperature or some correlate of temperature, seems to be the obvious variable'.

With little doubt, the diversity of diadromy in its various forms and wide variety of fish groups, probably precludes the production of simple general explanations for the evolution of this complex phenomenon.

There are two levels at which this question needs to be approached, these levels relating to the three groups of taxa discussed earlier in this chapter, where the phylogenetic occurrence of the various forms of diadromy was reviewed.

In that review:

1. Several largely phylogenetically primitive groups were shown to have

diadromy as a widespread and evidently fundamental characteristic of the group — the lampreys, sturgeons, the anguillid eels and various salmoniform families. In these groups diadromy tends to be characteristic of a high percentage of species in each family and to be of a fairly uniform character within each family (i.e. consistently anadromy, catadromy or amphidromy).

2. A second group of families was identified in which diadromy was found to be quite widespread and present in sometimes divergent phylogenetic lineages within families, and might be more variable in character — the shads/herrings, mullets, gobies, sleepers, etc.

3. In yet further groups diadromy was shown to be an intermittent and minority characteristic for each family - ariid catfishes, centropomids, temperate percichthyid basses, scorpionfishes, sculpins, flounders, etc.

It seems likely that the patterns of origin and evolution of diadromy in each of these informal groupings has probably been quite different and not necessarily uniform within the group — probably not!.

Simplest to understand is category 3 above, in which it seems that occasional species in otherwise non-diadromous families have extended euryhaline habits to become diadromous. Thus if there are catfishes (family Ariidae) that are properly described as anadromous, they are representatives of a huge group of fishes (Ostariophysi) that is almost exclusively and strictly freshwater, and are included in one of two families in that group which have invaded the sea with success (the other is the catfish family Plotosidae). The purportedly anadromous ariids have exploited both marine and freshwater habitats by the evolution of migratory patterns that involve shifts from one to the other at regular intervals. Some species, as in the cods (Gadidae), sand perches (Mugiloididae) and southern rock cods (Bovichthyidae), are diadromous representatives of otherwise exclusively (or almost exclusively) marine families. Diadromy in any of these groups would seem highly unpredictable.

In a few families, both freshwater or marine habitats are occupied by various species, and the evolution of species that exploit both is perhaps not so surprising, even if still unpredictable — as in the sticklebacks (Gasterosteidae), temperate basses (Percichthyidae), snooks (Centropomidae), gobies (Gobiidae), and sleepers (Eleotridae).

How diadromy evolved in any particular case is bound to be both complex and species-specific, and I know of no instance in which it has been explored. For the New Zealand torrentfish, (*Cheimarrichthys fosteri*, Mugiloididae), there is not the slightest clue from its life history or from confamilial species, as to how amphidromy developed, but since all its known close relatives are marine it seems to have successfully relocated its breeding habitat from marine to fresh water.

Diadromy could have evolved in these groups quite recently as there are few indications that there has been much, if any, speciation of the diadromous species. However, there is no reason why some could not be more ancient and

137

primitive survivors whose sister taxa are no longer diadromous, or no longer extant. In some of these families, e.g. the pleuronectid flounders and the temperate basses, it is likely that diadromy has developed independently in several geographical areas and from distinct phylogenetic lineages. There seems no reason to regard the starry flounder (*Platichthys stellatus*) of Pacific North America, the river flounder (*Pleuronectes flesus*) of the eastern north Atlantic, and the black flounder (*Rhombosolea retiaria*) of New Zealand as in any way closely related just because they are catadromous (or for any other reason, for that matter). Similarly, the temperate basses are anadromous in North America (*Morone saxatilis*) and catadromous in Australia (*Macquaria novemaculeata*), and there is no reason to hypothesise close common ancestry for these disjunct but confamilial species on the basis of their diadromy.

Explanation of the evolution of diadromy in group 2 above, in which diadromy is a little more widespread, is no more simple. Diadromy in these groups tends to be more variable in character and to occur in divergent phylogenetic lineages within families. If they are divergent, confamilial lineages, as is most certainly true of the gobioids, then we are viewing within-family phenomena comparable with the group 3 taxa just discussed, i.e. the appearance of diadromy in diverse taxa (but within families), although in some instances the evolution of species groups that are diadromous is evident. Examples are the *Sicydium– Sicyopterus* group which contains many diadromous tropical gobies, or the several diadromous *Gobiomorphus* in Australia and New Zealand. Anadromy is representative of several of the shads (genus *Alosa*) in the North Atlantic, and of some of the *Hilsa* group in the Indo-West Pacific.

The mullets demonstrate the strongest committment to movement in and out of fresh water from the sea amongst these groups of fish, but it remains uncertain how explicitly diadromous they are as a group. Life history details for most mullets are sparse, but it seems that several species of *Liza* and *Myxus*, as well as one or more *Mugil* and *Agonostomus* and a *Joturus* are, more or less, catadromous, some more than others. To what extent in these species a majority of the populations makes the migrations, to what extent it is at a fixed life history phase, or to what extent it is obligatory, has not been clarified for most of these mullets. There seems to be a continuum from highly facultative and opportunistic movements into fresh water at various life history stages in many marine mullets, as in *Aldrichetta forsteri* in Australia and New Zealand, to fairly regular and explicit migratory patterns in the North American *Agonostomus monticola* and the South African *Myxus capensis*. (The fact that the Australian/New Zealand *Aldrichetta forsteri* was once included in the genus *Agonostomus* is no more than a taxonomic coincidence.)

Perhaps this evident continuum is illustrative of how diadromous migratory patterns have evolved in many groups — euryhaline fishes venturing from one salinity medium through estuaries and into the other, with selection favouring migration and eventually locking the species into a more or less strictly diadromous life history strategy.

Most interest has been expressed in the evolution of diadromy in the larger and more primitive groups of diadromous fishes (group 1 above), and for these, a question that immediately prompts itself is whether diadromy is a primitive characteristic of these groups, or derived, or perhaps both. For some of these groups the universality of diadromy, as in the anguillid eels, points to diadromy (catadromy) being a primitive character of the group. For others, in which diadromy predominates, and in which the evolution of species can plausibly be interpreted as following land-locking of formerly diadromous stocks, again, diadromy is easily interpreted as a primitive character — as in the three families of lampreys. Hubbs and Potter (1971) considered the non-diadromous freshwater lamprey genus *Ichthyomyzon* as being more primitive than but not ancestral to the anadromous *Petromyzon*, while *Petromyzon* is also seen as ancestral to some other Northern Hemisphere lampreys. There is no way of knowing whether the extant *Ichthyomyzon* is a freshwater survivor of a formerly anadromous lineage, or whether anadromy in the petromyzontids began with *Petromyzon*. Anadromy is a predominant feature of both northern and southern temperate lampreys and a close common ancestry may probably be assumed (Hubbs and Potter, 1971). Anadromy may also be a primitive feature of the sturgeons.

Only in the salmoniforms, and to some slight extent in the anguillids, has there been much discussion of the way diadromy developed, and on the whole, this discussion is best described as very speculative.

It is in the salmoniforms that there is greatest variation in the characteristics of diadromy, and the greatest diversity of fishes that is diadromous. There is also the least indication of consistent patterns. All three forms of diadromy are represented (catadromy only 'just'), but a consistent theme of the eight salmoniform families is that, with very few exceptions, reproduction takes place in fresh water. Perhaps this is a pointer to the origins and primitive life history patterns of the group.

Reverting to the view expressed much earlier on in this book, that diadromy should be seen as a specialised form of migration, it is profitable, here, to consider briefly how migration itself evolved, although this is not developed as a major theme. Foster (1969) in reviewing this topic, suggested that migration developed slowly by the increasing geographical separation of various essential or beneficial environmental factors, like distinct breeding and feeding habitats. No doubt a variety of other mechanisms has been involved, one of which for diadromous fishes may have been initial random wanderings of euryhaline fishes that led them from the sea into riverine situations — or the reverse — with selection favouring wandering into more favourable rearing-/feeding habitats and eventually making it a regular, purposeful, and ultimately obligatory phase in the life cycle. This sort of wandering is possibly what can be seen in many of the mullets, today.

The question of the marine or freshwater origin of the salmonids was reviewed by Tchernavin (1939), who pointed to a long debate on this question, contributed to over many years by such illustrious ichthyologists as A. Gunther, F. Day, C.T. Regan, A. Meek, G.A. Boulenger, and others. Tchernavin argued this question at some length, finally concluding that a freshwater origin for the the salmonids is most plausible. Hoar (1976) more recently re-examined Tchernavin's (1939) arguments, finding that they still made sense, and that nothing seemed to have been discovered in the last 50 years that invalidated his hypothesis of a freshwater origin of the Salmonidae. The critical question for Tchernavin, and others, seems to have been that reproduction in salmonids is restricted to fresh water. He stated that:

If the salmon had originated in the sea one would expect [that] its eggs, alevins and parr would flourish in salt water, and [that] their negative reaction to a salt milieu greatly tends towards the supposition of their freshwater origin - the rule is that young stages are passed through in the ancestral home.

Various other authors have suggested that eggs/sperm are likely to be more conservative in their environmental tolerances and thus to be less adaptable to varying salinities. Thus Foster (1969) wrote of the general tendency among fishes to exhibit more evolutionary conservatism in the early stages (eggs and larvae) of the life history than in their later stages (juveniles and adults). He argued, on that basis, that egg and larval habitats provide clues to the evolutionary origins of groups. Foster's inclusion of larval stages in his hypothesis has its difficulties in that not infrequently, especially in the amphidromous species and especially in the southern salmoniform families, the egg stage is in fresh water and the early larval stages at sea. Furthermore intolerance to different osmoregulatory conditions can easily be imagined to be a derived character that develops rapidly in evolutionary time. But if the arguments of Tchernavin, Foster, and others are accepted, it follows that the reproductive medium is likely to reflect the original one. Applying this interpretation, the anguillid eels would be seen as originally marine (as are all other anguilliforms), while lampreys, sturgeons, and the various salmoniform families would originally have been freshwater in habit. If the habitat origins of the salmoniforms are examined much more widely than Tchernavin (1939) did (he looked only at salmonids), it is true that all the freshwater-inhabiting salmoniform families (Salmonidae, Osmeridae, Plecoglossidae, Salangidae, Galaxiidae, Aplochitonidae, Retropinnidae, Prototroctidae) are almost exclusively freshwater reproducers. The exceptions are three marine spawning osmerids and, marginally, one galaxiid. This, I believe, points to an ancient freshwater origin for the salmoniform families as a whole.

Thorpe (1982) reviewed present evolutionary trends within the Salmonidae and pointed to the fact that evolution is producing derived and entirely freshwater species by landlocking of formerly diadromous species, and he summarised this as indicating a trend for salmonids to evolve away from a marine/freshwater life to a purely freshwater one. Thorpe's view reflects a similar one presented by Regan (1911) who regarded salmonids as fish that may be regarded as marine fishes which are establishing themselves in fresh water. This trend Thorpe interpreted as indicating evolution from more marine to less marine habits and on this basis he suggested a marine origin for salmonids. However, this argument rests on a false inference from contemporary import- anc to historical priority (Gould, 1983), and there seems to be no reason to assume that the direction of evolution in existing salmonid species is an indicator of the direction of evolution in their Mesozoic/Tertiary ancestors. This is particularly true because any evolution leading from fresh water to marine conditions seems as though it may now be severely constrained by the evident inability of the eggs of existing fresh water species to survive in the sea. Even so, a very few salmoniforms have succeeded in evolving towards repro- duction in the sea. Thorpe's argument does not grapple with the question of where the species of salmonid reproduce, and there seems to me to be no rea- son why speciation has to follow a path leading in one direction. Evolution/spe- ciation by landlocking in freshwater impoundments is such an obvious and simple evolutionary/isolating mechanism that there is no doubt that it would be taking place whatever other evolutionary patterns were developing in the Sal- monidae. Speciation, involving the evolution of entirely freshwater species from a diadromous ancestry, can as easily be interpreted as a *retreat* from a par- tially marine life history as a move away from a primitively marine life history. Many freshwater-limited salmonid species, should, I believe, be interpreted as secondarily freshwater-limited species (Balon, 1968).

A freshwater origin for the salmonids is now fairly widely accepted. Thus Rounsefell (1962) wrote that 'Since all salmonids spawn in fresh water (pres- umably their ancestral home), the anadromous habit may have evolved grad- ually from population pressure and a higher survival of fish feeding in the sea'. Hoar (1976) regarded sea-going salmonids as having evolved from strictly freshwater trout-like ancestors, assuming 'that Salmonidae evolved in fresh water'. Rosen (1974) thought, on the basis of his proposed phylogeny, that sal- moniforms are therefore primitively freshwater fishes and that the common ancestors of the Salmonoidei and Osmeroidei evidently either breed or bred in fresh water.

This view might seem to be supported by the fact that the freshwater sal- moniform families are believed to be phylogenetically closest to the esoci- forms, the northern-temperate pikes, that are entirely freshwater in habit. However, this argument is rather tenuous as the Esociformes is a small group with only about 10 species, and its restriction to fresh waters in the present day may not indicate primitive habits or habitats. Margolis (1965) addressed this question from a distinctive viewpoint arguing that parasites of Pacific salmon (genus *Oncorhynchus*) support a freshwater origin for the group on the basis that a host, or group of hosts, will have a greater variety of parasites, particu- larly parasites peculiar to its in the habitat where it has lived and evolved for the longest time (Manter, 1955).

Conclusions about when in geological history diadromy evolved in these various groups are difficult to derive with any rigour, and depend in part on other conclusions about their phylogenetic relationships. The fossil record is of very little assistance. As was noted before, the strongly diadromous groups like the lampreys, sturgeons, anguillid eels and salmoniforms have their phylogenetic roots at the base of the vertebrate and fish lineages, and all of them are undoubtedly ancient groups. Romer (1966) considered that the teleosts began their expansion in the late Mesozoic and have been dominant since Cretaceous times. He reported salmon-like fishes at least as far back as the upper Cretaceous and *Osmerus* back to the Miocene, while the salmonid genus *Smilodonichthys*, of Pliocene age, was a very large, probably salmon-like fish. It is believed to have been a filter feeder, and this, with its large size, may indicate marine/pelagic habits (Cavender and Miller, 1972). David (1946a,b) described salmonid scales from as early as the Eocene of California. He considered that they came from marine sediments, and that the fish had lived wholly in the sea, not migrating to and from fresh waters. Studies of salmoniform relationships (McDowall, 1969; Rosen, 1974; Fink and Weitzman, 1982; Fink, 1984), although differing in some of the detail of their conclusions, all suggest that the eight freshwater salmoniform families that are diadromous, four in the Northern Hemisphere and four in the Southern Hemisphere (Figure 8.1), have close relationships that span the tropics at least once and perhaps twice or more. These relationships could conceivably date back to a time when the earth's continental masses were united in Pangaea during the Mesozoic. The progenitors of the northern families (Salmonidae, Osmeridae, Plecoglossidae, Salangidae) would have moved northwards with Laurasia and the southern families (Galaxiidae, Aplochitonidae, Retropinnidae, Prototroctidae) southwards with Gondwanaland, and then have dispersed around the Northern and Southern Hemispheres as these ancient continental masses redistributed the land around the two hemispheres. Since diadromy — in somewhat varying form — is a feature of all of these families, it might be considered to be a primitive feature of this group of salmoniform families that could have been characteristic of the group since they had a common ancestry in Pangaea as long as 180 million years ago.

The same argument can be made for the lampreys, which can easily be imagined to be just as ancient. Anadromy in the three lamprey families may date back to their common ancestry, and possibly on Pangaea in the Mesozoic, or earlier.

This view is in distinct contrast with discussions of the origins of diadromy by several other authors, and in several major taxa. The development of diadromy has frequently been interpreted as relating to the conditions that applied in the northern cold temperate as the ice sheets retreated northwards following the glaciations as recently as the Pleistocene. The theory is that highly diluted sea waters at the interface between the ice sheets and the sea

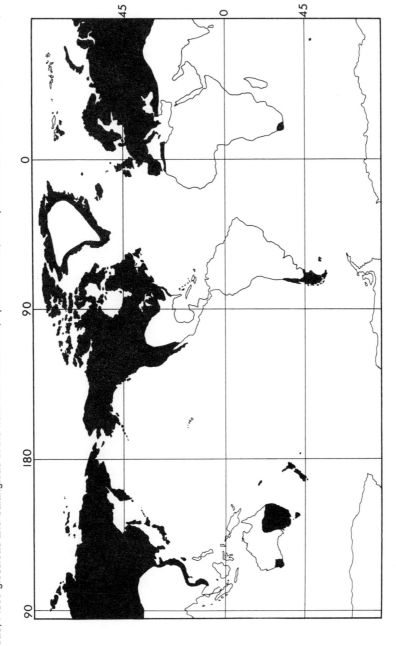

Figure 8.1: Amphitropical distribution in the cool-temperate zones of the Northern and Southern Hemispheres of salmoniform fishes (Salmonidae, Osmeridae, Pleco-glossidae and Salangidae in the north and Galaxiidae, Aplochitonidae, Retropinnidae and Prototroctidae in the south)

would have provided mesohaline conditions that would have facilitated a move by such fishes as trouts and salmons from an entirely freshwater life to a euryhaline and diadromous one. Nikolskii (1961) discussed a movement of ancestral salmonids from the sea to fresh water in Tertiary times and hypothesised a secondary return to marine conditions in migratory species during the glacial periods (but he did not specify when). He postulated convergent evolution of marine migratory habits in both the North Pacific *Oncorhynchus* and the North Atlantic *Salmo*, and considered that *Salmo* later invaded the North Pacific, possibly as recently as the Quaternary. He later (Nikolskii, 1963) argued that enormous masses of melt water significantly lowered the salinity of adjacent parts of the sea during the end of the Tertiary period and in Quaternary times. Balon (1968) also adopted this argument. Mottley (1934) suggested that *Oncorhynchus* and *Salmo* were separated and diverged by the penultimate glaciation, with evolution of *Oncorhynchus* in the Pacific and *Salmo* in the Atlantic, since that glaciation. Tchernavin (1939) had argued that:

> the migratory Salmonidae split off from purely freshwater species ... The ancestors of Salmonidae from the Eocene onwards were probably freshwater fishes and only during the glacial period, when all the recent species of this family already existed, did some of them acquire migratory habits. ... Probably this took place during the Glacial period when the low temperatures induced great changes in freshwater life ... when conditions of life in fresh water in the Northern Hemisphere were unfavourable and food scarce ... some of the Salmonidae acquired migratory habits of descending to the sea from exhausted rivers in order to feed.

This is a rather more extreme and also explicit hypothesis than that offered by Nikolskii (1961, 1963). Miller and Brannon (1982) considered that the initial emergence of gross patterns occurred in isolated North Pacific populations during the early Pleistocene, at which time they postulated the evolution of *Oncorhynchus*, although they recognised that the initial adaptation to long-term marine residence may have been a quite ancient one. Hoar (1976) also looked at a Pleistocene glacial period origin and believed that some of the Salmonidae acquired migratory habits of descending to the sea and that true migratory forms in different species of Salmonidae then gradually evolved. His hypothesis suggests that the more migratory species are derived and the non-migratory ones primitive. Vladykov (1963) wrote that 'the present genera and species of salmonid are of rather modern origin, appearing probably during the interglacial periods of the Pleistocene'.

The repeated and rather uniform evolution of anadromous migratory habits in a substantial number of existing salmonids, as Tchernavin suggested, is scarcely tenable, and there seems no reason to place such a recent date on the evolution of anadromy. Thorpe (1982) questioned such youthfulness of the

Salmonidae, with some justice — Behnke (1972) noted fossils of *Salvelinus* from the Pliocene and Miller (1972a) records *Salmo* from this period.

While some authors were discussing a Pleistocene origin for diadromous salmons, others have been exploring the likelihood that the retreat of these same Pleistocene ice sheets in North America and Scandinavia has been a prime cause in the development of species groups and subspecies of *Coregonus* (e.g. Behnke, 1972). While the two perspectives are not mutually exclusive, they seem scarcely compatible, and a Pleistocene origin of groups like the genus *Oncorhynchus* seems to me to be highly unlikely. And while some authors are hypothesising this Pleistocene origin for salmonids, others, e.g. Rosen (1974), have been writing about the involvement of a diadromous species like *Salmo trutta* in vicariance biogeographic events relating to the movements of continental masses, and of involvement of the southern temperate Galaxiidae in the early Tertiary fragmentation of Gondwanaland. Integration of these rather divergent perspectives seems to require considerable dexterity. Although a Pleistocene origin may still satisfy some as regards the evolution of the Salmonidae, and its anadromous migrations, it does not touch upon the occurrence of diadromy in various forms in other, highly diverse and related salmoniforms.

The suggested close relationships between the various Northern and Southern Hemisphere salmoniform families (McDowall, 1969; Rosen, 1974; Fink, 1984), and the fact that all eight families are predominantly diadromous, seems to discount any high likelihood that the evolution of diadromous habits by the salmons and trouts was in any way as recent as the Pleistocene. This is not to say that similar geological/climatic/environmental events in the Mesozoic could not have had a role in the movement of the primitive salmoniforms from an entirely freshwater life to a diadromous one (assuming that their origin was in fresh water). Diadromy in other groups, like lampreys, sturgeons and anguillid eels, may equally be a rather primitive feature that has very ancient origins in those groups. How diadromy evolved in the various fish groups, however, is not easily accessible to the generation of satisfactory and testable hypotheses. It is interesting to note that the superspecies of fairy shrimp, *Mysis relicta* (Mysidacea), is believed to comprise at least three sibling species in Scandinavia, and that although the present distribution of *Mysis* is said to have Pleistocene origins, biochemical studies indicate that they might have been distinct from as early as Tertiary times (Vainola, 1986).

9

Strategic Aspects of the Life Histories of Diadromous Fishes

LIFE HISTORY STAGE AT MIGRATION IN DIADROMOUS LIFE HISTORIES

The three forms of diadromy — anadromy, catadromy and amphidromy — represent distinct life history strategies, and within each there is substantial variation between species in how the strategy is carried out. These differences can best be illustrated by an examination of the various elements within the life histories. These elements in the life cycles of diadromous fishes can be isolated as the following stages:

1. Spawning and fertilisation;
2. Development and hatching;
3. Larval–juvenile feeding and growth; and
4. Sub-adult–adult growth and maturation.

Theoretically, diadromous migrations might take place between any of these ontogenetic phases, and in some instances, during them. The life history stage at which the osmoregulatory fresh/salt water transition occurs is obviously of critical importance, insofar as it is at this point in the life cycle that there is a period of osmoregulatory stress.

Migrating eggs

Although there are no evident explicit instances of 'migrating eggs', there are species in which there is a downstream movement of the eggs during development, in at least some individuals in the population. This has been described for the striped bass, *Morone saxatilis* (Talbot, 1966). Bagenal and Braun (1978) said that it is true of European and North American shads (*Alosa fallax* and *A. sapidissima*), but not of the alewife (*Alosa pseudoharengus*). The eggs of some of the shads of the Caspian Sea (*Caspialosa* spp.) were described as benthopelagic by Nikolskii (1961), and these, too, develop as they are carried

146

downstream towards or even into the sea. Lindroth (1957) considered that the eggs of *Coregonus lavaretus* are carried downstream, some lodging amongst the river gravels, but some being carried downstream into coastal waters before hatching. Some authors have hypothesised that the spawning of the mountain mullet, *Agonostomus monticola*, occurs in fresh water and that possibly the eggs are carried to sea, where the larvae are found (Gilbert and Kelso, 1972; Loftus *et al.*, 1984).

There are a few cases where the eggs are capable of tolerating a wide range of salinities. One is the estuarine spawning of the inanga, *Galaxias maculatus*; the eggs of this species have been shown to tolerate a wide range of salinities between fully fresh and fully marine waters, and to develop and hatch satisfactorily in either extreme (McDowall, 1978a). It has been shown that the eggs of the pink salmon, *Oncorhynchus gorbuscha*, may sometimes be deposited in estuarine circumstances (Hanavan and Skud, 1954) and possibly that the semi-floating eggs of *Morone saxatilis* may be carried into estuarine waters and develop successfully there. In the families Gasterosteidae (sticklebacks), Cottidae (sculpins), Gobiidae (gobies) and Eleotridae (sleepers) there is variation in the salinity at which the eggs develop, some species being marine and others freshwater. In general, however, there is strong adherence to a single 'traditional' osmoregulatory medium for egg deposition, fertilisation, and development through to hatching.

Migrating larvae

Many groups appear to be capable of migrating immediately or soon after hatching and prior to the assumption of feeding activity (between 2 and 3 above). This is true in at least some aplochitonids, galaxiids, retropinnids, prototroctids, osmerids, a minority of salmonids (but only after some weeks spent in the stream gravels after hatching), clupeids, gasterosteids and some of the heterogeneous array of perciforms (a mugiloidid, eleotrids, gobiids). In all of these, newly hatched larvae are known or thought to go to sea.

Migrating juveniles

Movements of well-grown juveniles between marine and fresh waters (in both directions) are widely present, and in fact seem to occur in all groups in which there is diadromy (between stages 3 and 4 above). The result of this migration can be either of two different strategies. It may be the initial outmigration in which the new cohort leaves the spawning habitat to move to a feeding habitat, as is true in lampreys, anguillid eels, salmonids, osmerids, sturgeons, clupeids, and others; or it may be the return migration of the juveniles from a feeding habitat to the spawning habitat, as in galaxiids, prototroctids, eleotrids, mugiloidid, etc.

Migrating adults

Similarly there are groups of fishes in which a migration follows growth to adult maturity (between stages 4 and 1 above); however not all diadromous fishes undergo migrations during the adult stage. Adult migrations occur in the lampreys, eels, sturgeons, salmonids, osmerids, retropinnids, some aplochitonids (*Lovettia*), gasterosteids (*Gasterosteus*), etc. In these fishes, too, the migration is a movement from a feeding habitat to a spawning habitat, but at this stage feeding and growth in most (but not all) species is complete, or nearly so. Gonadal development may be some way from completion, but depends in many species on transfer of energy from somatic tissues to reproductive tissues. In repeat spawning (iteroparous) species there is typically a dual alternation of habitats by the fully mature adult fish with each successive maturation of the gonads. The ripening fish cease feeding as the spawning period approaches and migrate into the spawning habitat. Survivors then return to the normal feeding habitat where they resume feeding and growing, in preparation for a further spawning. Growth after first spawning may be very restricted, especially in some salmonids. Size attained may approximate that at first spawning.

Combination of migratory stages

Bearing in mind that in all species there is a dual or reciprocal migration pattern — sea to fresh water and the reverse — the combination of stages within species at which migration takes place is also diverse. Fish may migrate initally as either eggs or newly hatched larvae, and return as juveniles or adults; or they may migrate as juveniles and come back as adults. It needs to be well recognised that in by no means all instances does diadromy involve the migration of mature to ripe adult fish (i.e. that amphidromy is much more widespread than is commonly recognised).

PASSAGE THROUGH ESTUARIES

Estuaries are areas of rapidly changing environmental parameters, which fluctuate both spatially and temporally at a rapid rate. For this reason, they are areas where aquatic organisms are subjected to considerable environmental stresses. However they are also transition zones, areas of intermediate values in such factors as salinity and temperature. For this reason they provide a place where migrating fishes can dwell for a period and allow their physiology to adjust during the freshwater/saltwater transition. The ability of fish to migrate between

Figure 9.1: Chronological development of seawater adaptation by juvenile chinook salmon, *Oncorhynchus tshawytscha*, measured as percentage survival for a period of 30 days after sharp transfer to the four different salinities
Source: After Wagner *et al* 1969

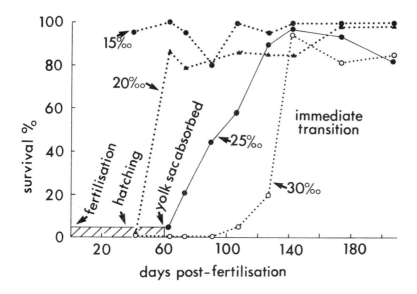

fresh water and salt water depends on their ability to cope with these stresses. In euryhaline fishes this ability seems to exist at most times, but in amphihaline species (Fontaine, 1975) it is periodic in occurrence and fish must migrate at specific, often limited, times. McCormick *et al.* (1985) studied a mixed population of brook char, *Salvelinus fontinalis*, in eastern Canada which contained both anadromous and non-migratory forms. They found that anadromous char examined at the period of peak seaward migration did not have a greater ability to osmoregulate in high salinity water than non-anadromous fish taken during August (a non-migratory period), at which time these authors did not expect the fish to have exceptional salinity tolerances. Ability to osmoregulate must develop at times appropriate to migration to and from the sea. This issue highlights the question discussed earlier of the fundamental and important distinction between euryhaline and amphihaline fishes. Extensive work on this phase of the life history of the Pacific salmons (Hoar, 1976) has shown that their ability to handle the transition from fresh water to the sea varies widely between species, and also within species between different growth stages (Figure 9.1), so that their ability to osmoregulate is not species- or time-universal. Hoar (1976) suggested that the ability of salmonids to osmoregulate in extended salinity ranges varies on a regular seasonal cycle. Survival of young Atlantic salmon, *Salmo salar*, also varies widely with age and size (Figure 9.2).

Figure 9.2: Survival of Atlantic salmon, *Salmo salar*, in sea water of various dilutions at different growth stages (ages and sizes)
Source: After Parry 1960

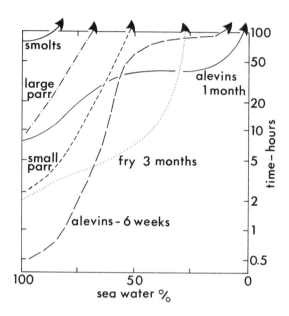

There is some, although not much evidence for the use of reduced salinity areas in estuaries for acclimation by migrant stages of various diadromous species. Certainly, the larval/juvenile phases of some of the Pacific salmons are known to settle in large estuaries for weeks or more, where they feed and grow before moving further to sea (Figure 9.3). Estuarine residence seems to be a feature of some stocks, e.g. chinook salmon in the delta of the Fraser Rivers but possibly not of stocks in others, as in rivers draining into the sea along the Washington and northern British Columbia coasts, where the river estuaries are much more limited in extent, and migrating fish must move rapidly from fresh water to the sea. There is nowhere for the vast numbers of smolt to reside in these estuaries. This is also a feature of chinook salmon populations in New Zealand, where the estuaries provide virtually no habitat for juvenile salmon to dwell, and thus to acclimate to marine salinities. Whether the habit of spending some time in estuaries is critical for some stocks of salmon to adjust their osmoregulation, or is because estuaries may provide a very rich and productive feeding environment for the young salmon, does not seem clearly elucidated. Dorcey *et al.* (1978) considered that:

> Estuarine marshes are often said to serve as critical regions for young salmon to generally acclimate to full seawater, thereby decreasing stresses of osmoregulation and increasing their survival. Support for this assertion

150

Figure 9.3: The place of salmon smolts in the intricate food web of the lower Fraser River estuary in western Canada
Source: Dorcey and Northcote 1978

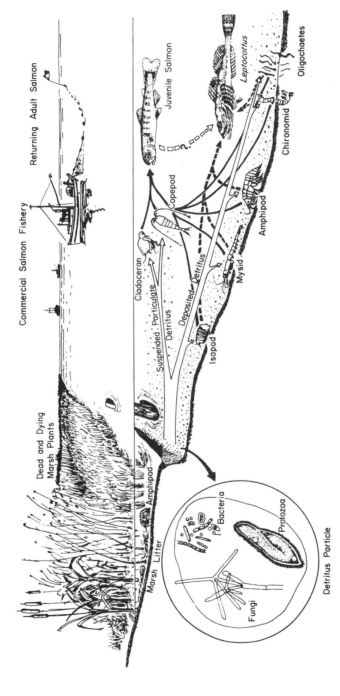

seems to rely largely on laboratory studies of salinity tolerance where survival abilities may be quite different [from] those expressed under field conditions.

These authors in assessing the role of the Fraser River estuary in the life cycles of salmon species there (Figure 9.4), suggested that there is wide variation in occupancy from 'doubtful' in pink salmon, to 'possibly up to a few months' in chinooks. In a review of the literature, they found that the juveniles of several *Oncorhynchus* spp. are unable properly to osmoregulate on direct transfer from fresh to sea water. They concluded that common assertions that

Figure 9.4: Varying migratory strategies of juvenile Pacific salmon, *Oncorhynchus* spp., as they move to sea from the lower Fraser River. A, smolts move straight to sea without estuarine residence; B, smolts pass through the estuary but linger in the deeper waters of the Straits of Georgia; C, some fish may occupy marshes and estuarine habitats before migrating offshore; D, other smolts may live in estuarine habitats, then deeper coastal waters and only then begin active offshore migration Source: After Dorcy and Northcote 1978

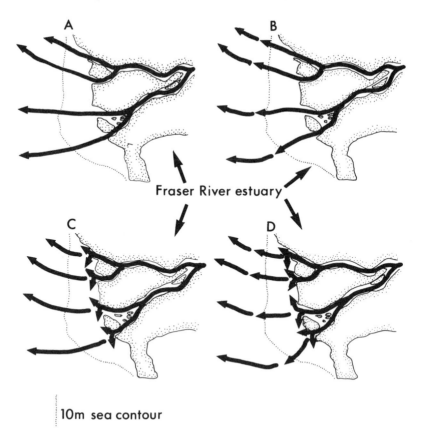

Fraser River estuary

10m sea contour

marshes and tidal flats are essential zones where young salmon may gradually acclimate to sea water would only seem to be supported in part by the available physiological evidence from perhaps two of the five species of Pacific salmon — coho and chinook. However, they point to the question of different stocks of the various species and the possibility that different stocks have differing abilities to osmoregulate. A similar use of estuaries for acclimation is likely in other salmonids. McCormick *et al.* (1985) believed that gradual acclimation to sea water significantly increases seawater survival of brook char and for this reason concluded that the estuary is an important site for acclimation of brook char, which ultimately permits their entry into sea water.

Upstream migrating glass eels (*Anguilla* spp.) are known to settle in tidal river estuaries for several weeks or more after leaving the sea, and before they move on upstream (Jellyman, 1979) although this has not been shown to relate in any way to a need to adjust osmoregulation.

In contrast, there are many examples in which movement between fresh and salt water is extremely rapid. This is likely to be true, for instance, of the downstream-moving, newly hatched larvae of some of the southern salmoniforms - *Galaxias*, *Retropinna*, *Lovettia*, and also of the New Zealand *Cheimarrichthys* and *Gobiomorphus*. In these fish, egg deposition is generally supratidal, though it may be many kilometres upstream, and as far as is known the larvae of these fish are carried straight to sea by the riverine and tidal flows. This is probably true of many amphidromous and some anadromous species whose larvae are extremely small at hatching. In one galaxiid (*G. maculatus*) the larvae may hatch successfully straight into sea water (or fresh water or any mixture of the two, McDowall, 1978a), while in some retropinnids hatching is just above the saline/tidal zone, and the hatching larvae move straight from fresh into saline waters. Similarly, sea-migratory adult eels (*Anguilla* spp.) are believed to move straight to sea without any period of acclimation in intermediate salinities.

Upstream migrants have a naturally greater opportunity to dwell in fresh water/sea water mixing zones at river mouths, especially by comparison with newly hatched larval fish that are swept out to sea in river currents. It is hard to quantify the extent of any transition period in upstream migrants. Nevertheless it is evident that there is little if any acclimation period in many of the species. In some river systems where coastal/lowland topographical gradients are relatively steep, there is little if any estuarine zone with intermediate salinities, so that even spatially there is little opportunity for acclimation. And, certainly, observations of the immense migrations of shoals of some species (*Galaxias*, *Lovettia*, *Retropinna*, *Oncorhynchus*, various osmerids, etc.) indicate that passage from the sea into fresh water is routinely accomplished without mortality in a matter of minutes. Theoretically, it is possible that some acclimation takes place in areas of reduced salinities at sea beyond the river mouths, but it is not known if this is actually a significant factor in acclimation. It is undoubtedly an interesting question, for which answers would be difficult, though the ability of various fish species to make a rapid transition can easily be tested experimentally.

Clearly we are dealing with a question important to the survival of diadromous fishes about which little is known. In the very intensively studied salmonids not much seems to be known; in other groups, virtually nothing is known.

LIFE HISTORY AS AN ADAPTIVE STRATEGY

There is a widely held view that life histories in animals are selected for and adapted to maximising the production of progeny (Schaffer and Elson, 1975; Stearns, 1977; Dingle, 1980; Miller and Brannon, 1982; Thorpe, 1982; Gross, 1985). Gross considered that most of the currently accepted models of the evolution of reproductive strategy are based on optimisation theory. This includes migration as one of the widely occurring life history strategies in diverse animal taxa. In evolutionary terms the persistence of migration needs to be seen in relation to the balance of advantages obtained from migration and the costs incurred by the population/species. Advantages include such aspects as increased food supply, avoidance of potentially harmful environmental conditions and/or a movement to more favourable ones, the occupation of habitats that have specific or specialised habitat requirements, and the availability of more living space. Costs of migration include mortalities resulting from migration itself, changed environmental conditions that may be intolerable (in diadromous fishes, specifically osmoregulatory stresses), higher predatory rates that might follow from lack of experience in new habitats, and the prospect of the migrants getting lost and being unable to return to suitable habitats. Northcote (1979) has reviewed the adaptive advantages of migration in fishes and found that 'Migration, particularly in freshwater fishes, often has been regarded as an adaptive phenomenon for increasing growth, survival and abundance, which may increase production ... migration seems assured as an adaptive feature of major significance in populations of freshwater fishes.' Dingle (1980) similarly wrote that 'Evolution of fish migration seems to have been driven by selection for more efficient adult feeding and growth (or greater winter survival) in one area but more successful breeding in another.'

Northcote (1979) explored the great variety of migratory patterns in fishes (primarily in relation to fish production) and concluded that perhaps we should not expect to find any single 'unifying principle' or explanation for the varieties of migratory patterns in freshwater fish populations. It was his view that migratory behaviour evolved independently and for different reasons in different instances. Or, as Dingle (1980) viewed the situation for salmonids no single mechanism can be expected to be universally responsible for the observed patterns.

Smith (1985) saw migrations of fish as their way of gathering energy in one portion of the environment and transporting it to other areas — migration permits species to utilise habitats that are only temporarily available to them. And

Northcote (1979) considered that animal migration in general may be regarded as a strategy to deal with areas of habitat that are subject to marked fluctuation and frequent periods of temporarily unsuitable conditions (Figure 9.5). Dingle (1980) held a similar view, claiming (though not documenting) that animal migrations tend to occur in environments or combinations of environments that are subject to marked temporal fluctuations. Cohen (1967) argued that migration involves the avoidance of the hazards of climate and food shortage during an unfavourable season in the breeding territory at the costs of the hazards of migration and survival in the wintering territory.

These seem unduly negative perspectives that do not seem wholly applic-

Figure 9.5: Summary of migratory patterns in fishes, focusing on the wintering (climatic), feeding (tropic) and spawning (reproductive) categories of Heape (1931). Each of these movements may also be osmoregulatory
Source: Adapted from Northcote 1979

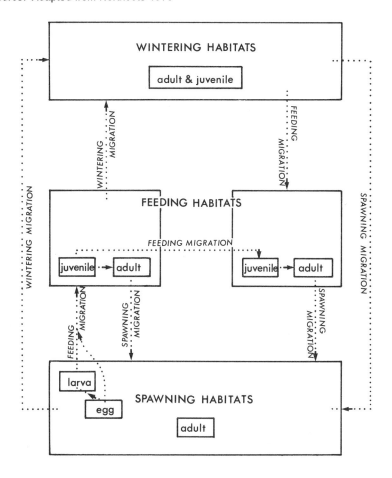

able in many instances of fish migrations. Fish migrations abound at nearly all latitudes, and although there is little doubt that in some species and some locations avoidance of potentially fatal conditions is an important issue, I doubt that it can be shown to be a general truth in fish migration. That conditions relating to migration are not necessarily unsuitable is demonstrated by the diversity of migratory strategies exhibited by various populations of a considerable range of migratory species and even of diverse strategies between individuals within populations of a species. Some individuals in many species will persist and survive over many generations in conditions from which others escape by migration. It seems to me that a more positive perspective can be adopted, in which migration is seen as a strategy that enables a fish species to occupy more favourable conditions for each of two or more life history stages.

DIADROMY AS AN ADAPTIVE LIFE HISTORY STRATEGY

Since diadromy is only a specialised form of migration it should follow that, in general, the adaptive/selective advantages of migration are applicable to diadromous species. There are bound to be some minor differences, since diadromy is a specialised form of migration, but the persistence of diadromy in many species in which it is to some extent facultative indicates that it does have selective advantages — or it would have disappeared. A large proportion of diadromous species has the ability to establish entirely freshwater or wholly marine populations. This being so, i.e. that easy alternatives are available, the long-term survival of diadromy as a life history strategy confirms, I think, that it is a successful one for many species. The opportunity to restrict life to either freshwater or marine habitats exists, but many species are facultatively diadromous, or have failed to exercise this option and remain diadromous.

Dingle (1980) considered that the evolution of diadromy occurred as a result of freshwater fishes moving into estuaries and eventually the sea, where they were able to grow larger and faster, and become more fecund. This is likely to be true for some species but the wide diversity of forms of diadromy requires that a broader variety of strategic patterns be investigated. Heape (1931) listed three key factors relating to migration as being trophic, climatic, and gametic; as discussed elsewhere (p. 22) diadromous fishes have the additional significant factor of osmoregulation.

If migration is to be adaptively beneficial, the costs involved must in general be less than the advantages that accrue. This may not apply in some (probably rare) cases in which there are specific factors or tolerances that make migration from one environment to another one mandatory; this could be true, for instance of the freshwater anguillid eels, in which it seems that marine spawning and the marine leptocephalus stage are mandatory in spite of the costs.

The structural patterns of the migrations of diadromous fishes are discussed in considerable detail elsewhere; these structures are so diverse that it is difficult to generalise briefly. However, various categories of movement and the life history stage involved may be summarised as follows:

A. Movement to sea may comprise:

1. eggs swept to sea prior to hatching;
2. newly hatched larvae carried to sea without any feeding occurring in fresh water;
3. movement to sea of larvae/juveniles after brief (a few weeks) to long-lasting (up to 3–5 years) feeding and growing in fresh water;
4. movement to sea of mature/prespawning adults which do not feed again.

B. Life in the sea, correspondingly, may comprise:

1. larval feeding to the juvenile stage (cf. A1 and A2 above);
2. larval feeding to sub-adult stage (cf. A1 and A2 above);
3. larval feeding to the mature adult (cf. A1 and A2 above);
4. juvenile feeding to the mature adult (cf. A3 above);
5. spawning, development and early larval/juvenile life (cf. A4 above).

C. Movement from the sea to fresh water may comprise:

1. migration of juveniles (cf. B2 above);
2. sub-adult migration (cf. B2 above);
3. spawning adult migration (cf. B4 above).

D. Life in fresh water may comprise:

1. feeding and growth from juvenile to adult and reproduction (cf. C1 above);
2. sub-adult to adult feeding and reproduction (cf. C2 above);
3. adult reproduction alone (cf. C3 above).

The adaptive significance of these movements is also far from easy to interpret for specific examples. Baker (1978) saw a paradox in diadromy in that some species moved from a life predominantly in fresh waters to feed and grow in the sea while others moved from a life predominantly in the sea to feed and grow in fresh water. McKeown (1984) found difficulty in understanding how catadromous and anadromous behaviour developed in the same drainage system. Apart from the fact that anadromy and catadromy do coexist in many river systems there is no evidence to show that they did develop in the same drainage system. And, if a common explanation is sought for the various forms of diadromy, then it is not surprising that it seems paradoxical and difficult to understand. Further, the situation is far more complex than either Baker or

McKeown seem to suppose. Both anadromy and catadromy are present in diverse forms, and this diversity needs explanation, as do the various forms of amphidromy, which they do not mention. However, I do not see that their coexistence needs to be an issue at all.

In terms of the categories listed above as expanded from Heape (1931), all these migrations are osmoregulatory; some in addition, are trophic; some are trophic and gametic. While some are undoubtedly also climatic, this is not nearly as easy to determine with any certainty.

Trophic strategy

Heape (1931) suggested that the primary reason for migration is related to food supply, and if this is so it is likely to occur when the optimal feeding habitat is spatially separated from the optimal reproductive habitat. Both Heape (1931) and Nikolsky (1963) have suggested that an absence of food may be a stimulus for outmigration, but evidence for this seems very limited. Slaney and Northcote (1974) showed that higher densities of rainbow trout in streams are proportional to food availability, but the extent to which there is a causal link between abundance and food availability is not clear. In general it seems possible that interactive behavioural mechanisms applying at other levels will establish both population levels and emigration patterns before food availability becomes a limiting factor.

If the migratory strategy is designed to optimise feeding it would follow that there would be selection to move away from a food-poor habitat early in life. However, few choices are that simple. It is easy to imagine that some of the trophic migrations have distinct advantages, particularly the migrations of larvae/juveniles from an often rather sterile and unproductive freshwater habitat where food items tend to be fewer, large and widely spaced, to a much more productive marine environment, where the food items may be more abundant, smaller and closer together. In some instances there is very early movement but in others additional constraints like ability to osmoregulate in changing salinities affect selection and their behavioural patterns. This may explain differences within Pacific salmons, with pink and chum, which are better able to osmoregulate, migrating at an early age; others like chinook, coho and sockeye are less able to osmoregulate, and move later. Sockeye have avoided part of this problem by choosing an alternative food-rich environment in lakes for early juvenile life. Movements to optimise these habitat uses are also likely to be interrelated with seasonal changes in the environments, so that a complex web of causal relationships is indicated. This becomes particularly evident when it is remembered that migration has considerable costs in energetic and survival/mortality terms, as well as advantages.

Migration to rich feeding habitats is reflected in rapid growth patterns, which is evident from growth rates as interpreted from scales. This was discovered long ago for salmon (e.g. Finlay, 1972). Widely spaced circuli near the centres of the scales are characteristic of fish that have moved to sea at a young age and have then grown rapidly, in contrast to closely spaced rings in those that have lived longer in fresh water and have grown more slowly (Figure 9.6). It must, however, be remembered that there may be compensatory losses. Migration to sea at a very small size may result in osmoregulatory mortalities and excessive predation. Thus there is a balance struck between:

1. Early migration/low feeding mortalities/higher osmoregulatory and predatory mortalities, and
2. Later migration/slower growth/higher feeding mortalities with lower osmoregulatory and predatory mortalities.

Data suggest that migrations to sea of larvae/juveniles of diadromous fishes tend to be more common at high latitudes where marine productivity is higher and more seasonal than at lower latitudes. Northcote (1979) suggested that movements of sockeye salmon to sea are timed to coincide with periods of peak marine plankton productivity.

Comparisons of related migratory and non-migratory species suggest that species may adopt alternative strategies to maximise production. In New Zealand both Galaxiidae and Eleotridae include species that are diadromous/migratory and others that are non-diadromous/freshwater-resident. The former species have numerous, smaller eggs which hatch into tiny larvae that go almost immediately to sea. This can be seen as an adaptation to the availability of highly productive, marine plankton comprising very small prey organisms, enabling rapid growth but high mortalities owing to such factors as osmoregulatory stresses, dispersal at sea away from land and freshwater habitats, and heavy predation/mortality there. The non-migratory species have fewer, larger eggs, produce larger larvae that are adapted to less productive, freshwater stream ecosystems, larger prey items, absence of osmoregulatory stresses, the need to maintain position in flowing waters, and lower predation mortalities in fresh water (McDowall, 1978a). Humphries (1986) has shown that in landlocked populations of the Tasmanian spotted mountain trout, *Galaxias truttaceus*, there are comparable trends, with larger and fewer eggs by comparison with diadromous populations.

Strategically, one of the interesting aspects of feeding migrations is 'Why do amphidromous migrants come back so soon, leaving a rich feeding environment at an early age and small size, to return to less productive fresh waters?' If life in the marine plankton is so advantageous from a trophic perspective, the quite numerous species that are amphidromous, in which the juveniles leave the sea at a very small size to return to fresh water, might seem anomalous. It is not easy to understand why such fishes, having survived the stresses of movement from fresh water to the sea at very small size, having found suitable marine waters in which to feed and grow, and having survived there for several months

Figure 9.6: Spacing of circuli on scales of chinook salmon, *Oncorhynchus tshawytscha*, reveal growth rates and is indicative of habitats occupied; arrows show the boundary between a period of presumed slow growth in fresh water (closely spaced circuli) and more rapid growth at sea (widely spaced circuli)
Source: NZ Ministry of Agriculture and Fisheries

subject to predatory pressures from other animals, then choose to leave that environment, to again make the osmoregulatory transition and return to fresh water.

In spite of considerable comment that much of migration is related to trophic advantage, there do not seem to be a great many examples where only one trophic habitat is occupied by diadromous fishes. This is true, however, of the Tasmanian whitebait, *Lovettia seali, which goes to sea on hatching and returns as the ripe adult. It is apparently true also of the various diadromous salangids, although data are sparse.*

Dual trophic habitats are described for diverse diadromous groups, and there is no doubt that equally diverse life history strategies are represented. Diadromous lampreys (Petromyzontidae, Geotriidae, Mordaciidae) behave as filter feeders in fresh water for several years, but growth is slow. It accelerates rapidly once the ammocoete larvae metamorphose and move to sea, where they feed as fish parasites for a year or two. It seems likely that there are restrictive growth-rate and size limits upon filter-feeding juvenile lampreys, that are reflected by the much smaller size of non-anadromous lamprey species (Potter, 1980). Anadromous populations take advantage of the abundance of large-sized fishes upon which to prey. More rarely, such predation may take place in large freshwater lakes (Davis, 1967; Maitland, 1980b), and anadromy does not occur.

Some anadromous fishes, like retropinnids, immediately move to sea on hatching, feed principally at sea, but in addition the mature adults may continue to feed after returning to fresh water to spawn (Eldon and Greager, 1983).

Many anadromous fishes, like shads (Clupeidae), trouts and salmons (Salmonidae), and others, initiate trophic life as larvae in fresh water following hatching. The duration of this freshwater, trophic life is widely variable between species with close affinities, and even in separate stocks within species. Thus some pink salmon go to sea virtually without feeding (although they have spent some weeks in stream-bed gravels between hatching and migrating), while others may feed for a few weeks as they move downstream. At another extreme, sockeye salmon tend to migrate into productive lakes to feed for a year or more before moving to sea (Figure 9.7). In almost all Pacific salmons the fish then occupy a second trophic habitat, feeding at sea for several years before maturing and returning to fresh water, to spawn. In a few salmonids, e.g. sea-run brown trout, there is also some feeding during the return, upstream-migration of the pre-spawning adults. In yet other salmonids, there is a seasonal alternation between freshwater and marine feeding habitats over a period of several years, both prior to first spawning and subsequently between spawnings.

The freshwater anguillid eels also adopt a diverse trophic strategy, although at life history stages that are the reverse of what is typical of salmonids. Larval feeding and growth are marine and there is a migration from the sea to fresh water by small juveniles, which then grow and mature in fresh water over a very

Figure 9.7: Comparison of the development of seawater adaptation in three salmonids measured as the percentage survival in seawater (salinity 30ppt) for 30 days; the horizontal blocks indicate periods at which these species are observed to migrate downstream
Source: After Wagner *et al*, 1969

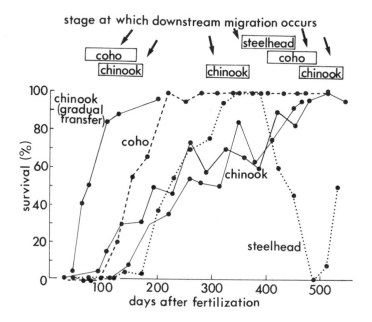

long period (large freshwater eels may be more than 60 years old before they mature and go to sea to spawn, - Todd, 1980). Again, from a trophic perspective, it is hard to see the selective advantages, for anguillid eels, of metamorphosis and leaving the food-rich environment of the sea and moving into fresh water. Dingle (1980) had difficulty with the strategic value of catadromy in eels and concluded that:

> Since the breeding site of eels is unknown ... speculation on the selective forces driving this system is largely futile ... the long life cycle with the freshwater phase lasting several years remains a fascinating problem in evolutionary biology ... the eel problem is left for what it is, an intriguing mystery.

Possibly catadromy in eels could be related to that fact that they have reached a stage where metamorphosis from the highly aberrant and specialised leptocephalus stage to the glass eel is physiologically 'necessary', though such a suggestion explains nothing either about how catadromy evolved or about how it is selectively advantageous. The juvenile/adult form of eels is also distinctive, a distinction that Gosline (1960) thought might be attributable to their habit of wedging themselves through small holes in coral or rocks. The eel metamorphosis involves a profound change of ecological niche occupied and it can only be assumed that anguillid eels have found a similar and favourable crevice-type habitat to occupy in fresh water, and that for this reason a movement to a freshwater environment is selectively advantageous. It is notable that there are few fishes that have successfully (or as successfully as anguillids) become adapted to the crevice habitat that eels occupy in either fresh water or the sea. Distribution of anguillid eels shows numerous species to be present in the rich and diverse tropical fish faunas, especially in the Indo-Pacific, and this alone indicates that the life history strategy is a successful one — they seasonally penetrate these complex ecosystems and become established in them. Perhaps anguillids have effectively taken the slender, elongate and highly sinuous form and predatory habits of the primarily marine anguilliforms into fresh water in a way that selection and evolution within other, primarily freshwater groups has failed to achieve.

A further strategy, as discussed above, and as occurs generally in amphidromous species, may involve the movement of newly hatched larvae from fresh water to the sea, without feeding. Brief life (one to several months) occurs at sea, where there is feeding and growth, and is followed by a return migration to fresh water of juveniles, which continue to feed and grow to maturity in fresh water.

Strategically it can be imagined that this is advantageous for getting the tiny larvae an early start in growth in a favourable growth environment, but it is not as obvious why these juveniles return to a freshwater environment. In some species the return may be related to the adoption of a benthic habit (as in amphidromous gobiids), but this is not always true (as in galaxiids, some of which remain pelagic, even in fresh water, e.g. the inanga, *Galaxias maculatus*). An alternative strategy to a return to fresh water would be for the fish to become adapted to life spent entirely in the marine plankton, and this is a strategy that seems to have been adopted by some gobies (Miller, 1973a). With such variability in trophic behaviour at diverse life history stages in both fresh water and the sea, it is difficult to imagine any common strategy as Northcote (1979) pointed out, and it seems more likely that natural selection has resulted in a series of strategies which, in diverse ways, have adapted species to diverse habitats and behavioural patterns, and all of which function to increase production in some way. Northcote (1979) suggested that for some species survival is apparently greatly enhanced by the fish reaching a certain minimum size before migrating to a secondary feeding habitat, and there is little doubt that this could be true in, for example, the various Pacific salmons. Larger size is achieved, in part by large eggs, in part by growth of the larvae in the redds based on egg yolk after hatching, and in some species, by feeding in fresh water prior to moving to sea. For some of them the ability to osmoregulate has been shown to be size/age related — early migrants suffer heavy mortalities. But in other groups, particularly the amphidromous species, minute size is obviously no barrier to making the freshwater/marine transition. Miller (1984) found that, amongst gobies, minimal egg size is found in riverine and stream-dwelling freshwater species whose tiny larvae are swept downstream to grow in coastal waters before ascending to the parental habitat. Again, there seem to be no general rules.

Climatic strategy

Not all migrations that appear to be movements from one feeding habitat to another are necessarily trophic migrations. Northcote (1979) pointed to what Heape (1931) described as climatic migrations, in which fish leave one habitat and move to another, to escape potentially hazardous conditions. Thus interacting with feeding migrations are climatic ones - in which species depart from unfavourable habitats and/or move to more favourable ones. It is very difficult to determine which of these is true in any instance, and they are not alternatives as both may apply simultaneously. Climatic migrations are most obvious in the high Arctic, where fish (salmonids in particular) leave the sea for the winter to escape from the extremely low temperatures. But in so doing, they then face the alternative hazard of icing of fresh waters. Bakshtansky (1980) described how attempts to establish pink salmon in eastern Arctic Russia failed because of the impact of winter icing on the salmon redds. Studies show that fish in fresh water during winter may avoid such hazards by moving into the deeper water of lakes, or into streams fed by underground springs which are less cold (McCart and Craig, 1973; Craig and Poulin, 1975). However, climatic migrations should not be seen just as movements away from hazardous conditions, but also as movements to enable the occupation of more favourable ones.

It is of interest that in some species the eggs may be deposited in stream-substrate gravels and develop below ice during the winter and early spring. The eggs develop during the winter at a rate that results in hatching taking place during the spring thaw, at which time stream habitat is increasingly becoming available to the young fishes. Depression of development rate by low temperatures could be interpreted as an adaptation to preventing hatching from occurring too soon, before the ice has gone. Possibly the distinctive habit of salmonids of living for several weeks in stream bed gravels, is also a similar adaptation extending the period in the spawning gravels well into the spring thaw. One of the particular costs of such climatic migrations involves the movement of fish away from productive marine habitats to freshwater habitats where there is little if any opportunity to feed over the winter. Thus in some species, which leave the sea for the winter, there is little or no growth until the following spring (Mathison and Berg, 1968; Moore, 1975a; Northcote, 1979).

In the tropics, the problems are different and migrations tend to relate to the seasonal monsoon rains. Upstream spawning migrations of the hilsa (Clupeidae) occur during the wet monsoon season of India and other parts of Asia

(Pillay and Rosa, 1963). At such times rivers are full, facilitating migration. The upstream feeding migrations of the chame, *Dormitator latifrons*, are reported to take place at times when the monsoon rains flood the land, making food available in higher quantities (Chang and Navas, 1984)

These examples reveal fairly clear cases of climatic reasons for migration, but they occur at the extremes in life history strategies in which the climatic role is easy to see. There may be far more numerous migrations for climatic reasons that are not obvious owing to ignorance of the environmental tolerances or environmental optima of the species involved. There seem to be few instances where a climatic migration is suggested for temperate species, but one is the anadromous brook char, *Salvelinus fontinalis*. It is believed that in northern cold latitudes, this fish is typically anadromous, leaving the sea to spawn in the autumn. In more southern latitudes it leaves the sea in the spring to avoid warm summer sea temperatures (McCormick *et al.*, 1985), and further south again it is confined to higher elevation, cool catchments, and the fish do not migrate to sea at all (Kendall, 1935).

Spawning strategy

Many diadromous fish migrations are quite obviously and explicitly related to the occupation of spawning habitats. These movements are designed to bring very dense concentrations of fish into specialised habitats of limited area, where the populations themselves could not be sustained. Several issues are involved — there is normally insufficient space, there is likely to be insufficient food, and the habitat may be unsuitable. Examples include the well-known salmonids that move onto gravelly stream beds where there may be enormous concentrations and densities of fish, or the much less-known movements of inanga, *Galaxias maculatus*, to vegetated estuarine tidal flats where spawning takes place. There are many such examples; in fact it may be a fairly general feature of diadromy.

An alternative, compelling reason for reproductive migrations is the need for species to occupy habitats providing suitable salinities. The most extraordinary example is the anguillid eels, which leave fresh water and migrate vast distances at sea to reach marine spawning grounds. The population costs of such migrations are quite unexamined, but must be enormous, yet vast numbers of progeny arrive back at the rivers where the adults live, and from which they departed months or years beforehand. In spite of the costs, the strategy is obviously successful. Only a portion of the anguillid migration can be attributed to the need for spawning/egg development to occur in the sea. Theoretically, it could take place very soon after the mature eels move to sea, but they also migrate long distances at sea, to find well-defined spawning habitats. So obviously some other factors are implicated. Among these could be the need for the adult eels to spend some time in the sea, developing from near maturity to ripeness (though why this should require the vast distances is not known, and there is really no obvious reason why maturation could not be completed in fresh water prior to migration to sea). In large measure the egg and larval stages are at the mercy of oceanic currents, and they drift for long distances, so the location of the spawning habitat may be adapted to ensuring that the more or less helpless progeny find their way back to suitable freshwater habitats within prevailing ocean current systems. Associated with this, it is possible that the location of the spawning grounds could have ancient historical causes - the spawning grounds may have become increasingly disjunct from freshwater habitats through geological time, as during the dismemberment of Pangaea established historical spawning areas have become more distant from freshwater habitats, and as oceanic current systems have altered their directions and velocities. Thus, Tesch (1977) linked eel spawning migration patterns to the opening up of the Atlantic Ocean, and suggested that the spawning grounds become increasingly distant from their freshwater habitats as the continents parted and ocean invaded the gap between them.

One of the costs of reproductive migration is a movement of mature fish away from areas rich in food and of low population density, to areas low in food and with high population density. However, many, if not most diadromous migrants cease feeding at the onset of the spawning migration, so that losses of food availability are evidently a loss already 'taken into account' by the fish prior to migration. There are, of course, very substantial costs in the act of migration itself, particularly in view of this cessation of feeding. Migration is undertaken with no or little energy intake, and the fish must depend on energy reserves in fat deposits and body muscle. This tissue, in many species, must also contribute significantly to the final maturation of the gonads themselves. These costs are not necessarily significant in survival terms, as is indicated by the duration of the non-feeding/upstream migration in many anadromous species. An extreme example is probably the southern pouched lamprey, *Geotria australis*, which is believed to spend as long as 16 months between ceasing feeding as it leaves the sea and spawning in fresh water; the distance moved is not excessive (up to 200 km), but the period without feeding is certainly of long duration. This time is not required for the fish to reach the spawning grounds, as might be argued for the long-distance anadromous migrations of species like sturgeons, salmons, clupeids, and others. However, the cost of migration, maturation, and then finally spawning, is often death.

Another indicator that long-duration, non-trophic freshwater life is not excessively costly might be taken from the fact that a surprising number of anadromous species have dual stocks — some of these are summer migrants, and remain in freshwater, without feeding until winter spawning, and others are winter migrants, and spawn very soon after reaching the same spawning grounds (Berg, 1959). Northcote (1979) suggested that the summer migrants accumulate much more extensive fat reserves than winter migrants, allowing the former to survive through until spawning. If there was a significant net cost to production of progeny resulting from the length of the non-trophic fresh-water life, it seems inevitable that such life would be negatively selected, and disappear.

The intense concentrations of spawning fish on spawning grounds has significant implications for the behaviour and survival of the hatching progeny. Vast numbers of young are emerging into the vicinity of the spawning habitats over a very short time period. First, the spawning habitats themselves may have little or no suitability for early feeding/rearing of the progeny. In such instances, rapid outmigration is more or less mandatory. Even should the spawning and associated habitats be suitable for early feeding/rearing, so great are the numbers emerging that for reasons of both space/territory and food availability, outmigration, if not necessary immediately, rapidly becomes so. For example it is easy to see that recruitment from the enormous reproductive potential of salmon runs would utterly overwhelm freshwater habitats, even for the smaller juveniles, and thus offer no opportunities for growth to the large size of adult salmons — an opportunity that is realised by migration to sea to exploit the immensely richer and more spacious living environment.

For these reasons, a rapid outmigration of many or all of emerging progeny from spawning grounds is characteristic of many species that spawn in fresh water, both anadromous and amphidromous species. These migrations can easily be interpreted as an immigration to reach more favourable trophic habitats, but can, as justifiably, be interpreted as emigration from areas that are space limiting and thus may have little or nothing to do with feeding. Hayes (1987) described the spawning of brown and rainbow trout of lake origin in a small tributary of Lake Alexandrina in New Zealand. So great is the density of spawners that Hayes estimated that the survival to hatching is less than 3 per cent, with virtually all surviving progeny being derived from the last mode of spawning fish. Rapid emigration from this stream by the progeny follows emergence from the gravels. Unwin (1981) showed that for chinook salmon, spawning in a small stream in New Zealand, as much as 98 per cent of fry may leave the stream almost immediately on emerging from the redds, with only a few remaining to rear in the spawning stream (Figure 9.8). In both instances massive emigration results in much greater survival.

In various of the anadromous sturgeons, clupeids, salmonids etc. the trophic/space migration downstream towards the sea may be a slow, growth-related process, in which the young fish move downstream as their territorial/feeding demands require increasing individual space, lower population densities, and therefore more total population space. Strategies within species may include both an early emigration of the bulk of the progeny from the spawning stream and a slower, downstream feeding migration of others (Unwin, 1981). Ultimately, however, separate behavioural/physiological processes begin to operate and what seems initially to be a behavioural/space interaction is overtaken by physiologically mediated movements to sea that have nothing to do with behavioural interactions between individuals of a species.

Figure 9.8: Migration from spawning grounds in the tropical western Atlantic to freshwater habitats by the American eel (*Anguilla postrata*) and European eel (*A. anguilla*)

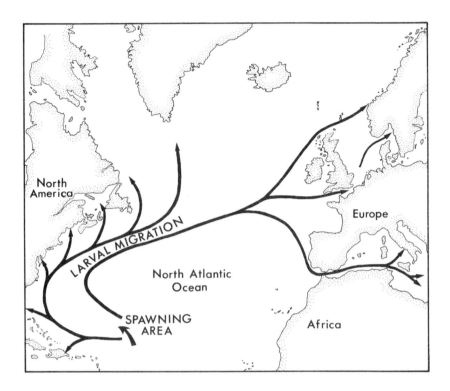

Ultimately the effect is to allow a growth rate and maximum size at sea that cannot be attained in fresh water. Observations of the considerable size and high densities of spawning adults of species such as Atlantic salmon and chinook salmon illustrate very clearly the fact that migration away from the spawning grounds for the bulk of the growth is mandatory for space reasons alone. This is supported by the fact that in species in which fluviatile resident stocks occur as an alternative life history strategy, the non-diadromous fishes are distinctly smaller than the migratory ones (see Chapter 13).

An interesting, alternative life history/growth/habitat strategy adopted by several salmonids, and perhaps by other groups, is the retention in freshwater, by anadromous species, of what have become known as 'precocious males'. These are small fish that never go to sea, but which mature, usually at an age of one year, in or near the spawning streams. These fish are at a huge disadvantage when it comes to reproduction in that they are unable to compete successfully on the spawning grounds for females with which to reproduce. In practice they do not compete, but instead become involved in the spawning activities of pairs of normal-sized adults, a strategy that Gross (1985) described as 'sneaking'. They hide among substrate boulders and dash into the spawning zone attempting to fertilise the eggs as they are discharged by the female (and apparently often do so successfully). Clearly it is more advantageous for this to be true of males than females, owing to the need for the production of a large bulk of reproductive products by females, but not males. And if the presence of precocious males has a selective/genetic basis (and there is no reason to believe that it does not), then the effective fertilisation of some ova by such fish is a mechanism of ensuring that there is a continued presence of some males on the spawning grounds to fertilise arriving females. Gross (1985) described 'disruptive selection' for dual sizes of spawning males in coho salmon, suggesting that very large males have an advantage in fighting for females while the tiny precocious males have an advantage in sneaking. The result is a bimodal length/frequency distribution, with intermediate males being too small to fight successfully for females but too large to hide and sneak (Figure 9.9). Precocious males are reported for several salmons and trouts — chinook and coho salmon, Atlantic salmon, Kamchatka and rainbow trout, and no doubt others.

One important strategic aspect of diadromy relating to deme survival is the colonisation of habitats that are new. Diadromy offers opportunities for the invasion of areas that become habitable when, for example, ice sheets retreat releasing streams and lakes, or when an obstruction to migration such as a rock fall, disappears. It enables the colonisation of distant island habitats, an aspect

Figure 9.9: Alternative life history strategies in the coho salmon, *Oncorhynchus kisutch*; large, 'hooknose' males which fight for the females and small 'precocious' males which become involved in fertilisation of the eggs by 'sneaking'
Source: Gross (1985)

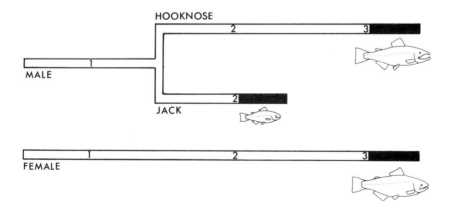

Figure 9.10: Emigration from spawning streams by fry and smolts of chinook salmon, *Oncorhynchus tshawytscha*, showing massive early emigration of newly emerged fry (induced by high population densities) followed by much reduced emigration of larger, growing smolts (that is more probably related to an urge to move to sea)
Source: Unwin (1981)

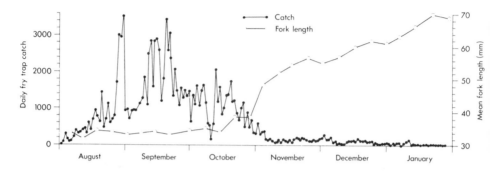

that is of biogeographical interest, and which is discussed in Chapter 8. The sea-migratory habitats, in both instances, enable the species to move to isolated habitats and colonise them. Examples are probably far northern land areas, like Greenland and Iceland, which may have been uninhabitable to fishes during the Pleistocene glaciations. Once ice sheets retreated, reinvasion of the habitats was possible for diadromous species, and is reflected by the fact that in both areas all the species present in fresh waters are diadromous.

The persistence of diadromy in fishes is heavily dependent on success in the return migrations to the spawning habitat. It must be obvious that in instances of failure or inability to return there is very harsh selection against diadromy — if the only survivors in the population are non-migratory variants, loss of diadromy is bound to be very rapid (i.e. immediate!). This applies where migration is prevented by the development of natural or man-made impoundments. The development of landlocked stocks is therefore very much encouraged by failure of return migration, as has happened in New Zealand with all three salmon species established there (*Salmo salar*, *Oncorhynchus tshawytscha*, and *O. nerka*, McDowall, 1978a). In only *O. tshawytscha* are there sea-migratory stocks as well as the landlocked ones. In all three species populations were probably based on original introductions of diadromous stocks. And although some landlocked *S. salar* were introduced from Canada (McDowall, 1978a), it seems most probable that the successful population was based on originally diadromous fish from Europe. This sort of selection pressure is the likely explanation of the reduction in the frequency of anadromy towards the equator in several cold-loving, northern hemisphere salmonids (Chapter 7).

10

Life History Deviation and Landlocking

Many of the fish that are diadromous establish 'landlocked' or exclusively freshwater populations. The mechanism of forming landlocked populations varies. Some populations are forced to become landlocked as a result of natural changes in flow patterns in the rivers they occupy, which prevent sea access; natural physiographic changes may alter flow patterns by stream capture of headwater catchments and thereby reverse or divert the direction of flow; tectonic changes or land-slips may impound streams behind natural dams. In other instances man-made impoundments prevent sea access. There are, however, examples where populations lose their diadromous migrations without any physiographical/topographical constraint — they exercise an apparently inherent capacity for some members of a population to not go to sea.

THE OUTCOME OF LANDLOCKING

The population consequences of the impoundment of a river system and the prevention of migration to and from the sea can be one of several possibilities.

Population extinction

In some species, no doubt, the population is unable to adjust to deprivation of sea access and it declines as time passes and the fish age and die. Judging by the fact that in various genera some species have successfully established wholly freshwater populations after being impounded behind a dam while other related and often sympatric species have not under the same circumstances, it would appear that the latter are unable to. Examples can be seen in New Zealand amphidromous species of *Gobiomorphus* (family Eleotridae). *G. cotidianus* has frequently established lake populations but the closely related species *G. huttoni* and *G. hubbsi* have failed to, yet there is no doubt that there have been opportunities for these last two species to do so when impoundments have been formed, either naturally or by man. Another gobioid, *Awaous tajasica*, from Central and North America failed to establish landlocked populations when artificially impounded in Puerto Rico, as did the mountain mullet, *Agonostomus monticola* (Erdman, 1984).

Landlocked stocks

In other instances, landlocked populations may become established and a lacustrine/fluviatile life history develops in which the lacustrine phase replaces and largely mirrors the former marine phase. This is widely true of many freshwater-limited lampreys, sturgeons, salmonids, osmerids, galaxiids, retropinnids, plecoglossids, etc.

Polytypic species

Some diadromous species are polytypic and their populations naturally comprise migratory and non-migratory stocks. This has been reported for many diadromous groups, as Northcote (1967) observed :

> Temperate freshwater fish, particularly the salmonids, appear to have retained a high degree of plasticity in their migratory behaviour. It is not uncommon to find in a single species, even in a single population, a certain fraction which is non-migratory, along with the migratory component.

Certain taxa appear to be much more facultative as regards loss of diadromy than others and in nearly all groups there are some species that are facultative and some that are not. This is well exhibited in the northern Pacific species of *Oncorhynchus*: *O. masou* and *O. nerka* seem adept at responding to landlocking, others like *O. kisutch*, *O. gorbuscha* and *O. tshawytscha* do so occasionally, but as far as I am aware, *O. keta* has never established a non-diadromous population. Rounsefell (1958) classified American salmonids as 'optionally' anadromous (*Salvelinus alpinus, S. fontinalis, S. malma, Salmo clarki, S. trutta, S. gairdnerii, S. salar*), 'adaptively' anadromous (*Oncorhynchus nerka, O. kisutch*), and 'obligatorily' anadromous (*O. tshawytscha, O. gorbuscha,* and *O. keta*) (Figure 10.1). However the experience with *O. gorbuscha* in the Great Lakes and with *O. tshawytscha* in New Zealand shows that these two could perhaps be re-classified as 'adaptively' anadromous. Similar classifications of species in other diadromous families could easily be produced. In the southern Galaxiidae, species like *Galaxias brevipinnis* and *G. truttaceus* are known to become landlocked very easily — whenever a river system becomes impounded (McDowall, 1978a; McDowall and Frankenberg, 1981). Others like *G. maculatus, G. fasciatus* and *G. argenteus* do so occasionally, but it has never

Figure 10.1: Degree of anadromy in various North American salmonids based on six
different criteria
Source: After Rounsefell (1958)

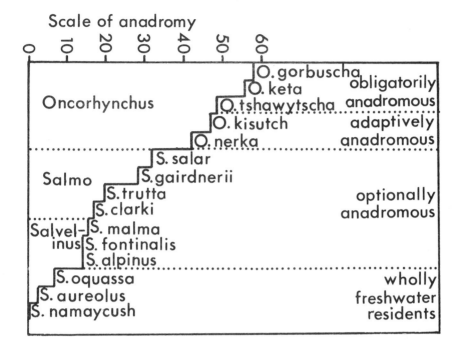

Scale of anadromy

0   10   20   30   40   50   60

been reported in *G. postvectis*. In the Retropinnidae, *Retropinna semoni* seems
to be facultatively euryhaline rather than diadromous, *R. retropinna* is anadromous but easily landlocked, there is one apparent instance of landlocking in *R.
tasmanica*, while *Stokellia anisodon* is obligatorily anadromous. The essential
point is that adaptability varies widely, with some species able to adapt and
other closely related species not able to.

Baker (1978) has claimed that whenever diadromy is lost, the result is that
the population becomes lacustrine and that a downstream migration always ter-
minates in a lake. Rounsefell (1958) was rather more tentative (and accurate)
in suggesting that freshwater forms of the anadromous salmonids normally use
a lake as a miniature sea. But it is by no means always true. A significant number of anadromous salmonids have non-anadromous, fluviatile populations, as
do some clupeids (*Hilsa*), cottids (*Cottus asper*), and gasterosteids (*Gasterosteus aculeatus*). In the brown trout, for instance, there may actually be alternation, or facultativeness in whether a particular individual fish is anadromous or
fluviatile in any particular year. The migratory status of some populations is
also clearly variable; Nordeng (1963) has shown that fluviatile Arctic char,
when transplanted to cooler, northern localities, became anadromous. Anadromous and non-migratory Atlantic salmon coexist in some waters of eastern

Canada and northwestern Europe, and Hutchings and Myers (1985) have shown that there are probably no behavioural isolating mechanisms that would prevent the two stock-types from interbreeding where sympatry occurs. They concluded that there is no reason to believe that the two types do not interbreed, unless spatial isolation during spawning can be demonstrated. Dingle (1980) believed that where streams have both freshwater and anadromous populations of a species the former usually occur in headwaters some distance from the sea while the latter spawn in the lower reaches. He cited sticklebacks as exemplifying this principle. This is another of those statements that is quite widely true, but for which there are common exceptions. For some species the statement is rather a self-evident truth, in that the migratory stocks are excluded from upstream habitats (above which non-migratory stocks exist) by barriers to upstream migration — the non-migratory populations inhabit impoundments that actually prevent the upstream migration. But in other situations, diadromous and non-diadromous populations are found to be sympatric. It has also been shown that anadromous sockeye salmon and resident kokanee may interbreed in British Columbian waters (Hanson and Smith, 1967). Johnston (1981) described sympatric resident and anadromous stocks of cutthroat trout, *Salmo clarki*, in coastal streams of the Pacific north-west of North America.

One interesting instance involves sympatric populations of *Retropinna retropinna* in Lake Waahi, Waikato River system, in New Zealand. In this lake there is a dwarfed, low vertebral number, resident population, which is joined each spring by anadromous stocks that enter the river mouth some 75 km downstream and make their way up into the lake (McDowall, 1979; Northcote and Ward, 1985).

Nevertheless whether or not a species is naturally polytypic, the replacement of an oceanic/pelagic, marine, feeding stage, by a lacustrine/freshwater feeding stage is widely present in many anadromous groups. This is true of nearly all groups — of some lampreys, sturgeons, salmonids, a plecoglossid, galaxiids, retropinnids, osmerids, clupeids, a gadid, percichthyids, gasterosteids, gobies, and eleotrids, and no doubt other families. The capacity to abandon diadromy rapidly becomes apparent when artificial transplants of species take place and the sea-going members of the population fail to return. Under such conditions there is extremely harsh selection against migration; if there are no ocean returns any residual stock in the lake or river receiving the transplant is based entirely on the minority, non-migrant deviants in what was regarded as an essentially migratory stock. Northcote (1967) has observed this in stocks of *Salmo gairdneri* above falls in British Columbia — the residual stock above the fall depends entirely on the non-migratory proportion of the population. The occurrence of resident and non-migratory fractions within otherwise diadromous populations of several other salmonids, as well as some osmerids was discussed previously in the detailed analysis of anadromy. This is also very clearly seen in *Salmo salar* and *Oncorhynchus nerka* in New Zealand. Introduced stocks of both species were based in diadromous source populations in

Europe (*S. salar*) and North America (*O. nerka*) (McDowall, 1978a; Scott, 1984), yet all that survived initially, and all that remains now some 70–80 years following establishment in New Zealand, are landlocked stocks, but stocks that were free to go to sea for many years (although both are now impounded behind dams). All the evidence suggests that in both species, some fish did go to sea but that none ever returned to fresh water. In the case of *O. nerka* only one stocking of fish was undertaken so that the surviving stock must be derived from the residual, non-migratory individuals from that single, original release. The survival of stocks of both of these species in New Zealand must have depended initially on the failure of some proportion of the population to migrate to sea. Existing stocks of these species are the progeny of such non-migrants. (Because the stocks are probably based on a restricted gene pool they would provide an interesting resource for study of the effect.)

Some situations are known where there is no physical barrier to migration but in which it appears that environmental conditions constitute an equally effective barrier to migration. Rounsefell (1958) attributed landlocking of sockeye salmon, *O. nerka*, in Canadian lakes in the southern (warmer) parts of the species' range, to the fact that during the summer migratory period, the surface waters, through which the fish would have to migrate to leave the lakes on their way to the sea, were too warm. He considered that the fish would not migrate through this thermal barrier. (He also observed that there are no populations of kokanee in the cold far north.) Flain (1982) has hypothesised the same situation for a lacustrine population of chinook salmon, *O. tshawytscha*, in a New Zealand lake. Warm surface water is suggested as forming a 'lid' on the lake through which the fish will not migrate (Figure 10.2).

Speciation

In some families it appears that non-diadromous populations have evolved and become distinct species that are reproductively isolated from either former or still-existing diadromous species. This is believed to be true of the paired species of lampreys discussed by Potter (1980). The brook lamprey, *L. planeri*, is interpreted as a non-migratory deviant of the anadromous river lamprey, *L. fluviatilis*, in Europe, and similarly, *M. praecox* is a derivative of the anadromous *M. mordax* in Australia. Potter cited several other similar instances. In a similar way some of the lacustrine salmonid species are interpreted as having an ancestry in diadromous stocks (Behnke, 1972; Miller, 1972a;). This phenomenon is probably widely true in all diadromous families, although phylogenetic analysis of species groups would be needed to establish explicit relationships and confirm specific examples. It seems to be widely assumed that much of the speciation in families like Salmonidae, Osmeridae and Galaxiidae is on the basis of geographical isolation of diadromous stocks by landlocking. This assumption is made on the basis that the early stages of the

Figure 10.2: Warm temperatures in the surface waters of Lake Mapourika (New Zealand) form a lid through which chinook salmon, *Oncorhynchus tshawytscha*, will not migrate. This may prevent emigration and result in an entirely freshwater stock
Source: Flain (1982)

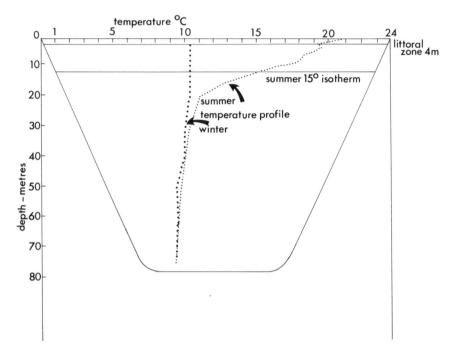

process can be seen in landlocking and morphological deviation with a great many diadromous species and in numerous lakes. It is also made on the basis that it would seem inherently more likely (simple) that populations more easily evolve from a highly facultative, euryhaline/osmoregulatory and migratory life cycle, to a non-migratory, stenohaline cycle, rather than the reverse. Further it can be argued that apart from the existence of diadromy, diadromous species tend often to be more generalised in both morphology and behaviour/habitat than the non-diadromous species. Morphological specialisation seems better developed in non-diadromous species (McDowall, 1970). Evolution seems to take species from a specialised condition with regard to migration and osmoregulation to a less specialised condition by the loss of diadromy, but morphology and some behavioural characteristics become more specialised in isolated fresh water stocks and species.

However, even if all these arguments are accepted as being generally true, there is a danger that assumption of their truth may unnecessarily lead to the further assumption that all evolution is from a diadromous to a non-diadromous condition. The argument could be largely true in the primitive, intensively diadromous groups like lampreys, sturgeons, salmoniforms, etc. In none of these groups is there evidence, nor has it been suggested, that diadromous species are derived from non-diadromous ones. But it seems unlikely that evolution is primarly from a diadromous to a non-diadromous condition in the perciform families, in which there are what appear to be one or a few highly distinctive diadromous species in each of otherwise wholly marine families. Outstanding examples are the southeastern Australian and Tasmanian bovichthyid tupong, *Pseudaphritis urvillii* and the New Zealand mugiloidid torrentfish, *Cheimarrichthys fosteri*. In these families, there is only the single diadromous species in what are otherwise entirely marine families, and both species would seem to be derived from a marine ancestry within their respective families. In both diadromy is most likely to be a secondary/derived condition. The primarily marine family Gadidae is similar, with only the diadromous tomcod, *Microgadus tomcod* (although there is a freshwater species in this family, also).

In some higher taxa in which diadromy is present, there seems to be little flexibility, either within or between species, the anguillid eels being the most striking example — the life cycle of the 15 species of *Anguilla* is consistently and undeviatingly catadromous. The southern graylings, as a family, are consistently amphidromous. A perhaps surprising example of inflexibility is the New Zealand torrentfish, *Cheimarrichthys fosteri*; although the only euryhaline/diadromous species in the Mugiloididae, it is invariably amphidromous, there being no entirely marine or entirely freshwater populations.

REVERSIBILITY OF LANDLOCKING

Reversibility of the loss of diadromy is a question that has seldom been addressed, the issue being whether landlocked stocks of diadromous populations can re-establish their euryhaline or amphihaline habits and their migratory patterns, should the opportunity for migration to sea be restored. Foerster (1947) examined this question with regard to kokanee stocks of the sockeye salmon, *Oncorhynchus nerka*, and found that smolts were able successfully to make the transition back to sea again, and that some mature adults, which were the first generation progeny of landlocked kokanee, did return from the sea. Thus the landlocking process was shown, at least in this instance, to be reversible. Rounsefell (1958) has shown that resident brown trout, *Salmo trutta*, can be derived from anadromous stocks, and vice versa, while Nordeng (1983) has found the same in Arctic char, *Salvelinus alpinus*. J.E. Thorpe (1987) described a situation where a stock of Polish brown trout, stream resident in waters 1600 km from the sea, was transplanted to coastal Polish streams. Slow-growing juveniles were found to smolt and go to sea. I know of no other cases where this question has been examined explicitly. However, in New Zealand there have been landlocked stocks of sockeye, *O. nerka*, since 1901 and of Atlantic salmon, *S. salar*, since about 1915. Attempts were recently made to transfer smolts of these stocks into sea water as a part of a study of their potential for commercial sea-cage rearing. In both species, there was a successful transfer, and stocks of both have been reared through to substantial size (Atlantic salmon to maturity) in sea water, after 20–25 generations exclusively in fresh water. The ability to osmoregulate in sea water has not been lost in that time.

CHANGES IN REPRODUCTIVE/LIFE HISTORY STRATEGY

When diadromous stocks are confined — either voluntarily or obligatorily — a change in the reproductive/life history strategy would also seem to be likely. Gross (1987) has explored the consequences of a change in life history strategy in diadromous and non-diadromous fishes and reached the conclusion that an examination of such important life history traits as egg size, fecundity, age at first maturity, and body size reveals no apparent significant differences between diadromous and non-diadromous stocks — on the basis of data on Salmonidae, Osmeridae, Clupeidae and Petromyzontidae. Looking more narrowly at salmonids he found few differences in life history traits for diadromous and non-diadromous forms within species, only larger body size at the same age of maturity.

Nevertheless, more explicit comparisons do reveal some interesting differences. There may be distinct differences in reproductive features between non-diadromous and diadromous species of close affinity, and between diadromous and non-diadromous stocks of a species. Landlocked lampreys are observed to grow to a larger size at the ammocoete stage and be larger at metamorphosis to the adult stage than anadromous ones, but because there is no post-metamorphic growth by the adult, they reach maturity and spawn at a much smaller size than anadromous species (up to 300 mm, compared with up to 800 mm). Hardisty and Potter, 1971c and Potter, 1980 have examined in some detail the implications of changing from an anadromous to non-anadromous life cycle, in what they have labelled 'paired species', i.e. sister species in which the derived taxon has abandoned anadromy for a wholly freshwater existence. In general this change has close association with one from a parasitic to a nonparasitic mode of life. The loss of anadromy is accompanied by loss of the marine trophic phase, so that a long period of potentially rapid growth is lost. This is compensated for, in some measure, by non-parasitic, non- anadromous species, like *L. planeri* and *M. praecox*, having a longer larval life, so that the latter are larger than their anadromous, parasitic sister species at metamorphosis. But the anadromous species feed at sea for several years while the non-parasitic species move directly to the process of sexual maturation, without further growth. As a result, anadromous species are normally much larger than non-anadromous species at sexual maturity. There are consequential differences in fecundity, which have several interrelated causes. Hardisty and Potter (1971c) showed that the actual number of oocytes laid down in the ovaries is greater in the anadromous species *L. fluviatilis* and *M. mordax* so that their potential fecundity is greater and has a genetic basis. In addition they observed extensive atresia of oocytes in the non-anadromous *L. planeri* and *M. praecox* which increased the differences between the two species pairs. Added to this were differences in absolute body size (Figure 10.3), resulting, overall, in substantial differences in actual fecundity between anadromous parasitic and non-anadromous, non-parasitic lampreys.

Hutchings and Morris (1986) carried out an analysis of life history traits in diverse salmonid species and distinguished between, at one extreme large, early maturing, semelparous individuals bearing few, large eggs in anadromous species, and the opposite suite of characteristics in small, iteroparous species that were non-anadromous. Mizuno (1960, 1963b) examined life history strategies of Japanese gobies and reported that *Rhinogobius brunneus* has amphidromous, fluvial and lacustrine stocks. The amphidromous fish were found to have many small eggs and tiny, free-swimming larvae, but the fluvial and lacustrine populations have fewer, larger eggs and benthic juveniles. The same was true of *Rhinogobius flumineus*, which he regarded as a landlocked derivative of *R. brunneus*. It has similarly been shown that the diadromous species of both *Galaxias* and *Gobiomorphus* in New Zealand have much smaller and more numerous eggs, producing more numerous and smaller larvae (McDowall, 1970). Humphries (1986) found that landlocked populations of the Tasmanian spotted mountain trout, *Galaxias truttaceus*, have fewer larger eggs than conspecific diadromous populations.

Blackett (1973), when comparing anadromous and resident stocks of dolly

Figure 10.3: Difference in size at metamorphosis of the anadromous lamprey, *Mordacia mordax*, and the non-anadromous *Mordacia praecox*, in an Australian river Source: After Hardisty and Potter (1971c)

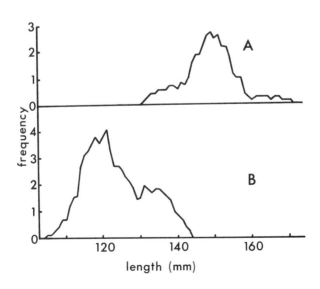

Figure 10.4: Differences in egg size and vertebral number in sympatric anadromous and freshwater-limited stocks of New Zealand common smelt, *Retropinna retropinna.* A, Presumed landlocked stocks; B, presumed anadromous stocks
Source: After Northcote and Ward (1986)

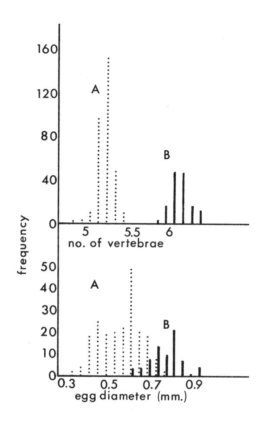

varden, *Salvelinus malma*, reported that the non-anadromous forms were smaller, matured younger, had fewer eggs per millimetre of body length, greatly reduced overall fecundity, and a shorter life span. Northcote and Ward (1986) reported reduced egg size in lake-limited populations of dwarfed common smelt, *Retropinna retropinna* in New Zealand (Figure 10.4). Another consequence of landlocking may be a shift in the spawning season, though there is presently only meagre evidence for this. Hagen (1967) showed a distinct shift in the period of spawning between the anadromous '*trachurus*' form of the threespined stickleback and the non-migratory '*leiurus*' form — a shift from June–August to March–June (Figure 10.5). And there are indications that a change in spawning season may occur in landlocked galaxiids. Pollard's (1971) study of a landlocked population of inanga, *Galaxias maculatus*, in Australia revealed a shift from primarily autumn spawning to spring spawning. Burnet, Cranfield and Benzie (1969) described shoals of 'mountain whitebait', koaro,

Figure 10.5: Differences in the breeding period of migratory stocks (*trachurus*) and non-migratory stocks (*leiurus*) in the threespined stickleback, *Gasterosteus aculeatus*
Source: After Hagen (1967)

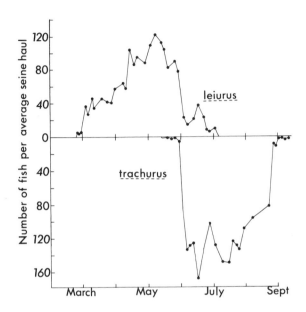

Galaxias brevipinnis, around the shores of lakes during the autumn, and runs of these occur into lake tributaries at this time of the year (McDowall, 1984c). In diadromous stocks these migrations are a spring phenomenon, and the difference in the timing of migrations possibly reflects a difference in the spawning season (autumn in diadromous fish but possibly spring in lacustrine stocks).

One of the most surprising and interesting consequences of loss of diadromy was described by Pollard (1971) for the direction of the spawning migration in the inanga, *Galaxias maculatus*, in Australia. Diadromous populations of this fish have a downstream spawning migration that is timed to bring them into estuaries at the high spring tides, at which time the marginal vegetation of the estuaries is inundated with water (McDowall, 1968a, 1978a). However, the landlocked population in Lake Modewarre in Australia has an upstream migration into lake tributaries, and spawning takes place during freshets, when stream bank vegetation is inundated. In both situations spawning takes place over terrestrial vegetation when this is flooded, and the eggs are deposited in sites that are out of the water most of the time, but the direction of the migration has been reversed to achieve this end in the landlocked population. Thus it appears that there clearly are differences in basic life history parameters between diadromous and non-diadromous stocks in many species.

CHANGES IN ABSOLUTE SIZE

A common consequence of landlocking is dwarfing of the fish. This is true both of species that are derived from presumed diadromous ancestry and of land-locked populations of diadromous species. It is characteristic, for instance, of non-parasitic, freshwater lampreys (Potter, 1970, 1980; Vladykov and Kott, 1979). The landlocked 'sebago' salmon, *S. sebago*, in eastern North America, and the kokanee strains of sockeye salmon, *O. nerka*, in the west are also known sometimes to be distinctly smaller at maturity and smaller relative to their age than their sea-going counterparts. It is true of introduced stocks of both these species, of landlocked chinook salmon, *O. tshawytscha*, in New Zealand, and of riverine rainbow trout, *S. gairdneri*, as opposed to sea-migratory steelhead. Sprules (1952) showed that anadromous Arctic char, *Salvelinus alpinus*, are much larger and more fecund than their landlocked counterparts. Mills (1971) found that anadromous examples were 3.5–4.5 kg in weight, occasionally reaching 9 kg, whereas lacustrine stocks might reach only 85 g and a length of about 15 cm. Scott and Crossman (1973) showed that landlocked populations in small ponds grew to only about 100 mm, but in Lake Ontario and the open sea they grew to 350 mm or more. Lenteigne and McAllister (1983) observed that the pygmy smelt, *Osmerus spectrum*, which is a landlocked derivative of the rainbow smelt, *Osmerus mordax*, grows more slowly, matures earlier and at a smaller size (less than 125 mm) than the rainbow smelt, and has a shorter life span. In New Zealand the dwarf inanga, *Galaxias gracilis*, reaches a length of only about 55 mm, compared with a common adult length of 95 mm and a maximum of 150 mm in the inanga, *G. maculatus*, the diadromous species from which it probably derived (McDowall, 1978a). This rule is probably generally true of osmerids in North America and Europe, of the plecoglossid ayu in Japan, of landlocked retropinnids and eleotrids in New Zealand and Australia, etc. Eiras (1983) reported that landlocked shad, *Alosa alosa*, matured at a max-mimum of about 500 mm, compared with anadromous fish which grow to 700 mm; the anadromous fish also lived longer.

Even though this is a common phenomenon, it is not a 'rule'. It is untrue, for instance, of stocks of various formerly diadromous fishes such as lampreys, alewives, and various Pacific salmons that have become established in the Great Lakes of North America. It was apparently always untrue of lake stocks of the Atlantic salmon in Lake Ontario, which once reached 22 kg (Webster, 1952). Comments above on the rainbow smelt show that Lake Ontario fish are roughly equivalent in size to marine ones. Perhaps the vastness of the Great Lakes creates a growth-favourable environment comparable with the open ocean for these species. It is untrue, sometimes, of brown trout — foundation stocks of brown trout in Lake Pedder (Tasmania) regularly grew to more than 12 kg for several years following their introduction, until food resources be-came depleted. In Lake Crescent, where spawning opportunities are limited, food remains abundant and the brown trout there still reach a large size. Rogan

(1981) described growth of lake-limited chinook salmon to 2.5 kg at two years and 5 kg at three years of age in a small Australian lake, and these sizes are not greatly different from sizes attained at sea at these ages. Pollard (1971) studied a landlocked population of the inanga, *Galaxias maculatus*, in Australia and showed that it reaches a total length of 170 mm, and has growth rates comparable with sea-migratory stocks. Growth to 60 mm at six months is similar to the growth in that period in sea-migratory stocks.

In at least some of the amphidromous species size at maturity in landlocked stocks is no different from that of diadromous fish. Since in both types most growth actually takes place in fresh water, between juveniles leaving the sea and their reaching maturity, this is not surprising, if differences in sizes attained are a result of differing productivity and food availability in the growing environment occupied. The appropriate comparison is of the size of the juvenile fish at the time they leave the sea. Data are sparse, but for at least the koaro, *Galaxias brevipinnis*, and the common bully, *Gobiomorphus cotidianus*, in New Zealand, the size at which the juveniles leave the marine plankton and enter fresh water seems to be greater than the size at which lacustrine fish leave the lake plankton and migrate into the neritic/benthic environment or into the lake tributaries. In the growth phase of the koaro in New Zealand following their occupation of fresh water, this species grows just as large and possibly just as quickly in its landlocked populations as it does in amphidromous ones. However, both the landlocked and the normally amphidromous populations inhabit small, rocky, forested streams that may not be especially productive, so growth differences are not to be expected if growth is related to the availability of food. In lakes adult koaro may inhabit the lakes themselves, where growth and the size achieved may exceed those typical of stream populations, e.g. in Lake Chalice (Meredyth-Young and Pullan, 1977). Growth rates and size achieved vary within species from population to population, and have also been shown to vary within a population from year to year — a dwarfed population of *Retropinna retropinna* in a New Zealand lake grew to near normal 'diadromous' size when the biota of a small lake was extensively disrupted by the removal of aquatic macrophytes, and the retropinnid population density greatly reduced (Mitchell, 1986; Figure 10.6). Growth rates and size reached vary widely in landlocked populations of *Galaxias maculatus* in Australia, New Zealand and South America (Pollard, 1971; McDowall, 1972; Campos, 1974). Svardson (1949, 1952) discussed dwarfing in Baltic Sea populations of *Coregonus* and showed that there is a profound influence of environment on fish size. Fish from 'dwarf' stock sometimes grew much larger when transferred to another lake. Reduced size of landlocked chinook salmon in Lake Ontario has been attributed to a restricted diet in an inferior food environment (although more recent results do not seem to indicate inferior growth). It could generally be true that growth potential of landlocked populations of otherwise diadromous species is not inherently different from that of diadromous populations. In some instances, at least, it seems to be simply a case of growth rates

Figure 10.6: Change in size of landlocked stock of New Zealand common smelt, *Retropinna retropinna*. When population density was reduced, a much greater size was achieved, indicating that differences in size between diadromous and non-diadromous stocks may be related to food limitations
Source: Mitchell (1986)

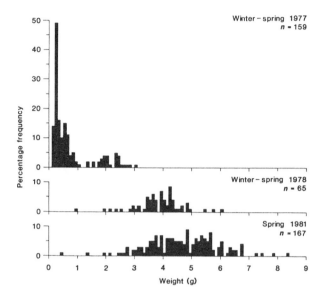

controlled by food availability.

CHANGES IN BODY PROPORTIONS AND MERISTICS

In addition to changes in absolute size attained, there are other changes in morphology. Hardisty and Potter (1971c) reported the eye to be larger in the anadromous *Lampetra fluviatilis* than in the non-anadromous *L. planeri*, and found that the former also had a larger sucking disc; these differences seem likely to have a functional connection to the parasitic habits of *L. fluviatilis*. Oral teeth are sometimes reduced in non-parasitic lampreys, like *L. mitsukurii*, again probably for functional/adaptive reasons. Differences were found in myomere counts in some species pairs but not in others.

Landlocked stocks in some groups, e.g. *Retropinna* (McDowall, 1979), routinely have larger heads, larger eyes and fewer vertebrae than diadromous stocks. Common smelt, *R. retropinna*, in New Zealand has 58–63 vertebrae in diadromous populations and 53–58 in brackish lake populations. There is a clinal increase in vertebral number, from about 49–54 in warmer, northern lakes, to 54–59 in southern, cooler ones. Sympatric and anadromous, lake-limited stocks in one lake have 58–63 and 48–54 vertebrae, respectively (Figure 10.4). Differences in the number of lateral line scale rows are comparable to differences in vertebral number. These meristic differences can probably be attributed to the effects of differing salinities and temperatures during early stages of development. Differences of a similar sort are present in *Galaxias* species. Diadromous koaro in New Zealand have about 58–64 vertebrae (with north–south cline), but there are only 53–58 in northern (warmer) lake populations, increasing to 58–62 in southern ones.

Other morphological changes appear in landlocked stocks. It is well known that the number of lateral scutes along the body of the threespined stickleback, *Gasterosteus aculeatus*, varies with the salinity of the environment. Anadromous populations have more scutes than freshwater populations (Wootton, 1976). McAllister and Lindsey (1959) noted that the prickly sculpin, *Cottus asper*, has a more intense covering of 'prickles' in anadromous populations than in inland, freshwater ones. Amphidromous populations of New Zealand 'bullies' (genus *Gobiomorphus*) have a well-defined series of laterosensory pores on the head (McDowall, 1975). Landlocked populations of one species, the common bully, *G. cotidianus*, lack these pores. It is of some interest that where pored fish of diadromous ancestry have been released into sympatry with lacustrine pore-less fish, the two stocks appear to have interbred and the hybrid population has a highly variable but intermediate number of pores. Thus the loss of pores has at least a partly genetic basis. Like non-diadromous stocks of *G. cotidianus*, entirely non-diadromous species of *Gobiomorphus* have no such pores.

One of the intruiging morphological changes that occurs with landlocking is exhibited by the sockeye salmon, *O. nerka*. Anadromous sockeye leave the sea with silvery coloration, but as they mature they become a brilliant red colour with a distinctly contrasting green head. The same colour change is observed in the landlocked stocks of sockeye known as kokanee. However, landlocked sockeye that are the direct progeny of anadromous parents — known as 'residuals' — fail to display this typical red and green coloration (Ricker, 1940; Foerster, 1968). In New Zealand, where there is an acclimatised, landlocked stock of sockeye, an intermediate situation prevails. This stock has been lacustrine for more than 80 years, and in general the coloration of the mature spawning adults resembles that of the North American residuals — they become darker, the females a greenish—olive and the males tending towards black. However, larger individuals in the New Zealand population — a small proportion that reaches a length of 300 mm or more — show something of the more characteristic red and green coloration of sea-run fish and kokanee.

11

Transportability of Diadromy

Because of their importance as game/food animals, diadromous fishes have been transported to many parts of the world where attempts have been made to establish populations. Three possible consequences of these attempts can be imagined:

1. Successful establishment of a diadromous population;
2. Failure of diadromous populations but the establishment of landlocked populations; and
3. Total failure.

Study of attempts to introduce species shows that all of these possibilities are represented, but that the establishment of diadromous populations in the target/release localities is in a distinct minority. Early attempts at transfers of diadromous fishes seem to have been made with the rather naive assumption that success was likely, but it was soon recognised that the task may be both complex and risky.

The New Zealand experience was, no doubt, little different from those in other countries. The early failure there in the late nineteenth century of both Atlantic salmon, *Salmo salar*, and chinook salmon, *Oncorhynchus tshawytscha*, prompted thoughtful fisheries biologists to give serious consideration to factors contributing to or against success. In particular, it was recognised that substantial releases of well-grown young fish might contribute more to survival and return, than smaller numbers of younger and smaller fish. In all parts of the world, and in more recent times, projects promoting releases have necessarily become increasingly aware of potential difficulties, as typified by Petr's (1985) proposal that *Lates calcarifer* (family Centropomidae) and *Tenualosa ilisha* (Clupeidae) be released into the Sepik River in northern Papua New Guinea. Petr recognised that the 'logistics of the transfer of migratory fish would have to be worked out as such fish have never been transferred to distant watersheds. There might be difficulties with their acclimatization'. Species for which there is good evidence of transplants, including many that are diadromous, were

186

listed by Walford and Wicklund (1973) and Welcomme (1981). In some instances it is not possible to be certain whether the source, or donor populations were actually diadromous, and in some instances they could have been land-locked populations of normally diadromous species. However, as there was commonly a clearly expressed intention of establishing diadromous populations in the recipient country, there is a very high likelihood that the fish transferred were of diadromous stock.

RECREATIONAL AND COMMERCIAL FISHERIES

Most historical examples of transfer are of salmonids and in most cases, particularly in the early years of transfers, the motivation was for the establishment of recreational fisheries for anglers rather than for the development of commercially important fisheries — in spite of the commercial importance of many diadromous fishes like salmon, sturgeons, eels, etc. This has been true, for instance, of most of the very extensive trout and salmon introductions in Southern Hemisphere lands, such as Australia, New Zealand, Chile, Argentina, South Africa, etc. (MacCrimmon and Marshall, 1968; MacCrimmon and Campbell, 1969; MacCrimmon, 1971; MacCrimmon and Gots, 1979, 1980), especially in the early years. And although this has probably been generally true, it has not always been the case. Chinook salmon were introduced into New Zealand as early as 1901 with the intention of establishing a commercial fishery (McDowall, 1978a). There are other examples, e.g. American shad, *Alosa sapidissima*, transferred from the east to the west coasts of North America dating back into the early 1900s but increasingly in the past two decades, there have been transfers particularly of Pacific salmon species to a variety of countries, aimed at establishing commercial fishery resources. The Russians sought to establish Pacific salmons in the eastern North Atlantic, while United States and Canadian transfers have aimed at establishing them in the western North Atlantic and the Arctic Ocean. There were also attempts to establish Atlantic salmon in the Pacific northwest of North America. In the 1970s and 1980s, there has been considerable Japanese and American interest in establishing Pacific salmon in Chile (Lindbergh and Brown, 1982) and also in Kerguelen (Davaine and Beall, 1981a,b). There have been discussions about establishing both Atlantic salmon and various Pacific salmons in the Falkland Islands (Stewart, 1980), although it seems that no actual transfers have yet taken place. Petr (1985) suggested the introduction of *Lates calcarifer* and *Tenualosa ilisha* into the waters of Papua New Guinea to support subsistence fisheries there.

GROUPS SUBJECT TO TRANSFER

A review of transfers of diadromous fish species is, to a considerable extent, a history of the establishment of salmonid fisheries in southern lands, a history which seems to begin with attempts to establish Atlantic salmon in Tasmania in 1864. A desire to see salmonid angling in southern lands like Australia, New Zealand, Patagonian South America and southern Africa followed very soon after the European settlement of these areas, proposals to establish Atlantic salmon in New Zealand dating back to at least 1859 (McDowall, 1984b). The history is summarised by Welcomme (1981), while MacCrimmon and co-authors have extensively reviewed the movements of some salmonids, and have examined the occurrence and present status of several salmonids in most areas of the world (MacCrimmon and Marshall, 1968, *Salmo trutta*; MacCrimmon and Campbell, 1969, *Salvelinus fontinalis*; MacCrimmon, 1971 *Salmo gairdneri*; MacCrimmon and Gots, 1979, *Salmo Salar*). The questions critical to this discussion are: 'What species were transferred?'; 'Where did diadromous populations develop following introduction?; and 'Why?', or 'Why not?' Without wanting to pre-empt the following discussion, the answer, briefly, is 'Scarcely anywhere'.

Transfer attempts have actually been made for relatively few of the families within which diadromy has been recognised. I can find no explicit records of deliberate transfers to establish wild stocks for any of the Petromyzontidae, Geotriidae, Mordaciidae, Osmeridae, Salangidae, Galaxiidae, Aplochitonidae, Retropinnidae, Prototroctidae, Centropomidae, Mugiloididae, Gasterosteidae, Eleotridae, Gobiidae — in fact we are left, essentially, with Salmonidae, Clupeidae and Percichthyidae, with the vast bulk of the transfers being salmonids. Welcomme (1981) did list transfers of anguillid eels, but on the basis of a continuous stocking programme for aquaculture, and not as attempts to establish wild populations and runs. (In fact, a few New Zealand *Anguilla australis* escaped from captivity in Japan and were found in the wild, but there was never a wild population nor any attempt to establish one — Jellyman, 1977a). Thomson (1966) reported an unsuccessful attempt to transfer grey mullet, *Mugil cephalus*, from the Black Sea to the Caspian Sea. And, as noted above, Petr (1985) suggested the release of *Lates calcarifer* and *Tenualosa ilisha* into northern waters of Papua New Guinea (Sepik River), but so far this has not occurred.

The percichthyid instance comprises a successful transfer of a diadromous stock of the striped bass, *Morone saxatilis*, from the east coast to the west coast of North America. The transfer took place between 1879 and 1900 — initially into the Sacramento River — and was rapidly successful; this species is now widespread along the Pacific Coast of North America from British Columbia south to beyond the United States/Mexico border, where it is said to be very successful. Welcomme listed this species successfully transferred to South Africa, but it is not clear whether it is diadromous there; Setzler *et al.* (1980) noted attempts to establish striped bass in Hawaii, Russia, France, and Portugal, but did not indicate whether or not these attempts were successful. Maciolek (1984) said that they failed in Hawaii. Introductions to Mexico were successful

(Contreras and Escalente, 1984), but again it is not clear whether the acclimatised stocks were diadromous; at least some of the populations there are landlocked. The diadromous Australian bass *Macquaria novemaculeata* was taken from Australia to New Zealand in the 1880s but failed to become established (McDowall, 1978a), as did a transfer to Fiji (Lewis and Pring, 1986) and a transfer of the sturgeon *Acipenser ruthensis* from the Baltic to western Europe (Welcomme, 1981). One clupeid example involves the successful transfer of the eastern North American shad *Alosa sapidissima*. Like the striped bass, the shad had a natural distribution along the Atlantic coast of North America and was transplanted to the Pacific coast. Hildebrand and Schroeder (1927) reported many experiments in transporting shad to waters in North America in which it was not native, and they mention the Mississippi, various of the Great Lakes, tributaries of the Great Salt Lake in Utah, and, again, the Sacramento River in California. Only the last of these was successful (Figure 11.1). Shad were released into the Sacramento River in the early 1870s, spread rapidly both north and south from that area, and are now present, as anadromous populations, from Alaska to Baja California. They eventually spread across the Bering Sea to northeastern Siberia (Whitehead, 1985). A second clupeid transfer mentioned by Tinker (1978) is of an introduction of the North American gizzard shad, *Dorosoma petenense*, from the Atlantic coast of the United States to Hawaii, in 1958. Only freshwater-limited populations resulted from this transfer. This species was also taken to Puerto Rico, where it was successful in lakes (Erdman, 1984), and the same seems to be true in Mexico (Contreras and Escalente, 1984). In another clupeid instance, Berg (1962) mentioned an attempt to transplant the Caspian/Black Sea species *Alosa caspa* to the Aral Sea in 1930. He reported that the transplanted fish grew to adulthood, but failed to reproduce. The Australian freshwater herring, *Potamalosa richmondia*, was taken to Fiji, without success (Lewis and Pring, 1986).

There were several attempts to establish the ayu, *Plecoglossus altivelis*, in Hawaii, six consignments of a quarter of a million being taken from Japan; they failed to become established (Maciolek, 1984).

Transfers primarily of salmonids

All the other examples of deliberate transfers known to me are of salmonids and the listing comprises very extensive transfers of a diverse array of diadromous salmonids, including *Salmo trutta*, *S. gairdneri*, *S. salar*, *S. clarki*, *Salvelinus fontinalis*, *S. alpinus*, *Oncorhynchus gorbuscha*, *O. keta*, *O. kisutch*, *O. nerka*, *O. masou*, and *O. tshawytscha*. (Information in the following discussion on the countries into which species have been introduced, derives largely from Welcomme (1981) except where otherwise indicated.)

The brown trout, *Salmo trutta*, was a very early candidate, and was very widely introduced (Welcomme (1981) listed 28 countries, Figure 11.2).

Figure 11.1: The American shad, *Alosa sapidissima*, has natural range along the Atlantic coast of North America, but is established as anadromous stocks along the west coast, and also in northeastern Siberia
Source: after Whitehead 1985

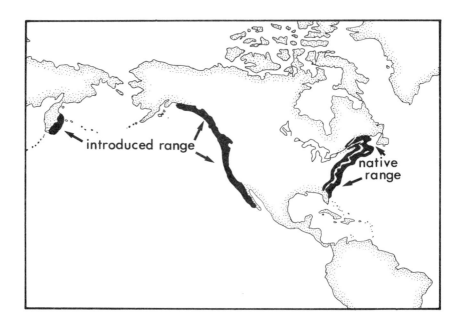

Although details are sparse, it seems that this species has become sea-run only in those areas where sea temperatures are sufficiently cool, as in Tasmania (but apparently not mainland Australia), in southern (but not northern) New Zealand, in the Falkland Islands (Stewart, 1980), and in Patagonian South America - Chile, Argentina, and Tierra del Fuego (Joyner, 1980a,b, Lindbergh, Noble and Blackburn, 1981; Zama and Cardenas, 1984). Recent results suggests that stocks in Kerguelen may be sea-migratory (Davaine and Beall, 1981b). There are also acclimatised sea-run stocks of brown trout in the maritime provinces of northeastern Canada (Crossman, 1984). As this species can be regarded as much a facultative marine wanderer as strictly diadromous, its success is perhaps not surprising. Certainly the fact that populations from within the natural range of the species contain both anadromous and non-anadromous individuals must have facilitated the relatively easy success in introducing this fish — the non- migratory individuals would have provided an insurance against weak returns, bolstering the spawning populations when returns were inadequate. Another factor in the success of brown trout is that it is believed to not move far away from river mouths when at sea, and this enhances the likelihood of successful returns.

The rainbow trout, *Salmo gairdneri*, has been even more widely introduced

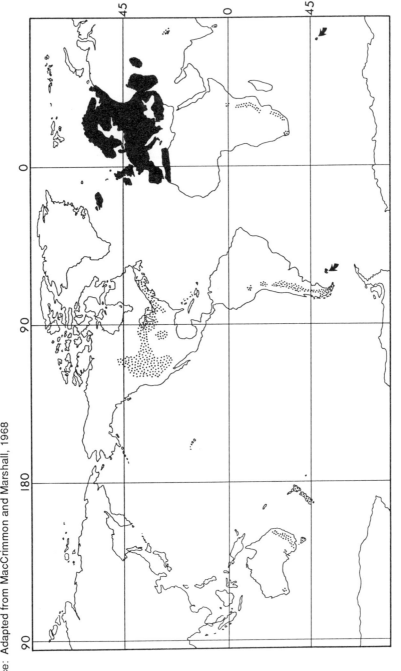

Figure 11.2: World distribution of brown trout, *Salmo trutta*: black, natural range; stippled, areas where naturally reproducing stocks have become established.

Source: Adapted from MacCrimmon and Marshall, 1968

(80 countries), also from an early date (mid-1800s). In many localities this species has failed even to establish self-supporting, wild populations limited to fresh waters, but is maintained only in captivity for stocking wild habitats for recreational angling by repeated hatchery releases, or for aquaculture purposes. However, also in many countries there are self-sustaining, wild populations, but in these the evidence for diadromy is, at best, meagre. There are very few well-authenticated reports known to me of 'steelhead' runs consisting of typical, large, sea-run rainbow trout outside the natural range of this species. Harrison (1963) evidently claims diadromy in southern Africa (Jubb, 1967) although Jubb himself did not corroborate the statement of Harrison, only reporting it. M.P. Bruton (personal communication) has advised that rainbow trout are anadromous in a confined area of the southwest Cape of South Africa, in an area where there are strong upwellings of cold, sub-Antarctic water; the fish leave the Erste River and enter the sea at False Bay, at about 34°S. Joyner (1980a) suggested that there is a sea-run of rainbow trout in the southern fjords, channels and sounds of Chile, but cites no data or authenticated reports of this. Occasionally similar claims are made of diadromy in New Zealand rainbow trout, but these too are unconfirmed and are presently regarded as improbable (McDowall, 1984b).

The Atlantic salmon, *Salmo salar*, was among the first candidates for transfer. Ten countries received stocks, beginning in 1864 (Tasmania), the most recent being Kerguelen in 1975 (Davaine and Beall, 1981a). In the few localities where the species has actually become established and self-sustaining in the wild (Argentina, Chile, New Zealand), there are no diadromous populations, and this is in spite of prodigious and enduring efforts to establish the fish. In New Zealand there were at least 25 importations of at least five million ova during a period of nearly 100 years (1867–1964). All that developed was a landlocked population (McDowall, 1978a). Joyner (1980b) discussed continuous production of Atlantic salmon in Chile for release, presumably from brood stocks from 1916 to 1938. A substantial egg take peaked in 1932 at 1,120,000. Again, only landlocked populations resulted. Attempts to establish *S. salar* in western North America failed (Hart, 1973). Stocks in Australian waters are maintained entirely by hatchery production.

The cutthroat trout, *Salmo clarki*, has had limited and mostly recent attention. There was an unsuccessful importation to New Zealand in the 1880s (McDowall, 1978a), and three other countries have tried to introduce this species in more recent times (since the 1960s), but none of these has produced diadromous stocks.

The Arctic char, *Salvelinus alpinus*, attracted minor interest (six countries since about the 1920s) with spasmodic success but no evidence of diadromy.

The brook char, *Salvelinus fontinalis*, has been much more fashionable (38 countries since 1869) with intermittent establishment, but no explicit evidence, or even hint, of success in diadromy. Populations recently established in Kerguelen in the sub-Antarctic Indian Ocean seem not to have moved across sea-gaps to reach unstocked river catchments and diadromy has not yet been documented (Davaine and Beall, 1981a,b).

There have been several major attempts to transplant pink salmon, *Oncorhynchus gorbuscha*. Berg (1977) and Bakshtansky (1980) described unsuccessful attempts to establish pink salmon in the Kola Peninsula/Barents Sea area of western Russia in the period 1933–39. Further transplants were begun in 1956 with no success. Then, beginning in 1959, pinks were released in large numbers — at least 200 million, with up to 60 million a year — between 1959 and 1964 (Bakshtansky, 1980). At least three million of the released smolts were derived from returning pink salmon adults. Initially there were good returns of fish to White Sea and Murmansk drainages — reaching a peak of 240,000 in 1973 — and numbers were still 165, 000 in 1977, but figures fluctuated very wildly from these large returns declining to virtually nothing. Bakshtansky (1980) stated that the fish spread eastwards across Arctic Russia as far as the Yenisei River, westwards across the North Atlantic as far as Iceland and Spitzbergen, and also southwards through the North Sea, some being taken as far south as France and the United Kingdom. Self-reproducing populations became established in Norway. Berg (1977) described catches in Norway of 20,000–25,000 kg in 1960, and from that time onwards, there was widely fluctuating abundance - peaks of 20,000 kg in 1965, 20,000-25,000 kg in 1971, and similar runs in 1973 and 1975. In spite of evident early success in northern Russia, Joyner, Mahnken and Clarke (1974) reported that the conditions in the rivers were too harsh for successful spawning and development, and that the populations soon declined. Bakshtansky (1980) found that the necessary 200-degree-days of heat for egg development could not be obtained before the water fell to its winter level of $0°C$. Ice might form to a thickness of 70 cm in the water flowing over the redds during the winter and as a result almost all the eggs often died. Although evidently unsuccessful in the long term, in far northern Russia, pink salmon seem to have become established in Norway. Berg did, however, caution that what had happened much earlier with pink salmon releases in the western North Atlantic — early success followed by failure — might still happen in Norway.

The events in the western North Atlantic are discussed by Huntsman and Dymond (1940) and Lear (1975, 1980), who described releases of pink salmon into various rivers in Maine. Small releases were made from 1906 onwards, the early ones consisting of about a million fry (1906–8), then much larger ones from 1913 to 1917 (29 million). Further releases, evidently from wild returns, were made between 1921 and 1925 (over two million). Runs of anadromous fish returned to the rivers for several generations following these releases, but they did not persist. Lear's (1975, 1980) accounts described transplants into the North Harbour River, in Newfoundland; five very substantial transplants were made between 1959 and 1966 (a total of over 15 million), with significant numbers of fish returning up to 1969. The last few years' runs were based on naturally produced progeny, but after 1969 the numbers declined and finally returns ceased in the early 1970s (Figure 11.3). Lear (1980) concluded that the experimental transplant of Pacific pink salmon from British Columbia to Newfoundland has obviously been a failure. He also alluded to releases of fry into Hudson Bay, in Arctic Canada in 1956 — about half a million during the winter and 225,000 in spring; this release was also a failure (Lear, 1980, Kwain, 1987).

Lear (1975, 1980) argued that failure in Newfoundland was due to predation on the fry by brook trout, eels and herrings, and to unsuitable donor stocks. As if discounting all but the last of these, he then suggested that better results might have been obtained with eggs from the western Pacific (Russian) rather than eastern Pacific stocks, on the grounds that the former may have migration patterns more compatible with current systems in the western North Atlantic. However, the fact that in at least two instances in the western North Atlantic some fish have returned from the sea suggests that the problem is more complex than that.

The chum salmon, *Oncorhynchus keta*, has similarly been released in western Arctic/eastern North Atlantic Russia; Bakshtansky (1980) alluded to releases into rivers of the Kola Peninsula during the period 1933–1939, from which only a few individuals returned and in little more detail reported on more than 50 million released during 1957 to 1964. From these only a few dozen returned. Chum salmon seem to have been even less successful than pinks. Recent stockings of chum into far southern Chilean waters have yet to show successful establishment. Releases into the large, semi-internal sea inside the Chonos Archipelago have resulted in some recaptures of adult fish, but evidence of sea–freshwater migrations of significant numbers of spawning fish, is

Figure 11.3: Returns of pink salmon, *Oncorhynchus gorbuscha*, to rivers along the east coast of Canada were ephemeral, and attempts to establish the species there in the 1960s and 1970s failed
Source: after Lear 1980

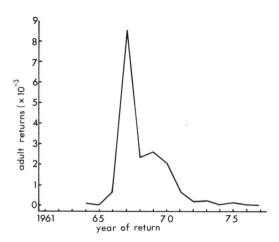

sparse. Nagasawa and Aguilera (1985) recounted releases of nearly 10 million chum into this area between 1980 and 1983, but recorded returns of less than 80 adult fish. The most recent report (Wurman, 1985), recorded successful returns from releases into Lago Llanquihue, but it is not clear that this is a sea-run stock. With regard to ocean ranching, he states that: 'Hitherto the number of returning adults has not been sufficient to justify a commercial fishery. However, the results have been encouragement enough to produce the necessary enthusiasm to carry on with the work in this direction'. Success seems highly elusive. An attempt to establish chum salmon in the Arctic drainages of Canada was also a failure.

The coho salmon, *Oncorhynchus kisutch*, has been released in a few European countries over a long period, but has not become established. An attempt was made to introduce coho into the waters of Maine during the 1940s, and Ricker (1940) reported that this induced ephemeral runs of a few hundred fish; these soon failed. Early releases in Chile (1904–1914) failed, as they did in Argentina (1905–1910, Joyner, 1980a). Releases more recently into Chilean waters appear to have generated some rather meagre returns but the long term results are still uncertain and it is too soon to forecast long-term success. Lindbergh *et al.* (1981) mentioned small returns of coho in southern Chile (Chiloe) but considered that insufficient would return to maintain the run properly. They described coho as a technical success but definitely not a commercial success. Davaine and Beall (1981a) noted releases of coho from Kerguelen in the far southern Indian Ocean, and stated that from a very modest release of only 5,000 alevins in 1978, there was a return of adults and natural reproduction in Kerguelen rivers. The returning fish could be those that have reared and grown within a long, semi-internal marine gulf at Kerguelen. Sagar (1986) thought that those which leave the gulf are likely to be lost. Long-term survival, in view of the ephemeral results elsewhere, must remain in doubt.

The sockeye salmon, *Oncorhynchus nerka*, was stocked in several European countries and in Chile and Argentina (early 1900s, Joyner, 1980a), without success, and also in New Zealand (in 1901; Welcomme, 1981, does not list this) where a landlocked stock has established and persists. Although there are some old, and rather tentative reports of a sea run in New Zealand (McDowall, 1984b, Scott, 1985), this seems unlikely to have been correct then, and is certainly incorrect now. Diadromy failed (McDowall, 1978a).

The chinook salmon, *Oncorhynchus tshawytscha*, was introduced many years ago into two European countries (but failed), and into Brazil, Argentina and Chile (where early releases in the early 1900s also failed). Joyner (1980a) mentioned reports of 'salmon-like' fish seen at falls in the Rio Uruguay, in Brazil, but could find nothing definite and no population was established. Rather surprisingly, one release was made in Hawaii in 1876, and again in the period 1925–1929 (Maciolek, 1984), but these, too, failed. Chinooks were taken to Australia, but there was no successful wild reproduction although hatchery stocks persist (Rogan, 1981), and New Zealand (1901– 1907) where a successful self-sustaining sea-run stock of fish developed very rapidly. This has persisted to the present day, primarily in the rivers of the east coast of New Zealand's South Island (McDowall, 1978a). Early attempts had been made in the 1870s in New Zealand rivers, but these evidently all failed — there are some reports of very large fish returning, — but it is certain that no consistent and significant run developed. Other attempts were made to establish chinooks in the rivers of New Hampshire, in the eastern United States. Hoover (1936) described transplants made in 1875 and in the 1930s, which were designed to restore salmonid stocks following the decline and virtual disappearance of Atlantic salmon. Hoover's investigation showed that some fish returned and spawned, but that the eggs did not hatch; obviously, no run eventuated. The results of more recent and still continuing attempts to establish *O. tshawytscha* in Chile (for commercial ocean ranching) are confused, though Joyner (1980b) described returns to a site on Chiloe in 1979 in 'gratifying numbers', and Lindbergh and Brown (1982) reported returns of adult salmon in the period up to July, 1981, as 'good' with chinooks and 'poor' with coho, and mentioned a continued return of both species through July, 1982. However, they also described the chinook returns as 'considerably less than anticipated'. Hopkins (1985) has noted returns to one Chilean facility on Isla de Chiloe, of up to 1 per cent of releases.

Little attention has been directed at masou salmon, *Oncorhynchus masou*. Nagasawa and Aguilera (1985) described the release of a small number of this species by Japanese interests into waters of southern Chile (Aisen), but without evident success.

In their summary of the process of introducing Pacific salmons, Childerhouse and Trim (1979) wrote that millions of chinook, pink and coho salmon eggs were planted in streams in Europe, Hawaii, Australia, New Zealand, Argentina, Chile, Mexico and Nicaragua, with even greater transfers being made from western North America to the streams of eastern Canada and the United States. Of all the tranplantations of Pacific salmons they found that only the chinook and sockeye that were taken to the South Island of New Zealand can be said to be successful, though the sockeye stocks are landlocked. This can only be described as a very small return on a very large investment.

OVERALL SUCCESS IN DIADROMOUS FISH TRANSPLANTS

On the basis of this review it must be concluded that success in the transplantation of diadromy in salmonid fishes is both highly erratic, and most often elusive. Obvious, long-term success is evident only in *Salmo trutta* (in several southern lands), *Oncorhynchus tshawytscha* (certainly in New Zealand for many years and perhaps recently in Chile), *O. kisutch* (perhaps recently in Chile), *O. gorbuscha* (possibly recently in western Russia and south into Scandinavia), and *O. keta* tentatively in western Russia and Scandinavia, and perhaps also in the Gulf of St Lawrence in eastern USA. It may be that this last stock is being sustained by diffusion of fish from the Great Lakes. Added to these are the two non-salmonid transplants — striped bass, *Morone saxatilis*, and American shad, *Alosa sapidissima* — from eastern (Atlantic) to western (Pacific) North America. Otherwise the only results have been the establishment of wholly freshwater-resident stocks, or, more often, total failure.

REASONS FOR SUCCESSES AND FAILURES

Reasons for the successes and failures of various attempted transplants have been the subject of extensive discussion. This question has caused both considerable disappointment among those keen to see species established and equal perplexity among those interested in explaining successes or failures. For most species diadromy appears likely to be a highly adaptive phenomenon that involves close integration of the species and its life cycle with the environment in which that life cycle takes place. Problems in developing the complete diadromous life cycle could conceivably occur in one or more of several phases:

1. In fresh water;
2. During migration from fresh water to the sea;
3. At sea;
4. During migration from the sea to fresh water.

Fresh water

It seems that in relatively few instances has there been any difficulty providing suitable conditions in fresh water for the appropriate phase of the life cycle — either in the wild, or in managed hatcheries built for the purpose where the eggs are tended and the young fry fed for a time prior to release. One suggested instance of problems in fresh water was the harsh winter conditions in the White Sea rivers in northern Russia, which are regarded as preventing the sustained presence of pink salmon in the area (Bakshtansky, 1980). It is interesting that those seeking to establish runs outside the natural range of the salmon species have tended to focus much of their attention on locating suitable conditions for the freshwater rearing of the eggs and juveniles fish. Joyner (1980a) described this for the Chilean investigations in the 1960s, which concluded that the rivers — their gradients, water quality, temperatures, gravels, foods, predators, etc. — were all favourably disposed to the establishment of salmons. But nearly all of these can, initially, be substituted, with relatively little effort or cost, by hatchery rearing of the young fish. In contrast much less attention was focused on the suitability of oceanic conditions and the likelihood of the released salmon returning to fresh water. The same has been true in New Zealand. Later studies in Chile, conducted by the United States and Chile, did look at oceanic conditions, and reached the conclusion that these were favourable for generating a return of fish.

Haedrich (1983) argued that one of the constraints on the existence of diadromy, and therefore its success or failure in transplanted stocks, is the availability of space (habitat), and this may be true in some instances (though hard to prove). He pointed to the failure of *Salmo salar* to become established after transplantation to British Columbia (Scott and Crossman, 1973), and his reasoning may be correct. However *S. salar* has also failed to establish diadromous populations in many other areas following transplant (Australia, New Zealand, Chile, Argentina, Falkland Islands, Kerguelen, MacCrimmon and Gots, 1979). The reasons for these failures are unlikely to include shortage of space as the faunas of these southern areas are extremely impoverished and other salmonids (that are not diadromous) have been extraordinarily successful in most of these lands. Joyner (1980a) described how, in Chile, the trout exploded into ecological niches left vacant by the devastation to freshwater life wrought by the Pleistocene glacial advances, and such a description probably applies also to parts of New Zealand and southern Australia. *Salmo salar* and *S. gairdneri* have failed to establish anadromous populations in New Zealand in the same waters in which *Oncorhynchus tshawytscha* succeeded. Furthermore, McHugh (1967) considered that striped bass, *Morone saxatilis*, and American shad, *Alosa sapidissima*, may be more abundant in their transplant habitats of the eastern Pacific, than in their native western Atlantic, and thus more abundant in those habitats where Atlantic salmon failed to become established. The reasons for the failure of diadromy are undoubtedly much more subtle and diverse.

Migration to sea

There seem to be few inherent reasons why in a transplanted situation migration to sea from fresh water should differ much between localities. If there is a need for the downstream migratory stages to linger in brackish river estuaries to acclimate to changing salinities before moving to sea, then the lack of suitable estuarine waters at river mouths could be a limiting factor in successful transplantation. However, comparison of the very extensive brackish estuarine areas at the mouths of important salmon rivers, like the Fraser River in British Columbia, with the rivers of the east coast of the South Island of New Zealand (where chinook salmon have become established) does not support this view. Many of New Zealand's salmon rivers where chinook salmon have become established), have virtually no slack-water, brackish/tidal estuaries at all. It seems to me to be a fair assumption that, in general, fish have been successfully hatched and reared for some time in fresh water (often in hatcheries) and that they have made the transition downstream and to sea successfully.

Life at sea

There is diverse although not extensive evidence that in many instances there has been survival and growth of the fish at sea. One example is the capture of a large, adult Atlantic salmon at sea off the New Zealand coast, in 1893 (McDowall, 1984b), in spite of the failure of sea runs of this fish to develop there. Another is the taking of mature coho salmon in the fjords of southern Chile, in association with largely unsuccessful ocean ranching ventures there (Zama and Cardenas, 1984, Nagasawa and Aguilera, 1985). There is no reason to suppose that the ability of the fish to feed and grow at sea in the transplant locality is a significant contributor to failure.

Returning to fresh water

In general, where transplants have failed, there have been few or no adult spawning fish returning from their oceanic life to fresh waters. The adult fish seem to fail to navigate accurately on the return migration, or perhaps encounter conditions at sea, like warm temperatures, through which they are unwilling to migrate. However, it is not simply a question of navigational failure, in all cases. There is a series of instances in which, following initial releases, there has been a meagre run back to fresh water followed, eventually, by failure, e.g. pinks, chum and coho along the Atlantic coast of North America. These initial returns show that some fish have successfully navigated at sea to enable them to return to fresh water, but the eventual failure of these runs suggests that there are too few fish that do successfully navigate to maintain stocks on a continuing basis. Perhaps the early returns are a result of a few fish that, for some reason, have not strayed far from the river mouths, and that the fundamental problem is one of navigational failure. Or, alternatively, it could be a case of random migration at sea resulting in a small number of fish sometimes making their way back to fresh water. There may be an absence of any inherited component in the return of these few fish, which results in erratic returns, and eventual failure.

Return from the sea involves several features, the principal one being oceanic navigation during the migration bringing the fish back from their often far distant oceanic feeding grounds. It is this point that is widely regarded as the weak point in the cycle, the point at which the cycle is broken.

This issue has been explored and discussed by several workers. It seems essentially a question of the ability of the fish at the marine stage in their life cycle to navigate accurately, and thereby make their way back to the mouth of the home river. In a transplant situation, the current systems and the various navigational cues are likely to be different from those within the species' natural ranges. The factors that control long-distance oceanic migration in diadromous fishes are, although extensively discussed, poorly understood. Stewart (1980)

199

has hypothesised that the movements of salmon at sea are controlled by large gyral current systems in the oceans, and that these gyres carry the fish away from the coastlines and then ultimately back to land again. Stewart suggested that:

> it may eventually be found that their journeying in the sea is affected by different factors than those affecting their movements inshore and upstream. It may even be found that their movements at sea are affected by physical forces such as ocean currents which return them to a point where they are influenced by discharges from their natal streams. [He] came to the conclusion that the absence of integral land masses in the Southern Hemisphere, and the lack of suitable oceanic gyres developing as a result, was probably the main reason for the failure of almost every attempt to establish sea runs of salmon there.

Stewart attributed the existence of Northern Hemisphere salmon to the presence of gyres, but did not explain the failure of Northern Hemishpere transplants, e.g. the failure of pink and coho salmon releases within the range of Atlantic salmon along the eastern seaboard of North America, or of Atlantic salmon along the western seaboard of North America where, according to his own theory, Pacific salmons are returned to the coastline by such oceanic gyres. Stewart's hypothesis seems superficially plausible, but he makes little attempt explicitly to relate his hypothesis to observations of salmon species distributions and the oceanic systems within which they disperse.

Gyral systems in the oceans are a product of west–east oceanic current systems generated in temperate seas by Coriolus forces (that relate to the rotational movements of the globe). The west–east current systems are interrupted and diverted north or south by large north–south aligned continents. Whether the fish actually move in relation to these gyral systems is unproven. And, inasmuch as what is known of the pattern of movements of Pacific salmons in the northern Pacific Ocean varies quite widely in relation to the different species, and in relation also to the rivers to which the salmon are returning, it seems to me unlikely that the movements of salmon at sea are related so simply to oceanic currents.

Scott (1985) has reviewed the general question of transplantations of diadromous salmonid fishes from the Northern to the Southern Hemisphere, and examined the general hypothesis that the probability of establishment of a complete migratory pattern is inversely proportional to the distance of migrations of the parent stock — simply, that species that migrate only short distances at sea are more likely to succeed when transplanted — a view that has been quite widely canvassed but never explored in detail. Scott (1985) assumed that the key factor in success is related to oceanic migration and navigation, and also that migratory behaviour at sea is an inherited character. He classified a series of salmonid species according to the distance of their migration:

1. Long-distance migrants — *Salmo salar, S. gairdneri, Oncorhynchus nerka, O. gorbuscha, O. keta*, and *O. nerka*.
2. Variable distance migrants, with some stocks distant and others closer inshore — *Oncorhynchus tshawytscha* and *O. kisutch*.
3. Largely coastal species — *Salmo trutta, S. clarkii, Salvelinus fontinalis*, and *S. malma*.

The only possible test for this hypothesis is to examine what has happened in practice, and there are relatively few examples to review.

The widespread success of *Salmo trutta* — a coastal migrant — certainly is consistent with Scott's hypothesis. The equally widespread failure of *S. salar* — a long distance migrant — is similarly consistent. Following the failure of *S. salar* to become established in New Zealand as an anadromous species, after substantial efforts (about five million ova were introduced over a period of many years, McDowall, 1978a) some fish of Baltic origin, which were interpreted as being relatively short distance oceanic migrants, were introduced into New Zealand. This introduction also failed, and this is, perhaps, inconsistent with Scott's hypothesis. *S. gairdneri*, another long-distance migrant, has generally failed to establish sea-migratory populations (in Australia, New Zealand, Argentina) and this, again, is consistent with Scott's hypothesis. However, there is some evidence that *S. gairdneri* is sea-migratory in southern Africa (M.P. Bruton, personal communication) and also in Chile (Joyner, 1980b).

Oncorhynchus tshawytscha has succeeded completely only once — in New Zealand, primarily in rivers of the South Island east coast — but it has also failed over other and extensive areas of New Zealand in spite of attempts to establish populations. It has also failed in Australia, Tasmania, Argentina, and also in Chile — at least until recently; even now, success in Chile seems to be based largely on hatchery releases into a largely enclosed coastal sea (Golfo de Ancud), and long-term success cannot be assured yet.

Other failures include introductions of *O. nerka* to New Zealand and South America, *O. gorbuscha* and *O keta* in Chile, all of which are long-distance migrants, and of *O. kisutch* in Chile, a middle-distance, but somewhat variable species.

The example most inconsistent with Scott's (1985) hypothesis is *Salvelinus fontinalis* — a coastal migrant which, as far as is known, has become well-established in many southern lands and yet has not become diadromous in any of them.

The success of chinook salmon in New Zealand, because of its evident uniqueness, demands more detailed examination. Self-sustaining runs occur in rivers along the east coast of the South Island, from the Waiau River (about 43°S) south to the Clutha River (about 46.5°S, McDowall, 1978a). Runs are not large, numbering thousands or tens of thousands, but are nevertheless well developed. There are smaller and more intermittent runs in the Wairau River

201

(about 42.5°S), to the north of the main centre of distribution. On the South Island's west coast there are much more meagre runs (hundreds only) in a series of systems (Okarito, about 43°S, to Moeraki, towards 44°S); in all of these systems there are lakes not far from the sea. Occasional fish stray into other west coast rivers that lack lakes, but self- supporting runs do not seem to occur more widely.

On the east coast, where the substantial runs occur, Eggleston (1972) pointed out that the runs occur into rivers the mouths of which are enclosed within the Southland Current, of primarily warm Tasman Sea water, which passes around the southern tip of the South Island and then runs in a northeasterly direction, along or away from the coastline of the South Island. Between the current itself and the coastline is a wedge-shaped area of sea, in which Eggleston suggested the salmon occur that return successfully to the rivers; those salmon which stray beyond this area, it is suggested, fail to return. To the north, this area of sea is bounded by a subtropical convergence, and the northern limits of the distribution of chinooks in New Zealand are probably established by the location of that convergence — runs in the Wairau River, at the northern fringes of the range of chinooks, possibly occur at times when the sub- tropical convergence has a northerly position, and may not occur when the convergence is to the south. The southern limits of the distribution seem to be set by the point at which the Southland Current diverges away from the coastline itself. There are no obvious temperature or other habitat reasons that might exclude chinook salmon from this southern area. Uchihashi, Iitaka, Morinaga and Kikkawa, (1985) have elaborated a similar hypothesis, suggesting that the Southland Current forms a front through which the salmon do not migrate — or if they do, they do not subsequently recross, — to return to fresh water.

If Eggleston's hypothesis is correct, it could be argued that chinook salmon have succeeded as a result of the oceanic current system off the east coast of the South Island of New Zealand confining them to a restricted area of coastal sea. It is not so clear why chinooks have succeeded on the West Coast, although the populations are so meagre that it can scarcely be described as a success. Releases of millions of young salmon into one of the west coast rivers (the Hokitika, a little north of 43°S) during the 1920s and 1930s failed to produce a run into that river, although strays from these releases are the source of what stocks did become established there. The fact that populations persist only in West Coast rivers with lakes suggests that possibly the meagre returns of sea-run fish are just a few survivors of sea-run individuals in populations that are essentially lake-limited — without the lake populations the sea runs may rapidly fail — if the lack of runs in lakeless river systems is any indication.

If the ocean current hypothesis is correct it does not explain why sea-run populations of sockeye salmon, Atlantic salmon, and rainbow trout — all of which were released in the same river systems — failed to establish in New Zealand. This gives some credence to Scott's (1985) hypothesis that long-distance migrating stocks are less easily transplanted.

What chinook salmon do in the sea off the east coast of the South Island of New Zealand is little known. Flain (1981) has shown that salmon are present in coastal seas up to 48 km offshore and in waters up to about 60 m in depth. They occur frequently in harbours and deep bays, especially during the first year or so in the sea. Flain argued that the salmon probably move north with the ocean currents as juveniles, and then migrate south again against the currents as returning adults, prior to entering the spawning rivers. However, he later (Flain, 1983) argued that if the salmon behave like their parent stocks, they would move south on first going to sea, against the current, and then subsequently move north with the current on returning, and have to turn left to enter the spawning rivers. Data on the straying of tagged salmon, released from the Rakaia River close to the middle of the chinook's latitudinal distribution in New Zealand, suggest that at least those fish that stray move widely in both northern and southern directions. On the basis of what is known about the New Zealand populations, little can be stated with any assurance about what contributed to success in transplanting diadromous species, apart from the suggestion that it seems more likely to be successful with short-distance migrants than long-distance ones.

Instances of successful transplants in the Northern Hemisphere — pink salmon to northwestern Europe, ephemeral successes with pinks and coho to the east coast of North America, and lasting success of two non-salmonids, striped bass and shad, from the east to the west coast of North America — seem no more informative, other than that again, the two non-salmonids are regarded as short-distance migrants.

Instances of failures also provide little, if any information on causes (Figure 11.4). Joyner (1980b) suggested that the failure of attempts with Pacific salmon in Chile (since their beginning in 1905) was due to the Humboldt Current carrying the fish north into waters that are too warm for their survival. However, this seems little different from the oceanic situation in New Zealand, where there was success.

It seems that, especially in diadromous fishes that travel long distances at sea, there are bound to be highly adaptive navigational/behavioural patterns and cues to which the fish are responding, and therefore that with major latitudinal and longitudinal and oceanic current-system changes, transplantation is a hazardous — very much a 'hit-and-miss' affair, — in which success cannot be predicted.

Ricker (1954) made the point that success is likely to be limited if homing has a strong hereditary (rather than environmental) component, but since then it has been shown that, at least in the freshwater phases, homing is strongly environmental. However, if the migration that brings the fish back to coastlines and to the proximity of river mouths does have a strong hereditary component, and a few fish do return to a release location, then it would seem that there would be very powerful and rapid selection that would reinforce the hereditary component of successful returning; this should herald rapidly growing success.

Figure 11.4: Ocean currents in areas of the world where attempts have been made to transplant anadromous fish stocks
Source Flain 1983

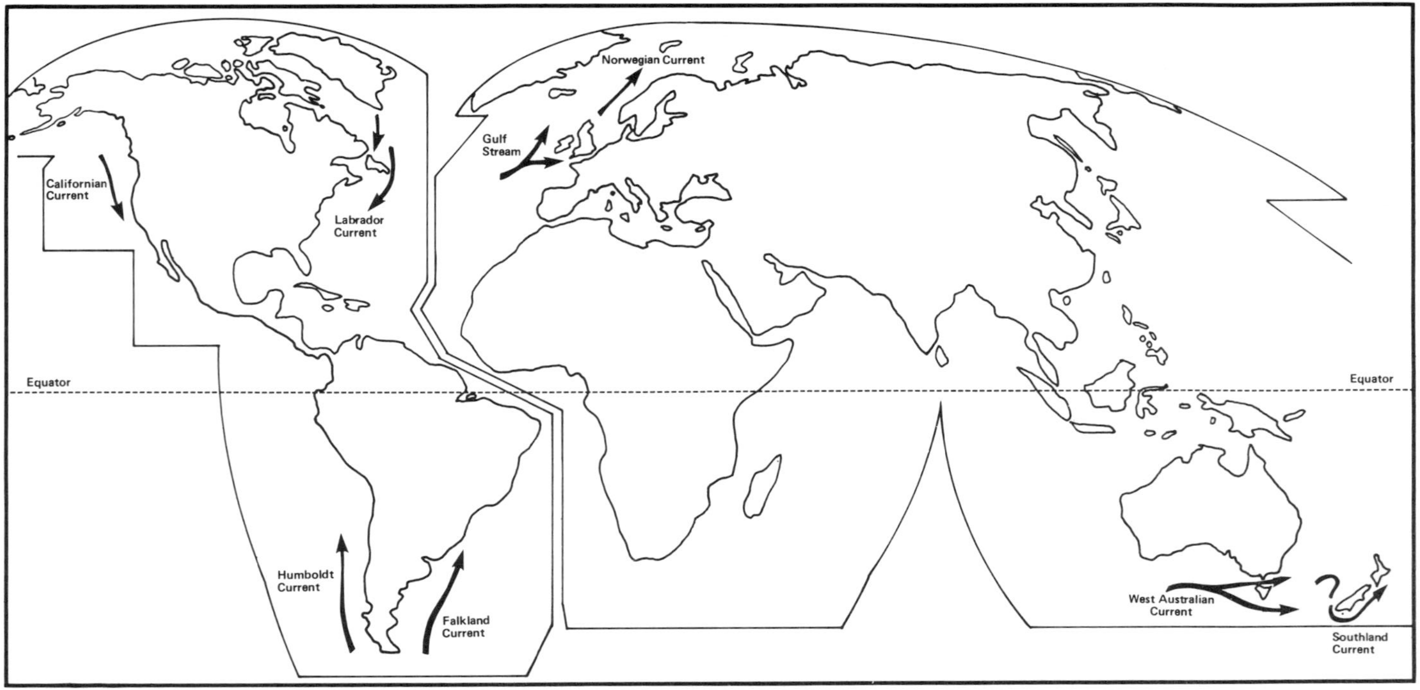

Here, it is important to note the comment made earlier, that meagre returns may be the result of random wandering of salmon in the ocean, in which case the returns of a few fish would herald nothing at all!

A likely reason for the general failure of salmon runs to develop is that the navigational cues that bring the fish back into freshwater habitats after several years at sea are either lacking, incomprehensible to the fish, or misleading to them in foreign waters, so that the fish are unable to relocate and return to fresh water. 'As can be seen, the establishment of seagoing runs of salmon is a complex matter' (Stewart, 1980).

There are two interesting instances where diadromous species have been transplanted accidentally and at least one of them successfully. These apparent successes seem ironic when the effort expended in many other species, which resulted in failure, is considered. Both species are of Oriental origin.

The Japanese goby, *Acanthogobius flavimanus*, appeared in the waters of San Francisco Bay, where Lee *et al.* (1980) said that it is primarily estuarine but ascends short distances into streams and lakes at low elevations. The Japanese sea bass, *Lateolabrax japonicus*, has recently been reported from near Sydney, Australia (Paxton and Hoese, 1985), and as yet it is known only from two fish (370 and 390 mm S.L.), and it is presumptuous to assume that the species is established there. In both instances transfer is believed to have resulted from the discharge of ballast water from ships that took on the ballast in Japanese or other northwestern Pacific waters.

12

Diadromous Fish and their Fisheries

As was mentioned in Chapter 1, various diadromous fishes have long been the target of fisheries exploitation by man. For as long as there are historical records, there seems to be evidence of man's exploitation of these fish. Archaeological remains and very early written records of man contain evidence of this. The occurrence of diadromous fishes in large and dense concentrations in both space and time has made them easily accessible to man for capture in large quantities, although before the advent of refrigeration there must have been attendant problems related to processing and long-term storage, especially for primitive man. In Europe, lampreys, sturgeons, and salmonoids of various sorts are all long-exploited and highly valued fishes providing traditional foods for mankind. And as European man explored and colonised the east, the Americas, Africa, and Australia and New Zealand, he routinely found aboriginal man in many of these areas exploiting and depending upon diadromous fishes as a source of protein. International fisheries production figures for the world's fisheries do not rank any diadromous fish species among the top 10 producers (Harden-Jones, 1981), and yet some of them constitute very important fisheries both at national and international levels. Salmons, in particular, fall into this class, and management of their fisheries has required the negotiation and maintenance of complex, international fishing/management agreements.

The species exploited vary enormously, from the minute, larvae/juveniles of some tropical gobies, to the huge sturgeons that reach lengths of 4 m and weights up to 1,500 kg. The minute species are generally eaten whole, *en masse*, while the larger ones are consumed more selectively for their flesh and sometimes their reproductive products. Diadromous fishes of almost all varieties are exploited as food by man.

LAMPREYS — FAMILIES PETROMYZONTIDAE, GEOTRIIDAE AND MORDACIIDAE

Lampreys were regarded as a gastronomic delicacy in Europe in the middle ages. Clearly there was a fishery for lampreys and a demand for them as food many centuries ago in the British Isles, and reports from Scandinavia also mention early use of lampreys (in both Sweden and Finland) as early as the fifteenth century (Sjoberg, 1980; Tuunainen *et al.*, 1980). Tradition has it that King Henry I of England died from eating too many lampreys — possibly the sea lamprey, *Petromyzon marinus*. At first glance it would be easy to regard such a report as apocryphal, but there is a plausible explanation. Not only did this English king apparently die from eating too many lampreys, but so too did Maori people in New Zealand, from eating the southern pouched lamprey, *Geotria australis* (Best, 1929). The explanation of these deaths possibly relates to the accumulation of bile pigments in the bodies of the adult lampreys when their bile ducts atrophy at ammocoete metamorphosis. Because the bile salts accumulate in the body of the adult lamprey, they are consumed when the fish are eaten. So it is possible that the consumption of greater quantities of bile pigments than the human body can cope with, caused death from bile poisoning.

In most parts of the world, lamprey fisheries have long since faded into obscurity. Bigelow and Schroeder (1948) wrote of a lamprey fishery in New England (USA) that has been 'scarcely more than a memory for 40 years past except locally and in a small way for home consumption or to supply the needs of biological laboratories.' They also mentioned the taking of ammocoetes in large numbers, by anglers, for bait. The decline of a once significant lamprey fishery in New England was attributed to the construction of impassable dams on the rivers, i.e. the decline was related to reduced availability rather than reduced interest — thus in 1847 several cartloads were caught daily for a considerable period after a dam was built on the Merrimac River. They quoted a catch of 3,800 lampreys in one night from below falls in the Connecticut River in 1840, but noted that by 1866 they had become almost extinct in the upper reaches of that river, where numerous dams had been constructed. Use of lampreys as food by North American Anuit people was recorded by Scott and Crossman (1973) and McPhail and Lindsey (1970); *Lampetra japonica* was evidently caught through the ice during the spring in the Yukon and other parts of northern Canada, primarily for dog food, but also for human consumption when other food sources were scarce. McPhail and Lindsey (1970) described how this species was raked out by hooked sticks from under the ice by the thousand in Alaska. Although now little-used for food in North America, it is smoked and exported for sale in Japan.

Although lampreys now attract little interest in the United Kingdom, they are exploited locally in small numbers, partly as bait for anglers (Wheeler, 1969) and continue to be exploited in some parts of Europe and Asia especially. *Petromyzon marinus* has little fishery significance in Finland, although it is rare there (Tuunainen *et al.*, 1980), but small numbers are taken for consumption in some other European countries like France, Germany, the United Kingdom, and also the United States. It is captured and salted in barrels in Europe (Scott and Crossman, 1973). Catches of the European river lamprey, *Lampetra fluviatilis*, in Sweden may be very large, and runs into Finnish rivers were estimated at half-a-million fish. Catches of more than 300,000 in some rivers in recent years are documented (Valtonen, 1980). Tuunainen *et al.* (1980) listed total Finnish catches reaching 3 million fish with 2 – 2.5 million (about 100 tonnes) being taken in recent years, all for human consumption. Catches in Sweden are given as between 4,774 and 27,300 kg for various periods between 1914 and the present (Sjoberg, 1980), roughly 100,000 to 500,000 lampreys. Nikolskii (1961) regarded *L. fluviatilis* as of great importance in the Neva River of Russia. *Lampetra japonica* is of commercial importance in Siberia (McPhail and Lindsey, 1970) and is taken for sturgeon bait in Russia (Scott and Crossman, 1973).

Berg (1962) related captures of *Caspiomyzon wagneri* reaching over 33 million in a year in the Volga River during early years of this century, but modern catches are likely to be much less than this vast figure.

In New Zealand there was historically an important fishery for the southern pouched lamprey, *Geotria australis*. The indigenous Maori people constructed elaborate timber weirs along the margins of the rivers that caught the adults as they migrated upstream. The fish were caught in baskets inserted into gaps in the weirs, and which were placed with their openings facing upstream. The lampreys, seeking the gaps to move upstream, were swept downstream into the baskets by the rush of the current through the gaps (Best, 1929). This traditional fishery has largely disappeared, although it is still carried out in some isolated localities (Todd, 1979). Lampreys are still taken by other means in New Zealand by both Maoris and Europeans, at localities where the migrating fish tend to accumulate during their upstream migration — primarily where migration is obstructed by natural falls or occasionally by dams. There is a demand for lampreys in New Zealand, by Maoris as a traditional and sometimes ceremonial food, and also from immigrants of northern European descent who were accustomed to lampreys as a food in their homelands. There are no catch statistics. In Australia Cadwallader (1985) noted that both *G. australis* and the Australian endemic, *Mordacia mordax*, are taken commercially in some coastal rivers of Victoria, but he provided no figures. Exploitation of the genus *Mordacia* in Chile has not been reported, to my knowledge.

World lamprey catches are recorded by FAO (1984) as 152 – 277 tonnes for the period 1980–1983, 63–195 tonnes of this being from the USSR, although they recorded 200 tonnes from France in 1980 only. Catches in the United States and Japan are not recorded. It is obvious from these various reports that figures are far from complete, rather erratic, and do not reflect the full extent of the world's lamprey fisheries.

No discussion of fisheries related to diadromous fish species could sensibly exclude reference to the invasion of the North American Great Lakes by the sea lamprey, *Petromyzon marinus*. The Great Lakes had for a long time been an extremely productive and important fishery for both the United States and Canada. The prime resource for the fishery was the lake char, *Salvelinus namaycush*,

Figure 12.1: Changes in productivity of lake trout, *Salvelinus namaycush*, in Lake Michigan, following invasion of the lake by sea lampreys, *Petromyzon marinus*. E, time of establishment of the sea lamprey; S, initiation of chemical control of lampreys; C, completion of initial phase of control
Source: Smith (1968a)

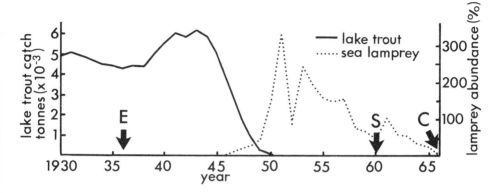

which yielded up to 7,000 tonnes a year. Other species were also highly productive, including diverse species of whitefish (*Coregonus*). In 1829, when the Welland Canal was constructed connecting Lake Erie to Lake Ontario, it became possible for the sea lamprey to invade the upper Great Lakes — Erie, Huron, Michigan and Superior. Landlocked populations of sea lamprey eventually became established in the lakes and then rapidly expanded there, preying heavily on the lake trout. As a result fisheries in the lakes collapsed. This began in Lake Huron about 1939, Lake Michigan about 1946 (Figure 12.1), and Lake Superior in the mid 1950s; fisheries in all of these lakes soon failed completely. Attempts to control lamprey numbers by poisoning streams with chemicals that affected only lamprey ammocoetes have been partly successful (and very expensive), but recovery of the lake trout in the lakes, and the fisheries based on them, has not followed in spite of considerable research and extensive plantings of fry into the lakes. Smith (1968a) described the comprehensive ecological disruption of fish communities in the lakes (Figure 12.2) which consisted of:

1. invasion of the lakes by lampreys;
2. consequent destruction of lake trout stocks and fisheries;
3. invasion of the lakes, also, by anadromous alewife, *Alosa pseudoharengus* (late 1940s);
4. declines in abundance of other major species native to the lakes, viz. burbot, *Lota lota*, and various whitefishes, *Coregonus* spp.;
5. subsequent domination of fish communities by alewives as populations of predators declined and food resources were exploited by the alewives.

209

Figure 12.2: Changes in the species composition of fish communities and fisheries in the Great Lakes of North America related to invasion of the lakes by sea lamprey, *Petromyzon marinus*, from a community and fishery dominated by lake trout, *Salvelinus namaycush*, top predator, to a situation where lake trout no longer feature in catches and much fishery production is from the alewife, *Alosa psendoharengus*, itself an invader of the lakes
Source: Smith (1968a)

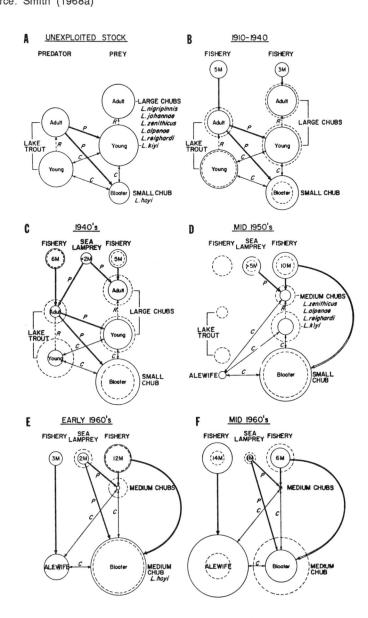

Subsequently several types of Pacific salmon, *Oncorhynchus* spp., have become established as landlocked stocks — chinook salmon, *O. tshawytscha*, and coho, *O. kisutch*, by deliberate strategy, and pink salmon, *O. gorbuscha*, by an evidently accidental release of fish intended for the Canadian Arctic (Kwain, 1987). Ironically it has been the pink salmon that have proved to be most successful, and have established expanding and self-sustaining populations that form the basis for fisheries in the lakes. Both chinooks and coho have also succeeded, with modest spawning runs developing into some rivers tributary to the lakes. These species have become the basis for both recreational and commercial fisheries, and seem also to have brought the huge stocks of alewife under some control by their intense predation. However, the lake ecosystems are profoundly different from those that prevailed prior to invasion and establishment of the sea lamprey in the 1930s. Vigorous efforts to re-establish lake trout in the lakes have been underway for many years, but success at persuading them to spawn in the lakes has been elusive (Foster, 1985).

STURGEONS — FAMILY ACIPENSERIDAE

The value of sturgeons to man has been recognised since earliest times, with both the flesh and reproductive products having culinary value. Sturgeon roe, in particular, is highly sought after, being the basis for caviar. In addition the bony skutes were once used as scrapers, oil in the flesh as a fuel for steamboats (in North America), and an extract from the swim bladder, known as isinglass, was used to clarify wines and to assist in the setting of jams and jellies. It is no surprise, considering all these uses and their very large size, that sturgeons were keenly pursued by man, but it was their value as food that attracted most attention. Their roe was smoked to produce the luxury gourmet food caviar which had its origins and has for long had its greatest production in Russia, particularly from fisheries in the Caspian and Black Seas where the enormous beluga, *Huso huso*, and several other sturgeons live. Kovtun, Bondarenko, and Rekov (1984) described sturgeon as the most valuable fish of the Soviets in the freshwater fish fauna of the Caspian/Azov Sea. Sturgeons have been taken through all history in Europe, Russia and eastern Arctic Asia, and are said to have been the basis for subsistence fisheries for North American Indians (MacCrimmon and Gots, 1980).

Because sturgeons are very slow growing and live to a great age, they are highly sensitive to overfishing, in spite of their enormously high fecundity — large female sturgeons may produce several million eggs. The great age of large sturgeons means that it takes a long time to recover from overfishing. It is difficult to assess how the sturgeon fisheries have changed over the decades, because they have been intensively and extensively exploited for such a long time and records are meagre.

Nearly all sturgeons seem to be valued for their roe, but the flesh is variable in character and quality. Greatest production of sturgeons is, and has probably always been, of Russian species, particularly of *A. guldenstaedti*, for which Berg (1962) reported catches of 340,000 fish in 1930, from the Caspian Sea. Numbers of sturgeon in Russian waters have evidently not been great for many years as annual catches of beluga, *Huso huso*, in the lower Kura River as averaged only 1,090 for the years 1829–1855 (Berg, 1962). About 4,000 were taken from the delta of the Volga River in 1924–1935; beluga numbers in the Kura River now average only a few thousand a year. The Atlantic sturgeon, *Acipenser oxyrhynchus*, has flesh of moderate quality but has high quality roe (Scott and Crossman, 1973), and has long been keenly sought throughout its range, being caught and exported from Canada to the United Kingdom as early as 1628. Its abundance has declined everywhere (Scott and Crossman, 1973). Catch peaked in the western North Atlantic in 1890, but had collapsed by 1905 and there are now only remnant stocks (Smith, 1985). A fishery persists along the Atlantic and Mexican Gulf coasts of North America, but at a much reduced level — catch in the United States is now only half that in the 1880s (Scott and Crossman, 1973). These authors nominated a catch of about 26 tonnes in Canada in 1962. McKenzie (1959) discussing sturgeon catch in the Miramichi River of New Brunswick, Canada (listing it as *A. sturio*) noted that they are not much wanted commercially; there was then no market near at hand and the fish were not valued locally as a food. In the Mexican Gulf, production in recent decades peaked at 26 tonnes in 1968, but last century up to 156 tonnes of meat and three tonnes of caviar were harvested. In Gulf rivers a viable fishery exists only in the Suwannee River of Florida (Huff, 1975). Capture of eastern Pacific white sturgeon, *Acipenser transmontanus*, fluctuated widely, reaching over 500 tonnes in 1897, but in recent years has been only 10–20 tonnes a year (Scott and Crossman, 1973). Wheeler (1969) set catch for the European sturgeon, *A. sturio*, as about 50 tonnes a year. Some species, particularly the western Atlantic shortnose sturgeon, *Acipenser brevirostrum*, are of small size and are less thus keenly sought , although the flesh is of good quality. *A. brevirostrum* is taken incidentally as a bycatch in other fisheries in eastern North America, although exploitation is constrained by the species being regarded as endangered. The Pacific green sturgeon, *Acipenser medirostris*, has inferior flesh and its roe is not used as food (Scott and Crossman, 1973); there is no fishery for it (Hart, 1973).

World production of sturgeons (both anadromous and non-anadromous) was reported to be about 28,000 tonnes for the years 1980–1983, by FAO (1984), nearly all of it being in the USSR, with small amounts in North America, central Asia and other Soviet bloc countries. About 70 per cent of world catches of sturgeon come from the Caspian Sea and vicinity.

All sturgeon fisheries have suffered greatly from overexploitation as well as from environmental effects, especially the construction of dams which hinder the migrations upstream to the spawning grounds. As a result the Russians, particularly, have developed technology for the aquaculture of sturgeons. Millions of juvenile sturgeon are being reared and then released into rivers to grow to maturity, for later exploitation. Such is the value of sturgeons to the USSR that population enhancement of stocks has developed in a major way. In 1985, the Russians released 10–15 million *Huso huso*, 130–135 million *Acipenser sturio*, and 290–300 million *A. stellatus*. They have a reported goal of catches of 50,000 sturgeon by the year 2000 (FAO, 1986). Similar technology is also being developed in North America (Rulifson *et al.*, 1982a; Dadswell *et al.*, 1984; Ceskleba, Avelallemant and Theulmer, 1985).

FRESHWATER EELS — FAMILY ANGUILLIDAE

Freshwater eels are the basis of fisheries over a very wide part of their range, comprising the North Atlantic and the Indo-west Pacific from Japan in the north, New Zealand in the south, the islands of Polynesia in the east, and eastern Africa in the west. Even though their serpentine appearance, sliminess, and tenacity of life following capture makes eels unattractive to some people, they are widely regarded as a highly nutritious fish, and as a delicacy by some. The behaviour and ecology of the 16 species of *Anguilla* are very uniform, so that although fishing methods and traditions may vary geographically, in virtually all areas the same life stages are being exploited while undergoing consistent behavioural patterns. Old records show that eels have been the target of exploitation by man for as long as there are surviving written records, and they continue to be exploited in virtually every life history stage. Sinha and Jones (1975) identified spears that were possibly used for taking eels as long as 7,700 years ago, and Mitchel (1965) reported the remains of eel weirs in Ireland that date back 3,000 years. There are evidently similar remnants of eel capture facilities in many places where primitive man settled near rivers and lakes that carried eel populations. Tesch (1977) considered that: 'We can be sure that eel fishing was practised in prehistoric times. Numerous archaeological finds indicate that fishes, including eels, were part of the diet of our ancestors.' Descriptions of eel behaviour and reproduction written by Aristotle about 2,300 years ago, and in particular his allusions to eels being viviparous and confusion between anguillid eels and gordian worms, are well known.

Eels feature strongly in the myth and legend of the New Zealand Maori people. Anthropologists have found the remains of eels in Maori middens that date from about the sixteenth century (A. Anderson, personal communication). Downes (1918), Best (1929) and others make it quite clear that the Maoris knew of eel migrations and that they made good use of their migratory patterns, to exploit them as food. There is no doubt that they were important, and they continued to be important following the European settlement of New Zealand. In contrast, there is little evidence that eels were captured or consumed by the Aboriginal people of Tasmania in Australia — in fact there is apparently little evidence that the Aborigines of Tasmania have exploited any freshwater fishes during the past 3500 years (W. Fulton, personal communication).

Regan (1911) described how elvers were caught in the United Kingdom in olden times with 'seives of haircloth, or even with a common basket and were fried in cakes or stewed and considered delicious ... hay ropes were suspended over rocky parts to help them on their ascent'. MacCrimmon and Gots (1980) mentioned the capture of the American eel by the Indian people of North America.

Eels are subject to exploitation by man virtually from the time they metamorphose from the marine leptocephalus larval stage and begin their migrations into fresh water as glass eels. The newly pigmented, small elvers are heavily harvested, as are the river-living, feeding, yellow eels and the downstream migrating, pre-spawning silver eels. Fishing methods are very diverse and involve baited pots, fyke nets, long lines, trawls, weirs and pound nets.

Nowadays, glass eels and elvers are exploited primarily for relocation. Some of this relocation is for establishment of eel populations in places from which the upstream migrating elvers are excluded by barriers to migration. Other relocations are to habitats where the eel densities are far below the carrying capacity owing to difficulties in upstream migration. Some relocation is for wild-rearing based on naturally existing foods, but in other instances there is supplementary feeding of the eels to increase growth rates.

Capture and carefully considered relocation of glass eels and elvers into depleted habitats makes it possible to minimise mortalities that inevitably follow from massive immigrations of young eels into restricted habitats at and above river estuaries.

In biomass terms, the major eel fisheries involve exploitation of river resident or 'yellow' eels and of migratory or 'silver' eels. Adult and sub-adult eels are taken for direct human consumption in almost all instances, although occasionally they are fattened for a time by artificial feeding. Eels are slow-growing and long-lived; as a result they enter adult fisheries at a relatively advanced age. Sinha and Jones (1975) suggested that males enter the fisheries at 6–12 years and females from 12 years onwards and they may live for 60 or more years (Todd, 1980). This makes eel fisheries difficult to manage and subject to decline over long periods when exploitation rates are too high. Because they are so robust, eels are often marketed alive, and are even shipped halfway across the world by air from Australia and New Zealand, to supply live eel markets of Europe and Japan. Major, traditional eel fisheries of very long standing have occurred in western Europe — Italy, Denmark, Sweden, the Netherlands, France, Germany, etc. — where the target of the fisheries has been the European eel, *Anguilla anguilla*. In general, figures for catch in several European countries show that yields from natural waters have not changed much since records are available, e.g. yield of the Danish fishery that catches in 1885 were not much different from those of today (Tesch, 1977). Catches in western Europe were in the vicinity of 1,000–2,000 tonnes in the period 1980–1983 (FAO, 1984). In the western North Atlantic the American eel has also long been heavily exploited, although Bigelow and Schroeder (1953) documented a decline in catch in the New England states from about 138 tonnes in 1919 to less than nine tonnes in 1947 in Maine.

Major eel fisheries occur in Japan; Kafuku and Ikenoue (1983) gave annual eel catch as exceeding 2,000 tonnes in recent years, but listed an increase in aquaculture production from 14,000 to 27,000 tonnes in the period 1972 to 1977. The FAO figures for the period 1980–1983 are up to 39,000 tonnes; however, a major portion of this is aquaculture production. The figure for Korea is 10,000 tonnes. These are the world's major eel fisheries.

Eel fisheries of any significance do not seem to have developed for countries/species in southeastern Asia and the countries bordering the Indian Ocean — there are few indications of eels fisheries for the diverse species in Indonesia, or in India and Africa. Jubb (1967), for instance, discussed angling methods for catching eels in South Africa but makes no mention of commercial fisheries. Whether there really are fisheries in these areas or not is unclear. If there are not, it is also unclear whether the failure to exploit eels is because of a lack of knowledge of existing resources, the absence of enough eels to generate real fisheries activity, or cultural issues connected with eels.

Although there has long been traditional exploitation of eels in New Zealand (both the shortfinned eel, *Anguilla australis* and the longfinned eel, *Anguilla dieffenbachii*), commercial exploitation is of more recent date. There were attempts to capture eels commercially and export them to Europe in the early 1900s but these did not last long owing, probably, to difficulties in storage and transport. Later eels were intensively exploited in a search for vitamins when supplies were threatened by the Second World War. But growth of a significant fishery and export industry began in New Zealand as recently as the mid-1960s. From these late beginnings, yields rose rapidly in the mid-1970s to make New Zealand one of the world's significant eels producers (Tesch, 1977; Jellyman and Todd, 1982). Even so, peak catch in the mid-1970s was less than 2,100 tonnes, and small by comparison with the Japanese fishery. Australia's eel fishery was even later in developing. A fishery in Tasmania for *A. australis* began in 1965 but it has not expanded to the size of the New Zealand fishery. A peak catch of about 94 tonnes in 1968 has never recurred owing to restrictive regulations (Sloane, 1982) and the catch now averges 20–40 tonnes. A bigger fishery is present in Victoria where catches of *A. australis* and *A. reinhardti* exceed 200 tonnes per year (Harrington and Beumer, 1980). Total world eel catch was between about 80,000 and 94,000 tonnes in the period 1980–1983 (FAO, 1984).

With depletion of both wild eel stocks and of the habitats that they occupy has come a move towards aquaculture of eels which has developed as a major means of meeting the world's demand especially in Japan. It probably can be traced back to early experiments with enhancing wild stocks by relocation of elvers and supplementary feeding/growing-on of wild eels. Owing to the complex and specialised nature of the life history and the long-lasting marine

215

larval stage, there is at present no technology available for eel culture based on captive reproduction and larval development. Aquaculture, therefore, comprises the rearing of juvenile fish taken from wild stocks (usually glass eels or elvers), in captivity by artificial feeding to a harvestable size. For this reason, migratory juvenile eels are taken from the wild in substantial quantities for transfer to aquaculture facilities. This is particularly prevalent in Japan, but also well advanced in Taiwan and in parts of western Europe, especially Italy.

Tesch (1977) in reviewing the development of eel aquaculture dated Japanese attempts at intensive, high density culture as early as 1880. However, he considered that only in the last few decades has aquaculture had a significant impact on eel production in Japan. Now, owing to both increased aquaculture production and reduced wild catches, aquaculture is by far the largest source of eels. Annual demand for elvers in Japan was estimated at 100 tonnes by Moriarty (1978). Some of this demand has been met from Europe (Deelder, 1970), and also New Zealand at times with indifferent results.

Attempts to establish eel aquaculture in New Zealand were based on the successful Japanese technology but have failed, and despite substantial investments of manpower and finance none of the half dozen farms established in New Zealand in the 1960s and 1970s is now operating (Jellyman and Todd, 1982)

TROUTS AND SALMONS — FAMILY SALMONIDAE

Salmonids are, without doubt, the most important diadromous fishes as regards fisheries. They have wide importance and high productivity in traditional fisheries, in recreational fisheries, in modern commercial fisheries, and in aquaculture. They are also important for acclimatisation, although success in establishing diadromous stocks in the new areas has been marginal (see Chapter 11). The natural range of salmonids meant that, traditionally, they were important to peoples in northern cool temperate and boreal lands. Elsewhere in this book, there has been discussion of harvesting of whitefishes and Atlantic salmons for many centuries in the Europe and Scandinavia, and of Atlantic salmon by Indian people along the Atlantic coast of North America and the earliest inhabitants of Greenland and Iceland. There can be no doubt that diverse diadromous salmonids were of critical importance as food for the Inuit people of northern Canada. It is well known that Indian tribes of North America's Pacific coast depended heavily on various diadromous trouts and on Pacific salmons for food, using a diverse array of traps, nets, and spears. Dunbar and Hildebrand (1952), for instance, have discussed the economic importance of Atlantic salmon to the Inuit people of remote Ungava Bay in northeastern Arctic Canada, and wrote of a commercial fishery involving the Hudson Bay Company dating from 1881. Atlantic salmon were taken by these people for both human and dog food, and for sale and barter. Netboy (1958) described how the first European explorers in the Pacific north west, at the beginning of the nineteenth century, found Indians catching, smoking and drying salmon in huge quantities, both for their own use and for barter with tribes further inland. Pemmican, a traditional Indian food, is dried, pulverised salmon. According to Netboy, these salmon were the 'staff of life' for the Indians. The same was undoubtedly true of peoples of the western North Pacific, from the Bering Sea, south through Kamchatka to Japan. The occurrence of salmonids in such abundance, even though seasonal, must have had a profound effect on the ability of diverse human populations to survive in many of these cold and inhospitable areas, especially in areas of boreal Russia, Asia and North America that abut the Arctic Ocean.

Fisheries for Atlantic salmon, *Salmo salar*, have undergone profound changes over the past few hundred years. In Europe, heavy harvesting, combined with serious deterioration of habitat quality, has led to major losses of fisheries. In England there was once a time when serving salmon to indentured workers was limited by regulation to avoid abuse of a freely available food. In more recent times it has not been possible to catch a salmon at all in the River Thames. Norman and Greenwood (1963) dated the last salmon to be caught in that river as 1833, and had hopes that should the River Thames be 'miraculously purified' these fish might run up it once more. Miracle or not, huge efforts were invested in cleaning up the Thames and limiting effluent discharges into it during the 1960s, and in 1975, the first salmon in modern times was seen there. One salmon does not constitute a fishery revived, but it is a beginning.

A serious deterioration of the North American Atlantic salmon fishery is also well documented. McKenzie (1959) found that a William Davidson had settled on the banks of the Miramichi River as the first British settler by 1764, and for many years took about 400 to 500 tonnes a year. At that time, quantities of salmon were 'perfectly prodigious'. In the early 1800s Nova Scotia exported 8,000 barrels of pickled salmon a year (Anderson and Bremer, 1976). Intensive harvesting of Atlantic salmon both at sea and in the rivers has seriously stressed fisheries of the Atlantic coast of North America. Even by the time of Canada's confederation (1867), there were no Atlantic salmon left in Lake Ontario (Anderson and Bremer, 1976). This stress was greatly increased by the propensity of the early European settlers for constructing dams on the rivers up which the spawning fish migrated. As a result of these combined impacts, salmon fisheries along this coast have, for decades, been only a shadow of what they once were. In McKernan's (1980) opinion 'that the species still persists at all is something of a modern miracle'.

Intensive efforts have been invested in protecting and restoring these salmon fisheries — by international agreements on harvesting at sea, by limiting recreational harvesting of the fish during the spawning runs, by eliminating barriers to upstream migrations to the spawning grounds, by hatchery releases as compensation for other losses, and by restoration of deteriorating habitats following pollution and deforestation. Most of this progress, made in the past

50 or so years, has probably been lost in the last decade to a single new threat — the acidification of rivers and lakes of the Atlantic northeast of North America. There, as well as in northwestern Europe (especially Scandinavia), industrial air pollution has been returned to the land in rainfall, which finds its way into rivers and lakes. Where these flow over ancient granites and other hard rocks, the substrate is unable to neutralise the acidity generated by these sulphur and nitrogen compounds, the pH of the water declines, and fish populations collapse. Only where parent rocks are limestones can pH be maintained at levels that allow fish populations — and their fisheries — to survive. Acid rain is the environmental phenomenon of the 1970s but concerted action to deal with it has begun only in the 1980s and we may have yet to see the worst of its impacts on natural environments. This is a sad fate for what Wheeler (1969) described as one of the most valued food fishes in the world.

Even so, important fisheries for Atlantic salmon remain, and production was assessed at about 11,000–17,000 tonnes in the period 1964–1976, mostly from Canada, Denmark, Greenland, Norway and the United Kingdom (McKernan, 1980). However, in many ways, one of the chief values of the Atlantic salmon today is to recreational angling — perhaps not in direct financial terms, but rather as a source of human recreational pleasure and the economic spin-off generated by angling, tourism/travel and associated activities. In North America and western Europe, salmon fishing is regarded as the finest of angling and is a source of great enjoyment to those with the financial resources to engage in it.

The esteem with which the Atlantic salmon is held made it a prime candidate for aquaculture, and this fish is the basis of a huge industry. Most production is from Norway, with lesser amounts in other parts of Scandinavia and the United Kingdom, with some also in eastern Canada and elsewhere. Production is almost entirely from sea cages, although experimental ocean ranching has been carried out in eastern Canada. Introduction of Atlantic salmon into other countries has been discussed in Chapter 11; although no anadromous stocks of Atlantic salmon have been established in any of these recipient countries, there is some sea cage rearing of the species developing in both Tasmania and New Zealand. Annual world production of Atlantic salmon is given as 17,000–27,000 tonnes for the period 1980–1983 (FAO, 1984), including 4,600–17,600 tonnes from Norway and most of this from aquaculture.

By contrast with the Atlantic salmon, the similar and closely related anadromous brown trout, *Salmo trutta*, is almost exclusively a target of recreational anglers — a species of great renown but one which has virtually no significance to commercial fisheries or aquaculture.

This is not true, however, of the rainbow trout, *Salmo gairdneri*; sea-run steelhead are taken in significant numbers amongst and in addition to the eastern Pacific salmon fishery, and freshwater rainbow trout are also of great importance to recreational angling in many countries. However, the most important production of rainbow trout is from aquaculture (trout farms), where fish are captive-reared entirely in fresh water. Production in 1980–1983 was 43,000–55,000 tonnes, especially in Japan but by 1985 the figure had risen to 180,000 tonnes for Europe alone (FAO, 1986).

Fisheries for Pacific salmons (*Oncorhynchus* spp.) are vast. They have a history in North America that is associated with early European settlement of western regions. Early explorers, seeing the Indians harvesting salmon rapidly recognised their commercial potential. Many Indian tribes were actually relocated onto reservations where access to the fisheries was more restricted, and in the 1860s the new settlers began to catch and can the fish. From a small beginning, canning production rose to about 14,000 tonnes in 1889 from the Columbia River alone, with 39 canneries operating there (Netboy, 1958, 1980). Initially chinook salmon, *O. tshawytscha*, were taken; fishing spread quickly to coho, *O. kisutch*, chum, *O. keta*, and sockeye, *O. nerka*. Pink salmon, *O. gorbuscha*, were later to be exploited. Catching of the fish, which began in the rivers, spread down to the coast, into coastal waters, and eventually the open ocean. From 1875 to 1920, mean annual catch was about 14,000–18,000 tonnes.

As this fishery developed, in addition to the massive intensity of exploitation, there were serious habitat problems. Dam construction hindered migration upstream and downstream, and in spite of heavy investment in fish passes and other mitigation measures, passage difficulties were not overcome. Land management and water abstraction caused deterioration of habitat quality and reduction in water flows. Slowly, a seemingly inexhaustible resource was in decline, and then in trouble. The massive Pacific salmon fisheries of the Pacific northwest declined to the extent that, as with Atlantic salmon, international fishery agreements on how, where, and by whom the fisheries could be exploited were negotiated between various competing countries. Management of the United States salmon fisheries was made much more complex by a court ruling allocating a substantial proportion of salmon production in some river systems to the Indian people (the Bolt amendment). The Canadian Government moved in the late 1970s and 1980s to invest over $300 million in a salmon enhancement programme, and huge hatcheries were built both there and in the United States to try and augment the stocks. Many river-specific stocks were found to be dangerously depleted, some of them extinct. Attempts to restore some depleted stocks revealed the presence of genetically adapted local stocks in various river systems, and suddenly there were no simple answers to what turned out to be very complicated questions. Restoration of depleted stocks by releases of fish from river systems was not as straightforward as it seemed. Ocean ranching developed and sea-cage rearing of Pacific salmons was experimented with. In spite of these problems, the North American Pacific salmon fishery remains a very large and valuable one.

There have been similar historical developments in the northwestern Pacific Ocean — in Siberia and Japan — although details of changes in these fisheries are less easily documented. It seems that they, too, have been seriously overfished and that, especially in Japan, environmental deterioration has been even worse than in North America. In Japan and Siberia, vast enhancement programmes, especially with pink and chum salmon, have been instituted. In both areas annual releases of more than a billion smolts have occurred in recent times. Whereas in North America a real effort was made to restore and re-establish fisheries based on return migrations of the spawning fish to fresh water, in Siberia and Japan much harvesting is based on oceanic fisheries.

Considerable effort has been expended in establishing anadromous Pacific salmon fisheries in southern hemisphere lands. Outside their natural range, Pacific salmon occur as anadromous populations only in New Zealand (Chapter 11). Exploitation there is almost exclusively by recreational angling, although some salmon are taken by commercial trawlers at sea as a by-catch. Ocean ranching is developing rapidly as is sea-cage and freshwater-pond rearing of both chinook and sockeye salmon (Figure 12.3). Sea-cage rearing of salmon is also developing rapidly in Chile.

Fisheries based on Pacific salmons produce varying quantities, depending on species. In the period 1980–1983 (FAO, 1984) chinook salmon yielded about 18,000–25,000 tonnes and coho 30,000–45,000 tonnes, both primarily in the northeastern Pacific (United States and Canada). There is a small Japanese fishery for masou salmon, *O. masou*. The biggest producers are sockeye, 112,000–163,000 tonnes; chum, 167,000–185,000 tonnes (both primarily eastern Pacific - United States and Canada — and Japan) and pink, 167,000–260,000 tonnes (all areas). Aglen (1980) listed total production for all Pacific salmon species from all areas as 426,000 tonnes, in 1971.

Some other salmonids, even those having a lower fisheries profile, are nevertheless of considerable fisheries significance. All of the chars, genus *Salvelinus*, are of significance to angling, and some only to that activity in modern times, e.g. brook char, *S. fontinalis*. However, MacCrimmon and Gots (1980) noted a subsistence fishery for chars from prehistoric times and that they were a means for survival for native peoples of the Arctic more than 5,000 years ago. Scott and Crossman (1973) assigned slight commercial fisheries value to the dolly varden, *S. malma*. But the widespread and prolific Arctic char, *S. alpinus*, has for centuries had wide subsistence and some economic fisheries importance. They were used at least back to the twelfth century in England and other parts of Europe. These authors also point to their importance to Norsemen, and later the Inuits in the remote Arctic lands of Greenland, Iceland and the Faroe Islands. McPhail and Lindsey (1970) recorded their use by Inuit people for both human and dog food in Alaska — the former either fresh, salted, smoked, or dried. Scott and Crossman (1973) considered Arctic char to be the only fish of real economic importance to 'polar and central' Inuit people of Canada. Dempson and Green (1985) listed a catch in northern Labrador in the late 1970s of 175 tonnes a year, and this species was described as becoming a gourmet food in recent years in large cities of North America; some is also canned, such that Scott and Crossman (1973) feared overexploitation. No production is listed by

Figure 12.3: Sea cages are a rapidly growing technology for the production of salmon

FAO for recent years (FAO, 1984).

Whitefishes, genus *Coregonus*, are much less widely publicised as forming the basis for fisheries and yet are important in virtually all areas where they occur. Diverse species are involved, Arctic cisco, *C. autumnalis*, in Arctic Canada and parts of Siberia, lake whitefish, *C. clupeaformis*, and lake herring, *C. artedii*, in parts of Canada, Bering cisco, *C. laurettae*, perhaps in the far north of Canada. Least cisco, *C. sardinella*, are of slight importance in Canada but more so in Siberia. Scott and Crossman (1973) considered the lake whitefish to be the most important freshwater fish in Canada, although exploitation was largely of non-diadromous stocks in the Great Lakes. They regarded the broad whitefish, *C. nasus*, as esteemed for both human and dog food in the far northwest; McPhail and Lindsey (1970) classed it as possibly the finest freshwater table fish in Canada. Houting (or pollan), *C. lavaretus*, and European whitefish, *C. albula*, are important in western Europe, especially Finland.

Total catch of these various species amounts to many thousands of tonnes each year, e.g. European whitefish 8,000–9,000 tonnes per year, mostly in Finland; lake whitefish 10,000– 2,000 tonnes, primarily in the Great Lakes, and miscellaneous coregonids 24,000–33,000 tonnes, mostly from Russia (FAO, 1984).

The inconnu, *Stenodus leucichthys*, also belongs amongst the salmonids of

221

some importance to fisheries, though Scott and Crossman (1973) found its appeal to be far from universal, owing to its high oil content.

SMELTS — FAMILY OSMERIDAE

Smelts of the Northern Hemisphere family Osmeridae are widely exploited in sub-Arctic and Boreal lands, and probably always have been, although the value of the fish varies widely with both species and location. Bigelow and Schroeder (1963) reported that in the colonial period of North America, according to a Captain John Smith, American smelt were in such abundance that the native people caught them up the rivers with baskets like sieves. Perhaps most interesting is the fishery for the eulachon, *Thaleichthys pacificus*. This fish has long had an important role in the lives of the Indians of western North America. It was important to them as a food, but in addition it has a very high oil/fat content and the oil solidifies at normal ambient temperatures (unlike the oils of most fishes). Eulachon are so oily that when they are dried they can be lit and used as a primitive lamp — and they were widely used for this purpose by the Indian people. Some accounts mention insertion of a wick or taper down the throat of the fish, to carry the oil to the site of burning. Eulachon were once taken for the value of the oil alone, which was extracted by the Indians with a press, and was a barter item amongst various Indian tribes. Capture of eulachon in northern British Columbia is now reserved for the Indians (Hart, 1973). Much more eulachon was formerly taken for food, for man and his dogs, but most of that now taken is used for captive rearing of fur-bearing animals such as mink. Catches in the period 1941–1970 were 34 to 200 tonnes.

The rainbow smelt, *Osmerus mordax*, is the basis for fisheries in many places in North America, especially the Great Lakes, and also in the Arctic USSR. Rainbow smelt for decades supported very important fisheries with thousands of tonnes caught, especially in Lake Ontario. They were taken in the spawning rivers in nets of diverse types and were the target of 'ice fishing', in which a hole was cut through the surface ice of lakes to provide access to fish living beneath. They appeared in 'almost miraculous' quantities each year (McKenzie, 1959). Decline in abundance evidently began very early so that by the end of the nineteenth century there was a hatchery in New York State which was releasing many millions of fish a year to augment the runs. Returns were poor so that by about 1900 there were only enough to provide for releases of a few million (Bigelow and Schroeder, 1963). The related European smelt, *O. eperlanus*, was said not to be favoured in the United Kingdom, even though regarded by some as delicious eating (Wheeler, 1969). It is clearly important in the USSR, as Nikolskii (1961) reported that the species is reared artificially to boost production in the Neva River.

The longfin smelt, *Spirinchus thaleichthys*, has been little exploited (Scott and Crossman, 1973), this being attributed to its limited availability, in spite of its good flavour (Hart, 1973). In Japan *Hypomesus transpacificus* is important amongst the wide diversity of fishes eaten by Japanese people.

World production of smelt for the years 1980–1983 was between about 31,000 and 45,000 tonnes (FAO, 1984), most of this being taken by Canada, the United States and Russia.

AYU — FAMILY PLECOGLOSSIDAE

The ayu ranks as the most favoured freshwater fish in Japan, in spite of its rather modest size (about 300 mm and 250 g). Kafuku and Ikenoue (1983) attributed this to its special flavour and its beautiful, slender shape. Needless to say it is heavily exploited, being caught in fresh waters by diverse types of nets (set nets, cast nets, drag nets), by construction of traps/weirs across rivers, and with baited hooks and artificial flies. It has also been a target of the famed Japanese cormorant fishery, in which birds are trained to return fish they catch to their master. Okada (1960) said that this method of fishing for ayu can be traced back 2,500 years.

Annual production of ayu from wild populations has risen rapidly in recent years, from about 7,500 tonnes in 1967, to nearly 15,000 tonnes in 1979 , and contributes 20–30 per cent of the total fisheries catch from Japanese rivers (Kafuku and Ikenoue, 1983). Ayu is also known from eastern China and Korea, but information on a fishery for the species in these areas seems meagre or difficult of access.

Because of the great favour with which the ayu is viewed, there has been development of an aquaculture industry, for which production has grown from about 2,000 tonnes in 1968 to 8,000 tonnes in 1979 — now amounting to about half the wild catch. Aquaculture production has traditionally depended largely on the growing-on of ayu mostly captured from the wild, as fingerlings taken from freshwater populations. Some diadromous stock is also caught for aquaculture purposes but these fish have some difficulties with osmoregulation. In the past five years, technology for artificial maturation and the production of ova from captive stock has been developed.

ICEFISHES OR NOODLEFISHES — FAMILY SALANGIDAE

The very elongated, slender, small fishes of the family Salangidae are of considerable fisheries importance in the rivers and coastal seas of the Orient, especially in the USSR, China, and Japan. Osbeck (1762), in his translation of J.R. Forster's report on travels to China and the East Indies, wrote of these strange little fishes that 'the Pack-Fanny is the Chinese name of a long, transparent white fish which caught ... and being dried is boiled and eaten'. Fang (1934) referred to them being dried in great quantities, and said that they were also delicious fresh. Quantities caught are not well recorded, but Kawanabe, Saito,

Sunaga, Maki and Azuma (1968) noted catches of over eight tonnes from Lake Nake-umi and the vicinity, in Japan.

Very little information is available about these fishes and their fisheries, though clearly they are still caught in large numbers in modern times as trial exports to New Zealand from China were made in the early 1970s. Apparently they were not well received by comparison with New Zealand's indigenous whitebait (*Galaxias* spp.), as exports did not last for long. Nikolskii (1961) regarded them as only of 'secondary commercial importance'.

SOUTHERN SMELTS — FAMILY RETROPINNIDAE

There are old historical records of Maori exploitation of retropinnid smelts in New Zealand — probably both common smelt, *Retropinna retropinna*, and Stokell's smelt, *Stokellia anisodon*, although the two species were apparently not distinguished by the Maori people. Numbers of fish migrating into the rivers of New Zealand over the spring and summer may be vast and the fish were well-known to and heavily exploited by the Maoris (Best, 1929). Hubbard (1979) described one of the methods traditionally used and which still persists in some remote areas. Channels are dug in the shingle beaches of river margins into which the upstream migrating shoals are diverted. When there is a substantial number of fish in the channel a trap net is placed across the mouth and the fish are driven down into it. The fish might be eaten fresh but were often dried in the sun for storage and later use.

In recent years European man has begun to harvest these fish commercially, although the quantities taken are still small by comparison with the apparent size of the resource. The ripe fish are very oily and rich in flavour and are regarded as a delicacy by some. Their strong cucumbery odour is disliked by many people unfamiliar with their flavour when properly prepared. Some of the catch is frozen and some is sun or kiln dried, both products finding ready acceptance amongst the Asian community in New Zealand (McDowall, 1983). In Fiji and southeast Asia the Indian and Chinese communities use other small, silvery fishes, usually sun-dried, and known as 'ikan bilis' or 'natali', and the New Zealand retropinnids are an acceptable substitute for other fishes traditionally exploited. Small exports of the New Zealand product are made but there is a huge potential market owing to the decline in abundance of the species used in this way in other parts of the world.

The juveniles of the common smelt also migrate into the estuaries of some larger New Zealand rivers, and although they find less favour as a food than the *Galaxias* whitebait with which they are caught, they constitute a significant fraction of the New Zealand whitebait fishery (McDowall, 1984c).

There is little, if any, evidence to suggest that retropinnid smelts are exploited in Australia, although the Tasmanian smelt, *Retropinna tasmanica*, was a minor constituent of the Tasmanian whitebait fishery (based on the aplochitonid *Lovettia seali*) before this fishery was closed because of overfishing in the 1950s (Lynch, 1965).

SOUTHERN GRAYLINGS — FAMILY PROTOTROCTIDAE

The New Zealand Maoris were well familiar with the native New Zealand grayling, *Prototroctes oxyrhynchus*, and old reports describe capture of mature adults as they migrated upstream. They were taken in traps inserted in timber weirs built across the rivers (Best, 1929). In a country with sparse sources of readily available protein, such fish, even though of modest size (to about 300 mm), were inevitably an important food source.

When European man began to explore and settle in New Zealand in the mid-1800s, this fish was also highly favoured as a food, being commonly described as the only native species that was regarded as of much value for angling. Grayling were caught in large numbers both by anglers and by various netting methods, but their exploitation was short-lived as the fish soon became rare, and was extinct by the 1920s.

The Australian species, *P. maraena*, is not reported to have had any significance to the Australian Aborigines, although it was taken by the early European settlers, in much the same way as the New Zealand species. *Prototroctes maraena*, although now much reduced in numbers, still occurs in some of the less-modified catchments of southeastern Australia and Tasmania.

SOUTHERN WHITEBAITS AND GALAXIIDS — FAMILY GALAXIIDAE

The juveniles of five *Galaxias* species returning to fresh water from the sea in New Zealand constitute a significant fishery. They are the basis for the 'whitebait' fishery there, and are small, transparent fishes about 40–55 mm long (McDowall, 1984c). There are much smaller fisheries for the same stages of various *Galaxias* species in Chile, and they were once a minor part of the Tasmanian whitebait fishery (that was based on the aplochitonid *Lovettia seali*).

Before the arrival of European man in New Zealand (mid-nineteenth century) these fishes and their migratory habits were well known to the Maoris, and the fish were well exploited in various stages. The upstream migrating juveniles were taken in nets and weirs. Commonly the shoals were diverted into blind, watered channels along riverbanks, from which they were scooped in small woven baskets. They were taken in such quantities that they were preserved by various means. Often they were dried in the sun or on racks above fires for storage and later use, but some were cooked over hot stones and also evidently stored in some way.

The adults of these *Galaxias* species are in some instances of quite substantial size; the largest of them, the giant kokopu, *Galaxias argenteus*, reaches a

Figure 12.4: The Europeans soon adapted traditional methods to catch the fish more efficiently. Here metal gauze nets are placed in gaps along a gravel groyne constructed across the river channel
Source: NZ Ministry of Agriculture and Fisheries

size of 2.8 kg and 580 mm. Commonly it grows to 300–400 mm. Other species are smaller, having maximum sizes closer to 250 mm. But in New Zealand, a land with limited, easily available protein sources, these fish were keenly sought as food, being caught in traps, pots, nets of diverse types and with primitive hooks. Amongst these galaxiids, the inanga, *Galaxias maculatus*, is distinctive in having shoaling habits throughout its life and well-defined, lunar-related downstream migrations of the spawning adults. Prodigious numbers of these small fish (mostly 80–100 mm long) make this migration and, having fully developed ovaries and testes, they are rich and oily. Not surprisingly they were heavily exploited by the Maoris either for immediate consumption or, again, for drying and storage.

European man very rapidly recognised the culinary value of these species (Figure 12.4). In the early days of settlement of New Zealand, freshwater resident adults were sought and exploited. The first settlers found them a useful source of protein, and the early explorers found them to be equally palatable (McDowall, 1984c). Later, in times of poverty and hardship during the New Zealand gold rush years, adult galaxiids were a notable source of food for miners, who often encountered them when diverting streams and draining swamps to sluice alluvial gravels for gold.

It was the upstream migrating juveniles, however, that attracted the early

Figure 12.5: Catch in the New Zealand *Galaxias* whitebait fishery has fluctuated greatly throughout the period when catches have been documented

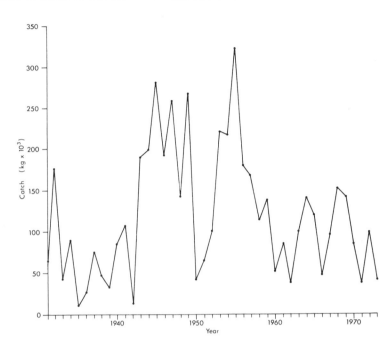

interest of the Europeans, and which have sustained a fishery in succeeding years. They were soon recognised as an easily available culinary delicacy. Initially the fishery was virtually New Zealand-wide, but deforestation, land development, swamp drainage, and other effects of human habitation, led to a significant decline in abundance. Substantial fisheries are now restricted to less-developed regions. Catch quantities have fluctuated very widely (Figure 12.5), reaching a recorded peak of 322 tonnes in 1955.

Although there is no doubt that, overall, catches have declined markedly in much of New Zealand, huge catches may still be taken. One fisherman in the mid-1970s took nearly a tonne of whitebait in one day's fishing. Catch is by no means sufficient to meet local demand, although in the 1930s to 1950s whitebait was canned in large quantities and exported to Australia. Once it sold for just a few cents a kilogram. Today demand, particularly from hoteliers and restauranters, is so great that the product can fetch $50.00 per kg on the retail markets.

A fishery for the migratory juveniles of *Galaxias maculatus* is also present in Chile, but is relatively small, Campos (1973) noting that catch in the Rio Valdivia in central/southern Chile was about 1.5 – 2.0 tonnes per year in the period 1968–1970.

Figure 12.6: The Tasmanian whitebait fishery, based on the aplochitonid, *Lovettia seali* , had very short-lived, high productivity in the late 1940s, but crashed, never to recover
Source: Fulton, 1984

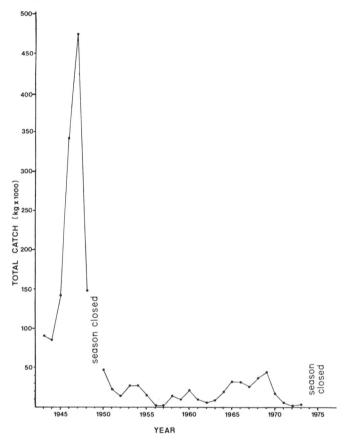

TASMANIAN WHITEBAIT AND PELADILLOS — FAMILY APLOCHITONIDAE

A fishery for the Tasmanian whitebait, *Lovettia seali*, developed during the early 1940s; prior to this time it seemed to be almost unrecognised as a resource of any value (Blackburn, 1950). The upstream migrating pre-spawning adults were captured by fishermen operating large, long-handled scoop nets from the river banks or from boats moored in mid-stream, mostly in estuarine areas. Catch rose very rapidly from the beginnings of the fishery, reaching a recorded peak of 480 tonnes in 1947 (Figure 12.6). However, the fishery did not last long, collapsing from this peak to only 1.5 tonnes in 1955 and little more than 1 tonne in 1972. Although Blackburn (1950) estimated a sustainable yield exceeding 300 tonnes for this fishery, it never really recovered from the collapse of the 1940s and 1950s and the fishery was eventually closed. It has not been reopened, although this is now being considered by Tasmanian authorities (W. Fulton, personal communication).

There are no apparent reports of a fishery of any significance for the Chilean/Argentinean peladillos, *Aplochiton* species, but Stewart (1980) described how *A. zebra* is taken by float fishing in the Falkland Islands, using sheep meat as a bait.

SHADS AND HERRINGS — FAMILY CLUPEIDAE

A diverse array of clupeid fishes contributes to important fisheries in nearly every ocean of the world — the herrings, shads, sardines, pilchards, etc. Many of them are entirely marine fishes, but there are also important fisheries that exploit various diadromous clupeid species, some of the fisheries being very well known.

Alosid shads (genus *Alosa*) of a considerable variety are involved in fisheries in the northern Atlantic. The two eastern Atlantic shads, the allis shad, *Alosa alosa*, and the twaite shade, *A. fallax*, are much reduced in abundance and are now scarcely common enough to be of much fisheries value. Wheeler (1969) described the allis shad as being of little value in Europe mainly because it is so scarce that only small local fisheries can exist to exploit it. Of the twaite shad he said that some are caught but it is not significant. Even so, catches of these two species are listed by FAO (1984) as about 1,000–2,000 tonnes (1980–1983), taken by various European countries.

Western Atlantic shads are more prolific, particularly the American shad, *A. sapidissima* and the alewife, *A. pseudoharengus*. The American shad was described by Scott and Crossman (1973) as large and meaty, a 'desirable' fish, taken both for pet food and for human consumption, fresh and smoked. Alewives were also taken for consumption fresh and smoked, with the ovaries sometimes used for caviar. These fish were extensively exploited by Indians prior to the arrival of European man, and were seen as a valuable fishery resource by early settlers. A substantial fishery soon developed. Rulifson *et al.* (1982a) regarded this fish as the most valued fish along the Atlantic coast of North America prior to the Second World War. It was held in such high esteem that when stocks became reduced as a result of dam construction, pollution, and overexploitation, moves were made to develop culture technology for the fish. Catch of American shad reached a peak of about 23,000 tonnes in 1896, but had declined to less than 4,000 tonnes by 1960 and less than a thousand tonnes by 1976 (Rulifson *et al.*, 1982a). It is taken primarily at sea by trawlers, but is also caught during its upstream migration to spawn. The alewife has also been a very important fisheries resource of very long standing in the United States and Canada. It, too, was taken by the Indians, for both food and fertiliser (Rulifson *et al.*, 1982a), and was the species involved in the first law applicable to the

229

fledgling settlement at Plymouth — a law to protect it from overexploitation. Rulifson *et al.* (1982a) noted that the alewife had been described as being of 'almost incalculable value in coastal areas'. Like the American shad, it has suffered serious decline from peak catches as recently as the 1960s. FAO (1984) listed recent catches of alewife of 18,000–23,000 tonnes (1980–1983). The American shad was much less productive, with 1,200–2,700 tonnes taken in the same period.

Reference was made earlier to the invasion of the Great Lakes of North America by the sea lamprey, *Petromyzon marinus*. Alewives also invaded these lakes by the same route. While the lamprey caused profound changes in the fishery ecosystem by preying on fish already there, the alewife invading an ecosystem disrupted by the lampreys established vast populations. Smith (1968b) recounted events relating to the first discovery of a few alewives in Lake Michigan in 1949–1952, followed by increases in abundance, and leading to their contributing 80 per cent of fish caught in experimental trawls by 1966. Alewives had so swamped the lakes that they replaced virtually all of the previously abundant species low in the trophic systems of the lakes. As a result they completely upset the fish community balance. In addition they became an offensive nuisance by making massive inshore migrations in spring. Mass mortalities followed, with rafts of dead fish being washed onto the beaches to rot. With releases of several species of Pacific salmon (genus *Oncorhynchus*) into the lakes, alewives have become an important forage species for predatory salmon and their abundance has declined. New balances in the relative abundances of various species in the fish communities in these lakes have yet to emerge.

The blueback herring, *A. aestivalis*, and the hickory shad, *A. mediocris*, also contribute somewhat to fishery catches along the eastern seaboard of the United States, primarily in the area between North Carolina and Florida. The blueback is listed by FAO (1984), but no quantities are specified, and for the hickory shad, the figure is very small, 34–44 tonnes. Scott and Crossman (1973) noted that the blueback is caught together with the alewife, which may explain the absence of catch data; Rulifson *et al.* (1982a) made the same comment, but noted that it has been a fishery item for a very long time, back into the 1700s. They described the hickory shad as essentially a bycatch taken with other species of shad.

There is also the American gizzard shad, *Dorosoma cepedianum*, which is taken by some fisheries (400–1,800 tonnes in 1980–1983, FAO, 1984), but the flesh is evidently soft, tasteless, bony, and of poor quality (Miller, 1960); it is used as a fertiliser and stock food (Scott and Crossman, 1973).

In the inland seas of central Europe — the Caspian, Black and Azov Seas - there are substantial shad fisheries, although FAO records indicate that yields in recent years are not huge — 2,300–5,000 tonnes (1980–1983) for all species listed. However, these species have for many decades made an important contribution to the fisheries of these inland seas. Old records suggest large yields.

Berg (1962) listed figures for the period 1936–1939 as follows: dolginka herring, *Alosa brashnikova*, 6,000 tonnes; blackspined herring, *A. kessleri*. 1,000 tonnes; Caspian shad, *A. caspia*, 19,800 tonnes. The tyulka, *Clupeonella delicatula*, yielded about 40,000 tonnes for this period (Svetovidov, 1963), but 47,000–122,000 tonnes in recent years (FAO, 1984). Berg (1962) described the Caspian shad as one of the most important fish species in the Caspian Sea, contributing about a half the total fisheries yield from that sea.

A further series of fisheries in the Indo-Pacific is based on a quite separate group of shads — the hilsa fisheries based on the clupeid genus *Hilsa*. These, too, are long-standing traditional fisheries, especially on the Indian subcontinent. The hilsa shad, *Tenualosa ilisha*, yielded 4,000–6,000 tonnes in Pakistan and Kuwait (1980–1983, FAO, 1984), the toli shad, *H. toli*, 1,500-2,000 tonnes in southeast Asia, and the kelee shad, *H. kelee*, 20,000–25,000 tonnes in India. There are probably additional hilsa shad fisheries in other areas and based on further *Hilsa* species, but they are indifferently documented. The local significance of the hilsa fishery in Pakistan is indicated by the fact that 5,000–7,000 fishermen are involved in the fishery in the Indus River, in Pakistan (Islam and Talbot, 1968.)

Like many diadromous fishes and their fisheries, that for the Indian shad fishery has been badly affected by development. Ganapati (1973) cited the construction of dams in the rivers to impound water for irrigation, which stopped the fish in their migration. Not only did this interfere with migration but it also rendered stocks prone to excessively heavy exploitation at locations where migration was impeded 'the barriers are so many and the fishing methods so efficient that the potential breeders are ruthlessly exterminated ... also floods no longer reach the river mouths to attract the fish and the rivers become obstructed.' Some of the shads of the genus *Ilisha* are also anadromous, although just which species fall into this category is difficult to determine as their life histories are poorly documented. The elongate ilisha, *I. elongata*, contributed 15,000–18,000 tonnes to the fisheries catches in Korea and China (1980–1983, FAO, 1984); the West African ilisha, *I. africana*, yielded 4,000–5,000 tonnes in West Africa.

And there are other fisheries for various clupeids which are listed by FAO (1984) as anadromous. Life history details of these are also poorly documented, but at least the following were listed: the Indian pellona, *Pellona ditcheli*, 1,500–2,500 tonnes, Malaysia; the double-armoured gizzard shad, *Clupanadon thrissa*, 10,000–14,000 tonnes, Korea; *Ethmalosa fimbriata* is also widely of fisheries importance in West Africa, but catches were not specified by FAO (1984).

ANCHOVIES — FAMILY ENGRAULIDAE

Many anchovies form the basis of fisheries, most of them marine and a few of them in fresh water. The Brazilian diadromous anchovy, *Anchoviella lepiden-tostole*, is the subject of a small fishery in one or more rivers (Nomura, 1962); production amounts to 'up to 1000 metric tons a month', during a season which Nomura specified as from 'October to March'. A fishery occurs in the Ribiera de Iguape, and possibly in other rivers. Consumption is primarily by a large Japanese populace in and near Sao Paulo, for whom it is salted and dried. The fishery was evidently started about 1935 by Japanese tea growers in the vicinity; Nomura (1962) regarded it as similar to the Japanese product 'iriko'.

CODS — FAMILY GADIDAE

The anadromous tomcod, *Microgadus tomcod*, migrates into eastern North American rivers during winter, and Scott and Crossman (1973) described it as a fish that is sought by both commercial and recreational fishermen of eastern Canada. Although it evidently has tasty and white flesh, it is used most extensively as an animal food. Recreational fishermen seek tomcod by bait fishing through the ice, especially at night. McKenzie (1959) commented that in the Miramichi River, when the commercial fishing season included the month of November, the upstream runs of spawning fish and the later downstream runs of spent fish were exploited. Peterson *et al.* (1980) recorded a commercial catch of about 200 tonnes in inshore marine waters of the Atlantic northwest. FAO figures for tomcod nominate catches of between 146 and 276 tonnes in the years 1980 and 1983; all of this was taken by Canadian fishermen.

PIPEFISHES — FAMILY SYNGNATHIDAE

Syngnathids, in particular any species that may be diadromous, do not contribute to fisheries of any sort.

STICKLEBACKS — FAMILY GASTEROSTEIDAE

It seems remarkable that there should be fisheries based on these tiny, bony and spiny fishes, and there is little detail to suggest that there is much of a fishery in modern times. However, there are reports of very extensive exploitation of sticklebacks in earlier years. Vast quantities were once taken in England (Regan, 1911); reference was made in Chapter 1 to very old English reports in which catches of up to 768 gallons a day were taken by fishermen over a considerable period. The fish were mostly taken for use as garden manure, although efforts were made to extract oil from them. During times of stringency in the Second World War the Dutch people caught sticklebacks to produce fish meal and to feed to their ducks (Baggerman, 1957).

Some quite substantial stickleback fisheries apparently still persist, as catches varying between about 1,000 and 28,000 tonnes of '*Gasterosteus*' spp. are listed for the years 1978–1982 by FAO (1984). Most of these fish were taken by the Russians, but small quantities also by the Danes. Scott and Crossman (1973) described how the ninespined stickleback, *Pungitius pungitius*, occurs in great abundance in some rivers of the Yukon where it is trapped by the native people and used as food.

SCORPIONFISHES — FAMILY SCORPAENIDAE

Several scorpionfishes have fishery importance, but the single diadromous species is not among them. Australia's bullrout, *Notesthes robusta*, although edible, is too small to attract interest (Merrick and Schmida, 1984).

SCULPINS — FAMILY COTTIDAE

Although species of *Myoxocephalus* (which seem not to be diadromous) are regarded as having some fisheries importance (FAO (1984) listed the species but nominated no catches in the period 1978–1982) the diadromous species of sculpin seem to be unimportant to fisheries. Nikolskii (1961) regarded cottids as being as yet of little commercial importance, but thought that many species which are very abundant should be exploited for the production of artificial manure, forage meal and vitamins. By contrast, McPhail and Lindsey (1970) knew of no uses for either *Cottus aleuticus* or *C. asper*, for either human or dog food, in Alaska.

SNOOKS — FAMILY CENTROPOMIDAE

The single diadromous centropomid, *Lates calcarifer*, the so-called barramundi (Australia), or giant sea perch, is a large species of very wide, though not necessarily great fisheries importance. The fishery for *Lates* seems to be a quite general one that has little direct relationship to migrations themselves. There are no apparent highly concentrated movements of this fish that render them more accessible to exploitation. However, Griffin (1987) showed that in the northern Australian fishery, catches build up as maturing males are moving downstream into river estuaries where the commercial fishery is concentrated. The species also has importance to recreational angling in many parts of northern Australia, but this is mostly upstream in riverine habitats for the smaller and younger fish. Quite substantial fisheries occur for *Lates*, with FAO (1984) recording figures up to about 3,000 tonnes in Pakistan, 450 tonnes in Malaysia, and about 14,000 tonnes in Indonesia. Australian catches are listed as up to

1,000 tonnes by FAO (1984), but Morrissey (1985) gave the figure for Western Australia alone as about 2,000 tonnes.

TEMPERATE FRESHWATER BASSES — FAMILY PERCICHTHYIDAE

These basses are generally large and mostly meaty fishes that are held in high regard for food. Relatively few of them are diadromous. Both the striped bass, *Morone saxatilis*, and the white perch, *M. americana*, are significant in fisheries along the east coast of North America. Catches of striped bass were listed as 785–2,075 tonnes and for white bass as 62–593 tonnes for the period 1980–1983 (FAO, 1984). Most of the fishing for these species is in the United States, where fisheries extend along the whole Atlantic coast. The striped bass has been one of the most popular and valued food and game fishes in this region since colonial times (Boreman and Austin, 1985). It is taken at sea by both commercial and recreational fishermen, and it is also important in coastal rivers, especially to recreational anglers. Mexican Gulf populations of striped bass are not anadromous but there is a significant angler fishery for the species in rivers of the Gulf. Populations have been heavily exploited, and also have been depleted by dam construction hindering migration to the spawning grounds, and by pollution. The species is regarded as so valuable to fisheries that there has been substantial interest and investment in population enhancement by hatchery releases. Nearly four million young striped bass were released into the waters between New York and Louisiana in 1977, with an additional four million in Texas alone (Rulifson *et al.*, 1982a). These authors wrote of proposals to release over five million in 1982. Introduced populations of striped bass in rivers of the North American Pacific coast are also now of importance to both commercial and recreational fisheries (although no commercial catch is registered by FAO, 1984). The white perch is less important, but is taken in much the same way and over the same Atlantic coast range as the striped bass.

Lateolabrax japonicus is a quite similar oriental bass, which is regarded in Hong Kong as one of the most popular food fishes, both to commercial fisheries and to recreational anglers. It is taken in winter during migrations from rivers to inshore marine habitats (Chan, 1968). Yield was over 10,000 tonnes in Japan and Korea in 1980–1983 (FAO, 1984).

The Australian bass, *Macquaria novemaculeata*, is a much smaller fish than the above species, and yet generates considerable interest amongst anglers in estuaries and lower reaches of rivers of southeastern Australia, where it is regarded as one of the best sport fishes. Some of this interest is in fishing for bass in freshwater dams, where there is artificial stocking in inland locations for fisheries purposes (Llewellyn and MacDonald, 1980).

MULLETS — FAMILY MUGILIDAE

Mullets are important food fishes throughout their range, especially in the tropics and subtropics. Thomson (1966) described them as forming the basis for thriving industries in several parts of the world, though to what extent mullets are diadromous is not clear (and was discussed in Chapter 6). The extent that the strictly diadromous (catadromous) species are amongst those involved in important fisheries is also not very clear. However, the grey or striped mullet, *Mugil cephalus*, is probably catadromous and certainly a very important food fish. Thomson (1966) listed world catches for the family as 49,000–76,000 tonnes in the years 1957–1962, but more recent figures are much higher — 178,000–213,000 tonnes for mullets excluding *M. cephalus*, and 32,000–37,000 tones for *M. cephalus* alone (1980–1983, FAO, 1984). Thomson reported that most catch came from the Mexican Gulf, but today the reported fishery is much more widely distributed, with large quantities taken from the seas around Peru, Pakistan, Thailand, Indonesia, Brazil, and Nigeria. Whether the evident increase between the early 1960s and the present reflects actually increased catch, or really is a result of better and more widespread reporting is a moot point; I suspect much of the latter. Catch of *M. cephalus* is widely distributed, including the tropical Atlantic, the Mediterranean, and the northwestern Pacific. The Caribbean hognose mullet, *Joturus pichardi*, which seems to be fairly definitely catadromous, supports a small fishery in Mexico and the vicinity.

Although mullets are important for fisheries, there is not much sense in which their diadromous habits are related to and of importance to their exploitation. They do not seem to be specifically sought during migration. Mostly mullets are taken in beach seines and gill nets operated along shores, in sheltered bays and harbours.

Because of the culinary value of mullets, presumably also because the technology is not too complex, and probably because mullet feed very low in the trophic structure of marine fish comunities, considerable emphasis has been placed on mullet aquaculture, especially in Asia and the Middle East. This may be in both saline and fresh waters. Sometimes mullet are grown in polyculture situations involving various Asiatic carps. Culture was based initially on the on-growing of wild-caught juveniles, but technology for breeding mullets in captivity has been developed in recent decades.

SANDPERCHES — FAMILY MUGILOIDIDAE

The New Zealand torrentfish, *Cheimarrichthys fosteri*, is only a small species commonly reaching a length of about 150 mm, but even so it was favoured as a delicacy by the Maori population of New Zealand (Best, 1929), at least until more easily available foods arrived with European man. Adults were taken as they moved out of rapids and torrents to feed in pools during the night, and were prepared for eating by the Maori people with considerable ceremony. Torrent-fish would not, however, have been available in the vast numbers typical of many diadromous fishes. They were not taken during their migration to or from the sea owing to the small size of the fish during such migrations (newly hatched larvae and small juveniles). Torrentfish persist in large numbers in New Zealand rivers but are no longer exploited by man to any significant extent.

SOUTHERN ROCK CODS — FAMILY BOVICHTHYIDAE

Although some marine bovichthyids make a minor contribution to fisheries cat-ches in Australia, this is not true of the catadromous tupong, *Pseudaphritis urvilii*, which seems to be of no fisheries importance.

GOBIES AND SLEEPERS — FAMILIES GOBIIDAE AND ELEOTRIDAE

That gobies and eleotrids should form the basis for fisheries is no doubt surpris-ing to those familiar with the small, stocky fishes that belong to these two families. However, some species are of modest size (reaching 400–500 mm) and both these larger species and many of the smaller ones make a contribution to fisheries production in diverse parts of the world.

Among the largest of these fishes is the Australian sleepy cod, *Oxyeleotris lineatus*, which reaches a length of 480 mm and a weight of 3 kg (Merrick and Schmida, 1984). This, like most of the other larger gobioid species, has not been shown to be diadromous. The 'chame', *Dormitator latifrons*, which is probably catadromous, is large enough to attract fisheries interest in parts of South America (Chang and Navas, 1984).

The Japanese have a remarkable ability for utilising a great diversity of fish products as food, and amongst these is the goby *Acanthogobius flavimanus*. Okada (1955) described it as the most delicious species amongst the gobioids of Japan, which is shipped live to the Tokyo market. Kawanabe *et al.* (1968) reported quite astonishing catches, of over 43 tonnes, from Lake Nake-umi and the vicinity, in Japan. In Africa *Glossobobius giurus* supports small fisheries in some African lakes (Bruton and Kok, 1980), and in Hawaii *Awaous stamineus*, in both the juvenile and adult stages, is the basis for a small fishery, some fish being marketed but most of it caught for sport and domestic consumption (Ego, 1954).

It is in the juvenile stages that gobioid fishes attract the greatest fisheries in-terest — the juveniles of a variety of genera (mostly amphidromous) are cap-tured by fishermen as they return to fresh water. These fisheries are most explicitly described from the Caribbean (*Sicydium plumieri*, Erdman, 1961,

1986) and the Philippines (*Sicyopterus extraneus*, Manacop, 1953), but are also reported from diverse other lands, including islands in the Pacific (Tahiti, Fiji, Samoa), Japan, and the Indian Ocean (Reunion) (Aboussouan, 1969; P. Ryan, personal communication). *Sicyopterus gymnauchen* was described as Sri Lanka's most delicately flavoured fish which is caught at river mouths and is known there as 'heen kongani' (Deraniyagala, 1937). However, the Hawaiian *Sicydium stimpsoni* is apparently not a species of fisheries importance, on account of its small size, although it is taken for bait (Tomihama, 1972). *Leucopsarion petersi* is taken as the anadromous spawning adults move into rivers from the sea (Akihito *et al.*, 1984), although details of the fishery are elusive. There is likely to be exploitation of similar migrations of various gobies in diverse parts of southeast Asia e.g. Indonesia. Truly astonishing numbers and quantities of these small fish are taken as they approach river mouths and then move upstream into fresh water.

In the Philippines diverse gobies have, for generations, been of great importance in fisheries — the so-called ipon fishery. Taylor (1919) described it as being of 'marked economic importance', there being 'no finer table fish' (Montilla, 1931). These tiny fish, usually less than 20 mm long, are taken by large traps, or in huge seine nets that may be 60–100 m long and 8–12 m deep, taking large teams of fishermen to haul them — Blanco and Villadolid (1939) suggested 50–75 people. The quality of the catch varies with location. When the fish are taken in the sea the quality is highest, but as the fish move into estuaries and fresh water, they become darker in colour and more bitter in flavour. After just a few days in fresh water they are of no value at all. Species included in the Philippines ipon catch were *Sicyopterus lachrymosus*, *Ophiocara aporos*, *Chonophoros melanocephalus*, *Glossogobius giurus*, *G. celebius*, *Eleotris melanosoma*, and *Rhyacichthys aspro*, about all of which very little is known (Herre, 1927). Only *Sicyopterus extraneus* is to any extent studied and understood (Manacop, 1953). The catch of these goby fry is handled in a variety of highly characteristic ways. Sometimes it is salted and allowed to ferment for days, or even weeks into an evil smelling substance which is greatly relished by many Filipinos (Taylor, 1919), the supernatant liquid being used as a sauce on rice and the solid 'mass' cooked, sometimes fried with tomatoes. Or it might be pickled in vinegar (Blanco and Villadolid, 1939); apparently the longer it is kept, the better it becomes! Montilla (1931) described how it is wrapped in banana leaves and boiled in water and vinegar. Clearly the Philippines ipon fishery produces a highly distinctive fishery product. Goby catches in the Philippines in recent years have reached over 6,000 tonnes (1978, FAO, 1984), but it is unclear if much of this production is still 'ipon'. These fisheries are evidently greatly diminished (E.A. Balyut, personal communication), and Philippines goby production may be largely adults of *Glossogobius giurus*.

A fishery for a related species, *Sicydium plumieri*, is well known from the Caribbean, the 'tri-tri' (Clark, 1905) 'ticky ticky', or 'seti' (Erdman, 1961) being one of the most important food fishes in the West Indies. The tiny fish

(20–40 mm long) were caught by the natives and 'eaten either boiled (whole) or fried into cakes. Although when cooked they bear a strong resemblance to maggots, they are very good eating, tasting something like whitebait' (Clark, 1905). They migrate in vast numbers; a column of these fish 10 cm wide, 2.5 rows of fish deep, passed on observation point at 300 per second (Erdman, 1961). Erdman calculated that an average migration lasting about 50 hours, would comprise about 90 million fish. With such vast numbers it is no wonder that they were exploited as a fishery by man. They are caught in the Caribbean in sacking nets on a wooden hoop, with up to 10 tonnes being taken in a single migration from some rivers in Puerto Rico (Erdman, 1961).

A somewhat similar fishery occurs along the Mexican Gulf coast of Central America — the tismiche — which consists of a 'diverse assemblage of post-larval fishes and crustaceans that appear in the [Tortugueros] lagoon ... twice a year, usually in May and December, at which times the local inhabitants exploit them for food' (Gilbert and Kelso, 1971). Gilbert and Kelso regarded these migrations as similar to those of the Caribbean seti and the Philippines ipon, but provided no further explanation of the fishery or the use made of the catch by the fishermen.

The amphidromous species of *Gobiomorphus* in New Zealand also undertake migrations, occasionally in very large quantities, though the fish themselves are very tiny, about 15–20 mm long. For this reason vast numbers are necessary to make up any significant quantity. Occasionally they are caught amongst the New Zealand whitebait fishery (*Galaxias* spp.); however, they are regarded with little favour by the fishermen who refer to the little fish as 'whalefeed' or 'Dan Doolin spawn' (McDowall, 1984c). They are not sought or marketed by the fishermen, but are usually just discarded. Their migration in substantial quantities seems very irregular and possibly for this reason they have never constituted a fishery. There is no evidence that the Maori people of New Zealand caught them, and certainly today they are not targets for fishermen.

World fisheries catch of gobies is given as about 7,456–13,631 kg in the period 1978–1984, with almost all of this (over 95 per cent) being recorded from the Philippines (FAO, 1984). How much of this reported production of gobies was from diadromous species is not known.

13

Distance and Persistence in Diadromous Fishes' Migrations

The migrations of some fish species are legendary, both for the vast distances up rivers that fish migrate and also for their persistence in struggling upstream past barriers that impede their movement. A quote earlier drew attention to the salmon which 'fling[s] itself into the air again and again until it is exhausted in a vain effort to surmount a water fall' (Haslar, 1966), and it is undoubtedly the various salmon species that have drawn most interest and have captivated the attention of people from all walks of life. Professional scientists, anglers, naturalists and others all marvel at the migrations of salmon. Photographs of salmon jumping at falls appear on picture postcards as characteristic features of some parts of the world. While this interest is justified it would be wrong to leave the impression that salmon are alone in their persistence in reaching some distant and perhaps inaccessible locations.

DISTANCE UPSTREAM

Earlier chapters have related the distances that some species migrate. These distances need to be seen in the context of both the size of the fish that are making the migrations and of the characteristics of the rivers up which they are migrating .

Having sought to divert attention somewhat from salmons, it must be admitted, in terms of sheer distance upstream, that the various salmons attain the greatest distances. Figures of up to 4,000 km are on record for chinook salmon, *Oncorhynchus tshawytscha*. It is probably no coincidence that this is the largest of the salmon species. It is also of some interest that for the Atlantic salmon, *Salmo salar*, Schaffer and Elson (1975) have suggested that the size and age of salmon migrating up various rivers in eastern North America is related to the difficulties experienced in the upstream migration. Stocks that spawn on far distant upstream spawning grounds and/or which have to find their way up difficult and swiftly flowing rivers, were shown to comprise larger and older individual salmon than those which spawned nearer the ocean and/or in rivers up

Figure 13.1: Differences in the sizes of chinook salmon, *Oncorhynchus tshawytscha*, between North America and New Zealand were shown to be due to the New Zealand fish migrating mostly at 2–4 years and not reaching 5–6 years as they do in North America

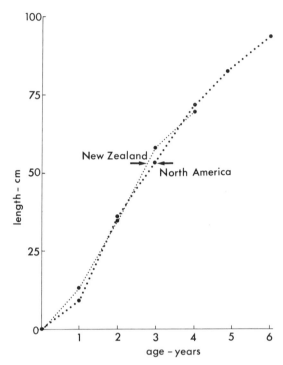

which passage is relatively less arduous. The interpretation is that large fish are stronger and more persistent swimmers and have better reserves of energy to achieve the upstream migration. They therefore have a higher chance of reaching the spawning grounds and are more likely to leave progeny, i.e. difficult rivers select for larger fish.

This question has not been examined for chinook salmon. However, there is some interesting, perhaps supportive evidence for Schaffer and Elson's (1975) hypothesis in what has happened to New Zealand's acclimatised stocks of chinook salmon. This species became established as an anadromous stock early in the twentieth century (McDowall, 1978a) but it was soon realised that returning fish were smaller than those in North America, from which continent they originated (Finlay, 1972). Study of New Zealand chinook salmon showed that they were primarily three- year-olds, with fewer four-year-olds, some two-year-olds, and only rare five-year-olds. None was older than five. By comparison, North American chinooks are frequently five, six or seven years old, occasionally eight (Figure 13.1). Thus the smaller New Zealand chinook salmon are also distinctly younger than their North American counterparts. It is possibly no coincidence that New Zealand rivers up which chinook salmon migrate are quite short, seldom more than 100 km from mouth to spawning grounds. In most of them, passage cannot be described as difficult. Possibly, selection for an optimal life history strategy, with regard to size and age at maturity and spawning, has led to a change in the age/size structure of chinook salmon in the New Zealand environment. Joyner *et al.* (1974) and Uchihashi *et al.* (1985) have suggested that the returning adults in New Zealand could be expected to be smaller on the basis that the oceanic system in which they move at sea is much shorter, bringing the fish back to the rivers sooner and at smaller size. However, it would seem that the critical issue would be age/size at maturity rather than the age/size at which the fish approach the coastline. Is there any reason why an earlier return would result in earlier maturity?

Although the greatest distances up which fish migrate into fresh water are achieved by salmons, various other salmonid fishes, as well as sturgeons and various clupeids also penetrate long distances inland. Examples include the Arctic Lena sturgeon, *Acipenser baieri*, which moves 1,300 km up the Yenisei River, the beluga, *Huso huso*, 1,000 km up the Volga River, and the kaluga, *H. dauricus*, 600–700 km up the Amur River, all of these in Russia. The omul, *Coregonus autumnalis*, is said to penetrate up to 1,500 km inland to spawn. While there are numerous diadromous clupeids, they seem to be less aggressive migrants than other families, though the hilsa, *Tenualosa ilisha*, is reported to penetrate more than 1,200 km upstream in some Indian rivers, and more than 700 km in Burma. Some alosid shads of the Caspian are vigorous migrants, *Alosa kessleri* penetrating 500 km inland and *A. pontica* more than 550 km (Whitehead, 1985). Even the chum salmon, *Oncorhynchus keta*, which has a reputation for having a relatively short migration (by comparison with the chinook — usually no more than 500 km) has been recorded 3,000 km up the Yukon River in sub-Arctic North America.

Much less well-known and reported are upstream migrations of adult lampreys, but the sea lamprey, *Petromyzon marinus*, is said to move over 300 km into rivers (Bigelow and Schroeder, 1953), the Pacific lamprey, *Lampetra tridentata*, over 400 km (Beamish, 1980b), and comparable distances are cited for southern hemisphere lampreys, like *Mordacia mordax* and *Geotria australis* in Australia (Potter and Strahan, 1968). Sjoberg (1980) mentioned a fishery for European river lampreys, *Lampetra fluviatilis*, 150 km up a Swedish river. Thus lamprey penetration of river systems may be considerable.

Migrations of the oxeye herring, *Megalops cyprinoides* — if properly designated a diadromous species — rank among the longest, with figures in excess of 700 km in the Sepik River in Papua New Guinea (Coates, n.d.).

From these high extremes there is a continuum of distances, with many species making migrations of 10 to 100 km, and many more moving only a few kilometres, often penetrating just upstream beyond tidal influence. Among the last are several 'classic' anadromous fishes like some populations of chum and pink salmon, *O. keta* and *O. gorbuscha*, as well as other much lesser known fishes — osmerids, salangids, retropinnids, eleotrids, etc.

Peak distances for species do not, of course, represent a norm, and to high-light this it should be noted that some of the most long-distance migrants like Atlantic and chinook salmon and the coregonid omul, are also known to have populations in which spawning takes place only a few kilometres above river mouths. Thus, huge distances accomplished in migration are not a 'species imperative', but rather opportunism by stocks of the species to utilise suitable habitats, even though difficulty of access results, possibly, in a great cost in energy consumption and perhaps mortality. Attention tends to be focused on the larger and better-known species with regard to spectacularly long migrations and this misrepresents, in some measure, what does take place.

Although small species cannot compete with distances achieved by salmonids, it is nevertheless known that upstream migrating juveniles of some diadromous species can attain considerable distances from the sea and elevations above sea level. These distances and elevations are generally achieved by amphidromous species, and in terms of life history strategy, this is an important distinction. The upstream migrations, for instance, of some diadromous galaxiids may be as far as 180 km, but are accomplished by small juvenile fish often less than 75 mm long that are feeding and growing as they move. The most penetrative galaxiid is the koaro, *Galaxias brevipinnis*, from Australia and New Zealand. This fish leaves the sea as a juvenile about 50 mm long, and within a day or two of leaving the sea (and only a few kilometres upstream from the sea) the migrants have begun to settle and feed in gravelly river riffles. They move upstream, feeding and growing, over a period of several months. The same is true of the various catadromous *Anguilla* species. The New Zealand longfinned eel, *Anguilla dieffenbachii*, attains distances upstream of 150 km, but this may take several years and be accompanied by extensive feeding. This sort of behaviour is reflected in the length/frequency distributions of populations of diadromous fishes from various locations along the length of a river system. There tends to be a broad range of sizes from juveniles upwards at lower elevations and a distinct tendency for the smaller, juvenile stages to be increasingly sparse in the populations with increasing distance upstream. The extent to which size (age) structure changes can be interpreted as indicating the time it takes the fish to penetrate river systems. This has been shown in two New Zealand amphidromous species, such as *Gobiomorphus huttoni* (Eleotridae) and *Cheimarrichthys fosteri* (Mugiloididae) (McDowall, 1965, 1973; Figure 13.2), and also in the Puerto Rico goby, *Sicydium plumieri* (Erdman, 1986).

PERSISTENCE

Another aspect of the migration of diadromous fishes that has attracted attention is their persistence and ability to make their way upstream past barriers of quite prodigious difficulty. Again, this attention has been focused largely on big salmon leaping repeatedly and often successfully at falls and foaming river

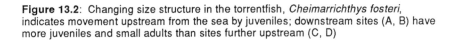

Figure 13.2: Changing size structure in the torrentfish, *Cheimarrichthys fosteri*, indicates movement upstream from the sea by juveniles; downstream sites (A, B) have more juveniles and small adults than sites further upstream (C, D)

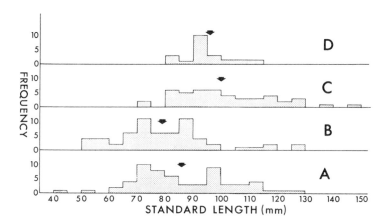

torrents. This is no surprise as the features of the rivers themselves are often of scenic attraction, and the additional value of large and beautiful fishes battling up such falls can only generate admiration.

Much less well recognised, again, are the abilities of often smaller and lesser known fishes to make their way past barriers. Very often these are the juvenile stages of amphidromous or catadromous fishes, amongst which the climbing migrations of eels are the best known. Up to a size of about 120 mm (weight about 5 g) elvers of *Anguilla* species are able to adhere to damp, vertical surfaces by surface tension (Jellyman, 1977b) and this enables them to climb such surfaces. To do this they leave the water flow, adhere to the damp substrate, and make their way up the surface by lateral, sinuous movements of the body, which is kept closely pressed to the surface. Skead (1959) described the movements of these small eels as:

> never the smooth, sinuous glide of a snake, but ... a series of jerky forward movements as the eel scrambles forward. The jerky, uninterrupted action is due to the necessity for the eel to keep some portion of its ventral surface attached to the climbing medium, while the other portion is advancing, and therefore out of firm contact.

There seems to be no theoretical limit to the height that the young eels can climb, and in practice vertical, or near-vertical surfaces of great height are climbed. Waterfalls of 60 m or more in height in New Zealand are climbed, as well as several dam faces that are up to 100 m in height. Graham (1953) described elvers climbing the face of the Karapiro Dam, on the Waikato River in New Zealand, as follows:

There was very little water coming over the spillway and the elvers appeared to prefer the concrete where there was no flow of water, only damp concrete. We saw young eels from three to five inches long — literally millions, trying to climb up the spillway of the dam — they must have climbed sheer rock and we saw thousands that had perished in the dust on the temporary bridge facing the spillway. We were told that what we saw was nothing compared with what struggled up during the night.

This ability is, no doubt, a general ability of migrating juvenile anguillid eels, and is reported in Australia by Merrick and Schmida (1984), who described the Australian species, *A. reinhardti* and *A. australis*, as able to 'negotiate enormous obstacles'; Sloan (1984a) described movement of *A. australis* up dam faces in Tasmania. Jubb (1967) found few eels in the Zambesi River above the Victoria Falls, but these falls are very high, and that any eels can climb past such an apparently formidable set of falls is itself remarkable. Elsewhere in Africa, Skead (1959) has described the passage of juvenile eels up a 30 m vertical concrete face alongside the outflow of a dam.

Predictably, anguillid eels are not alone in their climbing ability, which is shared in equal measure by some diadromous galaxiids. Best known of these is the New Zealand-Australian koaro, *Galaxias brevipinnis*, known by some in Australia as 'climbing galaxias' (Merrick and Schmida, 1984). Other New Zealand species, such as banded kokopu, *G. fasciatus*, and shortjawed kokopu, *G. postvectis*, have similar climbing skills, and as a result, some of these have very wide inland distribution above steep and very substantial barriers and falls. Populations are known above falls 60 m or more high. Climbing is accomplished in much the same way as was described for juvenile eels — the small fish, 50 perhaps 70 mm long, adhere to damp surfaces by surface tension; this is assisted in the galaxiids by their quite expansive and ventrally adpressed pectoral and pelvic fins, which both increase the ventral surface area of the fish and also offer some purchase against irregularities in the surface being climbed. Juvenile galaxiids climb damp glass surfaces with little evident difficulty. The broadly spread fins in *G. brevipinnis* have given rise to further common names — in Victoria, Australia it is called 'broad-finned galaxias' (Cadwallader and Backhouse, 1983) and in New Zealand 'elephant's ears' (McDowall, 1984c). Landlocked populations of *G. brevipinnis* are known from many inland and alpine New Zealand lakes and tarns, some approaching 1,500 m altitude. In some of these the access route to the lakes is far from obvious. Past geological events could explain invasion of some of these waters, but in others, it must have been by climbing very steep and obviously ephemeral overflows from lakes, possibly no more than seepages down rock faces below the lakes.

A fascinating example of persistence in migration was described at a velocity barrier designed to prevent the upstream movement of fish in a hydro-eletric diversion system in New Zealand. This barrier was constructed at the point at which a river diversion canal entered a lake. The barrier was constructed so that water flows very swiftly down a steep, smooth incline, the velocity of the water preventing fish moving upstream, However, migratory juveniles of koaro were observed circumventing this by leaping out of the water, adhering to the damp vertical concrete lateral walls of the structure, and wriggling up the damp concrete in the splash zone, above water level (E.J. Cudby, personal communication; McDowall, 1978a, Figure 13.3).

Adults of the southern pouched lamprey, *Geotria australis*, are known to surmount barriers to upstream migration of some magnitude. Where barriers are very difficult, they may work their way upstream along the margins of the stream flow. If this is not possible they can negotiate rocky falls by attaching themselves to rocks using the oral disc, beating the tail vigorously and driving the sucker forwards without detaching it from the rock. Jellyman (1984) suggested that the power of suction is so strong that a lamprey attached to a rock weighing up to 5 kg can be lifted clear of the water with the rock attached.

Climbing features in descriptions of several amphidromous eleotrids and gobiids. Maciolek (1977) discussed the juveniles of the goby *Lentipes concolor* climbing waterfalls 100 m high and series of falls up to 300 m high. Lake (1978) found that the Australian species *Gobiomorphus coxii* can climb up vertical slopes of damp concrete weirs using its large pectoral fins; juveniles of some New Zealand species of *Gobiomorphus* are also able to migrate upstream past low weirs. *Sicydium plumieri* is said to climb a dam 3 m high and reaches an altitude of at least 330 m (Erdman, 1986), and *Sicyopterus japonicus* is recognised as a skilled climber in Japan (Fukui, 1979); a related species, *S. stimpsoni*, in Hawaii, has a native name 'oopo napili, which means to cling, and presumably refers to the ability of this fish to cling to vertical surfaces, facilitating passage upstream over falls and similar obstructions (Tomihama, 1972). In Ceylon *Sicyopterus gymnauchen* is said to make violent attempts to ascend the side of a basin when water is poured down, suggestive of a propensity for climbing (Deraniyagala, 1937). A further Hawaiian goby, *Awaous staminius*, reaches an altitude of about 500 m only 13 km from the sea (Ego, 1954)

Persistence in downstream migration is less reported, but there are situations, again in anguillid eels, where fish migrating to sea have had to make their way across land to reach the sea. Obviously, when eels find their way across land into isolated water bodies, they must make their way back across land into fluvial situations if they are to return to sea. Another circumstance involves formation of sand/gravel bars across the outflows of coastal lakes, lagoons and rivers. Instances are well known in New Zealand in which the seaward migrations of eels in autumn have been interrupted by such sand bars (Hobbs, 1974); in such cases eels are known to migrate across the sand bar from lake to sea when, at high tide, waves break across the sand and keep it moist. A breaching of the sand bar is not necessary in induce this migration, and vast numbers of adult eels can be observed, at times, moving out of the water and across the sand.

Figure 13.3: The climbing skills of the koaro, *Galaxias brevipinnis*, allow them to migrate upstream past this velocity barrier on a diversion canal by jumping out of the water and climbing along the splash zone above the water flow (see inset)
Source: E. J. Cudby

LOCATION OF FRESH WATER

A feature of the migrations of fishes from the sea into fresh water that has captured little attention is their ability to locate extremely small streams that flow to sea across sandy or gravel beaches. Some such streams may be little more than seepages which virtually disappear into beach gravels; and yet migratory fishes are able to locate this fresh water and are observed to migrate into such streams.

The presence of adult anguillid eels in totally isolated water bodies is a widely reported phenomenon in many countries, and this has led to early conclusion that eels must breed in fresh water (e.g. Hall, 1905). However, it has also long been known that this involves the ability of eels to migrate across land. Graham (1953) described eels:

in a concrete water tank some chains from a creek ... This meant that these Eels had to leave the creek and traverse the long grass of an orchard and plantation and finally to crawl over the dry rubble which had been dug out to form the concrete tank on a small hill.

What prompts eels to leave water bodies and evidently roam around on land among damp grasses, and how they are able to locate other water bodies, are questions that remain unanswered; but that these things take place is quite clear.

14

The Conservation Status of Diadromous Fishes

Like all biological species, diadromous fishes have been affected adversely by the diverse activities of man — man as an exploiter of the fish themselves, and man as a co-inhabitant of the biosphere. Water, particularly flowing, fresh water, has always been seen by man as a very useful and valuable resource, but in many instances, the uses to which man has put these waters has been incompatible with their role as fish habitats.

Causes of decline in abundance of these fish species are varied and often cumulative. Fish populations have suffered from overexploitation as man has sought fish as food, and increasingly so as he has developed more sophisticated and effective ways of catching and preserving them. For many species overexploitation of the populations by man has been a fundamental cause of population decline, and has been well described. Mention has already been made (Chapter 1) of the fact that the concentration of diadromous fishes in space and time makes them particularly easy to harvest in large numbers. This same feature renders them very vulnerable to overexploitation. The ease with which they can be captured extends to an extreme where relatively simple weir/trap structures, which span a river mouth, can stop entire migrations of fish. Careful management of diadromous fish stocks at exploitation is needed to prevent catastrophic decline (such as was reported for the Tasmanian whitebait, *Lovettia* seali — Blackburn, 1950; see Chapter 12).

Natural fish communities have in some instances been severely disrupted by the introduction of exotic predators and competitors; the result has often been a reduction in the numbers of native species and their abundance in those habitats. Habitat deterioration is a further very general cause of fish population decline, the effects of which often combine with, and cannot be distinguished from, overexploitation. Fish habitats have been severely disrupted by changing land-use strategies. In particular this is true of deforestation leading to elevated water temperatures, increased flow variability, bed and bank instability, sedimentation, and other effects. Some species seem to require forest cover as a habitat characteristic for populations to thrive, quite apart from the secondary effects of forest removal on habitat quality itself.

Water pollution has been widely cited as a cause of habitat deterioration. It can result from discharges of waste from human population centres or industrial activities. Chemical waste discharges, especially from wood-based industries, have been particularly influential. Nutrient run-off from pastoral lands has resulted in eutrophication and thus changed habitat quality for fishes. More recent and insidious is the impact of acid rain, which has been felt in areas far removed from where the causal polluting discharges were released, and which has reduced the pH levels in many watersheds in Europe–Scandinavia and northeastern North America. Fish populations have been extirpated from many catchments where pH levels are so low that the fish can no longer inhabit the waters or reproduce there.

Construction of low barrage weirs and high impoundments has changed upstream habitats from flowing rivers to lakes, and downstream habitats have become subjected to changes in flow patterns. Often there are gross, short-term fluctuations in water discharge and level, and sometimes flows below dams cease, either temporarily or permanently. Water temperatures below dams may differ from those above them, or from those which prevailed before impoundment.

In rather fewer localities, but still a major cause of concern, has been water abstraction — the removal of water from waterways, and its diversion to metropolitan water supplies or for irrigation. Sometimes water flow is only intermittent, sometimes it ceases altogether. This has led to extensive habitat loss, and consequent population decline in some areas of the world. Manipulation of river channels by engineers and hydrologists leads to altered flow patterns, less diverse habitats changes to bank character, and loss of cover for fish.

These problems apply more or less equally to diadromous and non-diadromous fishes as regards occupation of habitats. The diadromous fishes are also particularly affected by some of the above, because their migrations are easily interfered with by changes in habitat qualities, and they pose some special and difficult conservation problems. Because they migrate up and downstream, often for very long distances, diadromous fishes are badly affected by anything that interrupts access to and from their various habitats. In its simplest terms, the problem is fish passage. This is customarily thought of in terms of the construction of impoundments across river systems which tend absolutely to prevent upstream movement of entire migrant populations (though a few fish like anguillid eels and some galaxiids and gobies can climb dams, see Chapter 13). But dams also drastically obstruct downstream movement of fish, if not entirely preventing it. And furthermore, impoundments are neither the only nor necessarily the most important impediment to fish migrations. Discharges of pollutants may deter or prevent fish from moving upstream or downstream through discharge plumes. Heated water, discharged from coal- or gas-powered electricity generators, can similarly create conditions through which fish will not migrate. Dewatering of river channels resulting from water abstractions is also fatal to fish migration. While these are the most obvious factors

Figure 14.1: Culverts, if not making allowance for fish passes, prevent upstream movement by diadromous and other migratory fishes

contributing to decline in diadromous fishes, other factors are also operative. Construction of culverts (Figure 14.1) and aqueducts to carry water over or under roads frequently prevents fish movement if the design of discharge structures does not take fish migration needs into account (and typically they don't). Flood gates are built at river mouths to prevent tidal water from flooding back across low-lying estuarine margins and these stop fish from moving upstream. In order to move past barriers, some fish need to be able to swim upstream along the stream bed amongst rubble for cover; others are better able to climb vertical surfaces but need a moist surface to move up, and may not be able to cope with overhangs. Some species are jumpers, but the height they need to leap may be too great, the pool below the jump may not be sufficiently deep, or the hydraulics of the pool may not allow the fish to generate enough

250

velocity to make the leap. Taking all these potential threats into account,. it is no surprise, then, that some diadromous fishes are a cause for concern amongst ichthyologists and conservationists. For some species, aspects of their life history strategies further accentuate the impacts of habitat changes. One such feature is that many diadromous fishes are obligatorily diadromous. When such fish are prevented from making their way to and from the sea, populations inevitably and rapidly become extirpated. Another feature is related to the vast distances that some species migrate up river. Obviously, the greater the distance, the greater is the probability that some barrier to migration will be encountered in the river system. This has been graphically displayed in such rivers as the Connecticut River in the eastern United States, and the Columbia in the west. Very early in the settlement of the United States, migrations of diadromous fishes like lampreys, sturgeons, salmons and smelts were truncated by dams constructed on the Connecticut River to provide power for mills. This prompted early but largely ineffective action to restore migrations by the provision of fish passes. In the Columbia River hydro-electricity dams truncated the huge populations of Pacific salmons; this, too, led to installation of fish passes for upstream migration. These were large, expensive and often of limited effectiveness, and also paid no heed to the need for the young salmon to move downstream again, to return to the sea. Fisheries managers discovered that getting fish upstream past dams to reach the spawning grounds, though difficult and expensive, only exposed the further problem of getting their progeny downstream again. This has proved to be both more difficult and more expensive.

Perusal of the fish literature on conservation, and on rare and endangered fishes, shows that virtually all of the potential impacts discussed above have played a role in contributing to the decline of diadromous fishes. It is perhaps surprising that the impacts have not actually been greater than those reported; even so, they have been serious enough to generate concern and to provoke vigorous action to mitigate against their impact. Examination of the International Union for the Conservation of Nature (IUCN) World Red Data Book for Freshwater Fishes (Miller, 1977) shows that only seven diadromous species are listed under their four categories, as follows:

1. Endangered — taxa in danger of extinction and whose survival is unlikely if the causal factors continue operating.

Acipenser brevirostrum (family Acipenseridae) — east coast of North America.

Stenodus leucichthys (family Salmonidae) — the nominate subspecies in Russia.

Prototroctes oxyrhynchus (family Prototroctidae) — New Zealand.
Lentipes concolor (family Gobiidae) - Hawaii.

251

2. Vulnerable — taxa believed likely to move into the endangered category in the near future if the causal factors continue operating.

Acipenser oxyrhynchus (family Acipenseridae) — east coast of North America.

3. Rare — taxa with small world populations that are not at present endangered or vulnerable, but are at risk.

Huso dauricus (family Acipenseridae) — USSR.

4. Indeterminate — taxa that are suspected of belonging to the first three categories but for which insufficient information is currently available.

Galaxias postvectis (family Galaxiidae) — New Zealand.

In addition, several non-diadromous subspecies of diadromous species, particularly amongst the Salmonidae, but also including a stickleback, were listed by IUCN (Miller, 1977) but are not relevant here. Thus, overall, only seven of about 175 full species listed in the World Red Data Book are diadromous.

The completeness of the Red Data Book as a listing of freshwater fishes in various states of endangeredness is obviously open to question, and the list itself is in need of frequent revision, a point made for Russian species by Pavlov, Reshetnikov, Shatunovskiy and Shilin (1985). They noted that the first edition included no Russian species, the second nine Russian species, and indicated their intention of adding an additional 31 species in the next edition of the book. For many regions of the world, no detailed analyses and listings of the faunas have been prepared. In some areas, although moderately well-known faunas exist, no contribution to the World Red Data Book has yet been made. An example is Australia, which has no fish listed at all. That this does not reflect the status of the fauna is exemplified by the holding of a conference in Melbourne, Australia, in August, 1985, to establish criteria for and prepare a list of Australian freshwater fishes that are in various ways subject to threat, endangered, etc.

In addition to the World Red Data Book listings of species, some countries have prepared official or unofficial listings of such species for their own faunas, in which categories for listing and criteria for inclusion may differ slightly to markedly from those used by the IUCN World Red Data Book. To some extent locally, rather than globally, oriented lists focus on the status of a species within the geographical perview of the region under examination, so that, for instance, the status of a species in Canada, as listed by McAllister, Parker and McKee (1985) may be different from that in the United States (Miller, 1972b). Nevertheless, local listings are legitimate descriptions of local status and are of value for that reason alone. The following review is based on various of these regional listings, and while undoubtedly incomplete, provides a perspective on the conservation status of the diadromous fishes of the world. The listing is in order of 'endangeredness'.

EXTINCT SPECIES

Only one diadromous species is regarded as extinct – the New Zealand grayling, *Prototroctes oxyrhynchus*, listed as Endangered by IUCN (see above), but in fact probably extinct for many years. There have been no authenticated reports from reliable observations since the 1920s (McDowall, 1978a). It was extremely abundant during the early years of European settlement of New Zealand (1850s–1870s), but by the 1880s it was being described as in decline and 'by no means common' in some parts. The last definite record was in 1923, when several were caught in a trap set to demonstrate the manner in which they had primitively been caught by the indigenous Maori people of New Zealand (Phillipps, 1923; McDowall, 1978a). By that time, Phillipps described the fish as 'rarely seen except in isolated streams and rivers remote from settled places'. There were occasional subsequent reports, but it seems that by the late 1920s this formerly abundant fish had disappeared, largely if not altogether, from New Zealand waters.

In spite of early recognition that the New Zealand grayling was in decline, it was not until 1952 that this fish was offered legislative protection, by which time it was far too late, and Stokell (1941) described its disappearance as 'a standing reproach on the administration of Wildlife in New Zealand and a monument to the indifference with which many natural resources ... have been treated'.

The sudden and unfortunate decline and extinction of the grayling has never been explained satisfactorily. There were early suggestions that the decline was due to the introduction of brown trout, *Salmo trutta*, into New Zealand (Rutherfurd, 1901). But Allen (1949) correctly pointed out that the fish had disappeared from waters where trout had not been introduced (although they may have very rapidly spread through the sea to most New Zealand river systems). The other likely cause would seem to have been changes that occurred as man began to clear the forests and establish pastureland. However, again, grayling disappeared from large areas that remained unmodified by man long after the fish had disappeared. Possibly the fish was already in a state of some decline when man intervened; perhaps its very specialised dentition and herbivorous feeding habits (McDowall, 1976) made it highly sensitive to habitat disruption. Removal of forest cover undoubtedly affected the occurrence of encrusting algae on rocks in the river systems and this may have been crucial. But little is known and no conclusive and satisfactory explanation seems possible.

Figure 14.2: The shortnosed sturgeon, *Acipense brevirostrum*, regarded in parts of eastern North America as endangered
Source: McAllister *et al* 1985

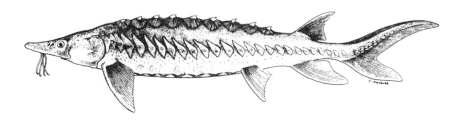

ENDANGERED SPECIES

The western North American shortnose sturgeon, *Acipenser brevirostrum* (Figure 14.2), was listed by Miller (1972b) as endangered. The United States Department of the Interior also listed it as endangered, and believed it to be seriously depleted throughout its range. It was classified as endangered by IUCN (Miller, 1977), was nominated as a 'runner-up' in a listing of the 24 most endangered of species of all classes, in the IUCN Newsletter No. 5, 1985, and was the only fish in that listing. However, up-to-date information suggests that this classification might be in need of review. Since the shortnosed stugeon was listed as endangered, several studies (e.g. Dadswell, 1979) have suggested that at least in some parts of its range this fish still occurs in substantial numbers and is, in fact, still taken in net fisheries along the Canadian–American coastline and rivers entering the sea there. Certainly, it is much reduced in numbers in mid-latitude states of the United States, but Dadswell (1979) showed that in the St John River in New Brunswick (Canada), it has a population of possibly 100,000 individuals including 18,000 adults. It could not, on that basis, be described as endangered.

Suggested causes of the decline of the shortnose sturgeon are varied, but include such factors as impoundments impeding migration upstream to spawn, habitat deterioration and pollution that accompanies increasing human population densities and industrialisation in the river catchments, and overexploitation of the stocks of the species by fishermen, either by targetted fishing for sturgeon, or as a bycatch in fisheries for other valuable species. Miller (1977) considered the decline to be caused 'chiefly by pollution but also probably from overfishing'. Sturgeons are highly valued for the flesh and for their roe, and as they are very slow-growing and long-lived fishes that are late maturers, they are very prone to overexploitation. Gorham and McAllister (1974) pointed, particularly, to river pollution connected with wood processing industries and municipal sewage discharges. Taubert (1980) noted that dams built as early as 1798 on the Connecticut River (USA) blocked passage by this and other migratory species, and reported that fish pass facilities built to assist fish passage in 1873, 1933, 1940 and 1951, all failed to achieve their objective. Another built in 1955 was successful, but Taubert found that only 26 sturgeon had been assisted upstream by the fish pass since 1955 (Taubert and Dadswell, 1980). Its effectiveness appears to have come too late to be of much value.

Dadswell (1984) concluded that:

> populations in each river system are larger than formerly believed. There is no evidence to suggest that Shortnose Sturgeon are increasing or decreasing. If properly managed, certain Shortnose Sturgeon populations could support small, gourmet-item fisheries. ... the population ... in the Saint John River is at, or near, the carrying capacity for the habitat available.

Although there can be no doubt that the shortnose sturgeon has suffered massive decline in numbers in some parts of its geographical range, its classification as endangered probably cannot be justified; Dadswell (1984) suggested that it 'could be removed from the US endangered species list in the next five years'.

Lentipes concolor is a small amphidromous gobiid fish from the Hawaiian Islands. It was listed as 'rare and endangered' by Miller (1972b) and 'endangered' by IUCN (Miller, 1977). It has disappeared from many of the streams in which it was once abundant; Miller (1977) showed that it was then present in only 6 per cent of the perennial streams of Hawaii. It had not been recorded from the island of Oahu for several decades. Miller stated that exotic fishes now dominate the faunas of these streams, and considered that stream modification and degradation had contributed to its decline. It is described as a species which shuns modified habitats (Maciolek, 1977).

VULNERABLE SPECIES

The Atlantic sturgeon, *Acipenser oxyrhynchus*, is a further western North Atlantic species described by Miller (1972b) as threatened in most eastern states of the USA, and as vulnerable on a world basis by IUCN (Miller, 1977). Miller considered it possibly extinct in Maine (USA), but McAllister *et al.* (1985) did not list it as being in any way under threat in Canada. Miller (1977) gave as reasons for reduction in its abundance, a combination of overexploitation, pollution and prevention of upstream migration by the construction of dams. Scott and Crossman (1973) said that numbers had declined seriously in the United States but thought that it remained as abundant as ever in Quebec (Canada) waters. They attributed decline to exploitation of juveniles, but also implied that a decline in interest in the species as a food has contributed to concerns that the fish is less abundant than it once was.

The Caspian/Russian sturgeon or kaluga, *Huso dauricus*, is also included in the category 'vulnerable' by IUCN. Miller (1977) attributed its reduced abundance simply to 'overfishing'. Although the population is probably only 'more than 2000 individuals' (Miller, 1977), Pavlov *et al.* (1985) pointed out that this species is not regarded as being under such serious threat by the Russians as it is not included in the 'Red Book of the USSR'. Nor did they suggest that it should be amongst 31 species that they proposed as additions to the USSR Red Data Book for later editions.

The inconnu, *Stenodus leucicthys*, has a very wide distribution in the northern cold temperate of Eurasia and North America. The nominate subspecies, *S. leucichthys leucichthys*, in the Caspian Sea, was described as vulnerable by IUCN (Miller, 1977) and its decline attributed to dams preventing its access to the spawning grounds. Numbers were thought to be less than 2,000, but again, this species was not regarded as endangered or threatened by the Russian authorities (Pavlov *et al.*, 1985).

SPECIES OF INDETERMINATE STATUS

In New Zealand the amphidromous shortjawed kokopu, *Galaxias postvectis* (Figure 14.3), is listed by IUCN (Miller, 1977) as 'indeterminate', a listing supported by the NZ Red Data Book (Williams and Given, 1981); this classification resulted from very limited knowledge of its abundance and natural history, and the fact that it was known only from small numbers of fish over a wide geographical range. Although it is a constituent of a fishery there is no reason to

Figure 14.3: The short jawed kokopu, *Galaxias postvectis*, a New Zealand species in severe decline, probably owing to the impact of land management practices on aquatic habitats

believe that it has suffered a serious decline in abundance as a result of exploitation. However, what is known of this species indicates that it is found only in unmodified and heavily forested catchments; it seems to be very vulnerable to habitat disturbance and modification, and its survival is clearly dependent on the protection of such habitats. Although knowledge of the species' natural history has not increased much in the past few years, intensive collecting in some highly remote and unmodified catchments has shown the species to occur in reasonable numbers over a quite wide geographical range.

SPECIES OF REGIONAL CONCERN

In addition to species listed in the IUCN international register of endangered fishes, some national agencies or individuals have prepared listings of species that are regarded as under some threat within their territories. Some of these accounts are both more up-to-date and more authoritative than the IUCN listing. The categories under which species are listed vary quite widely between accounts, so that the grouping of species according to categories is not easily possible and direct comparisons are difficult. Further, the term 'status' refers to that within restricted territories rather than on a global scale. Nevertheless, some discussion of the conservation status of various diadromous species is appropriate here, in addition to those species regarded as sufficiently under threat to be included in the international IUCN listing. Owing to the variable manner in which accounts have been presented, such species are listed here in a phylogenetic sequence.

Lampreys feature in some accounts, although most of them are non-diadromous species. Pavlov *et al.* (1985) included the Caspian lamprey, *Caspiomyzon wagneri*, in their list of 31 additional Russian freshwater fishes to add to the Russian Red Data Book but gave no details of the status of this fish, nor of reasons contributing to its decline.

Sturgeons have already featured strongly in discussion of status, and further species have had attention focused upon them especially by the Russians. One of these is the Atlantic sturgeon, *Acipenser sturio*. The Russian Red Data Book listed it, and Pavlov *et al.* (1985) regarded it as 'under threat of extinction'. They provided no explanation, although Wheeler (1969) described the Atlantic sturgeon as a 'rare visitor to northern European waters', where it has 'practically been exterminated'. In his list of rare British fishes Maitland (1979, 1985) included this species, but since it is not known to breed in British waters its rarity there cannot be attributed to a decline in abundance. A subspecies of the Sakhalin or green sturgeon, *A. medirostris*, was also listed but again no explanation was given by Pavlov *et al.* Russian ichthyologists have a strong tendency to recognise subspecies in their fauna, so that the significance of this listing is difficult to interpret. North Americans do not seem to share the Russians' concern for *A. medirostris* (Miller, 1972b; Scott and Crossman, 1973). A third

257

Russian species discussed was the ship sturgeon, *A. nudiventris*, which occurs in the Black and Aral Seas. Pavlov *et al.* (1985) described one subspecies as under threat. The white sturgeon, *A. transmontanus*, of the Pacific coast of North America, is said to be threatened in the inland states of Idaho and Montana (Miller, 1972b), perhaps because it can no longer reach these areas owing to the construction of dams in the major river systems. Miller (1972b) also listed the Atlantic sturgeon, *A. oxyrhynchus*, as threatened in most but not all Atlantic coast states of the USA, from Maine to Georgia.

The only anguillid that seems to feature in lists of threatened species is the American eel, *Anguilla rostrata*, which is a cause for concern in the distant inland states of the USA — Wisconsin and Dakota (Miller, 1972b) — again presumably because of access difficulties.

Among the salmonids one listed species is the Acadian whitefish, *Coregonus canadensis*, which occurs in Nova Scotia, Canada. McAllister *et al.* (1985) noted that its range is now very restricted, but it only ever occurred in a small number of catchments in southern Nova Scotia. Edge (1984) found that its habitat in one of these rivers had been affected by dams without fish passes constructed over the past 150 years, and in another it has been affected adversely for over 60 years. However, he regarded the most serious threat to this species to be the occurrence and impacts of acid rain. In addition it has been 'ruthlessly exploited' (McAllister *et al.*, 1985). It is now so rare (presumably both in the wild and in museum collections) that McAllister et al. (1985) asked that any moribund specimens be preserved and sent to natural history museums for study and long-term storage. Edge (1984) said that it is 'an endangered species requiring immediate attention in order to prevent its extinction'. He suggested that some rivers in Nova Scotia with 'good acid-buffering capacity ... might prove suitable for transplanting anadromous Acadian Whitefish in order to prevent their extinction'. However, one might ask why it has not already made itsN own way there through the sea.

Various of the salmonids were listed as threatened for some states of the USA (Miller, 1972b), without being a cause for general concern, e.g. Atlantic salmon, *Salmo salar*, from Maine to New York, brook char, *Salvelinus fontinalis* in Ohio, dolly varden, *S. malma*, in Oregon, cisco, *Coregonus artedii*, in Ohio and Pennsylvania (probably not diadromous stocks), and lake whitefish, *C. clupeaformis* in these two states and also New York (again, probably not diadromous stocks). Two British *Coregonus*, the whitefish, *C. lavaretus*, and the pollan, *C. autumnalis*, were listed as rare by Maitland (1979, 1985), and *C. lavaretus* was also a matter of concern in Sweden (Nilsson, 1975). Nilsson blamed pollution and overfishing.

In addition to these species, concern about salmonids relates primarily to diverse landlocked stocks/subspecies of diadromous species, for which there is wide concern with regard to quite a variety of distinctive stocks (Miller, 1972b; McAllister *et al.*, 1985; Pavlov *et al.*, 1985). Rainbow smelt, *Osmerus mordax* (family Osmeridae) are a cause for concern in the American state of

New Jersey.

Michaelis (1985a) discussed the diminutive Tasmanian aplochitonid, *Lovettia sealii*, in her account of rare or threatened species from the inland waters of Tasmania, but assigned it no 'rarity value'. Although once the target of a substantial fishery which collapsed dramatically under overexploitation in the 1950s, this species is not and has never been under threat of extinction. Consideration is presently being given to reopening the fishery, but recent comment indicates that this is unlikely in the short term (Anon., 1986).

Williams and Given (1981) added the giant kokopu, *Galaxias argenteus*, to the New Zealand listing in the World Red Data Book on Fishes (Miller, 1977), and classified it as under possible threat. The species has suffered very extensive habitat loss associated with deforestation, swamp drainage and the establishment of agricultural pasture; it has probably also declined in abundance on account of the introduction and widespread success of brown trout, *Salmo trutta*, in New Zealand (McDowall, 1978a).

The Australia grayling, *Prototroctes maraena*, has long been regarded as in serious decline (Lake, 1971; McDowall, 1976) and was listed by the Australian Council of Conservation Ministers as 'rare'. Michaelis (1985a) considered that its decline was related to habitat alteration and dam construction, and (1985b) noted that it has long been scheduled as a protected species in Tasmania. A meeting of the Australian Society for Fish Biology held in Melbourne during 1985 to address the question of the conservation status of Australian freshwater fishes gave the Australian grayling detailed consideration. On the basis of increased knowledge (deriving in some measure from the studies of Berra (1972) in a Victorian river), the species was removed from the list of Australian rare and endangered fishes.

The two shads present in British fresh waters, the allis shad, *Alosa alosa*, and the twaite shad, *A. fallax*, were both listed by Maitland (1979, 1985). The twaite shad is also threatened in Sweden (Nilsson, 1975). Wheeler (1969) described the allis shad as so scarce that only small local fisheries can exist to exploit it. Causes of the rarity of these species are probably mostly the construction of dams and impoundments that obstruct movements up and down stream. Several of the North American shads are regarded as under threat in some states — the Alabama shad, *A. alabamae*, in Missouri, the skipjack herring, *A. chrysochloris*, in South Dakota, and the American shad, *A. sapidissima*, in Delaware, New Jersey and New York. In southern Africa the freshwater mullet, *Myxus capensis*, is described as threatened by Skelton (1977, 1983). Jackson (1960) attributed its decline to harmful interactions with trout.

Both diadromous sticklebacks are regarded with concern in Massachusetts, USA, both *Gasterosteus aculeatus* and *Pungitius pungitius* being described as 'rare' (Miller, 1972b). The gadid tomcod, *Microgadus tomcod*, is threatened in the American state of New Jersey although not elsewhere (Miller, 1972b). Several Hawaiian gobies (in addition to *Lentipes concolor* — see above) were listed by Miller (1972b). *Awaous stamineus* is regarded as depleted, *L. seminu-*

dus as rare, and *Sicydium stimpsoni* as rare, the populations on the island of Oahu being the most seriously affected. Another goby, *Awaous tajasica* is rare in Florida (Yerger, 1978).

In addition to species about which international, national, or local concern has been expressed with regard to their conservation status, a large proportion of the diadromous species has sustained population decline as a result of a wide range of factors.

IMPACTS OF EXPLOITATION

Virtually any diadromous fish species that is the target of a fishery, from Newfoundland to New Zealand, has declined in abundance as a result of exploitation. This decline varies with location and the accessibility of the stocks, and their proximity to large population centres. However, even highly remote populations have also suffered severely in some cases. Decline varies with the cultural/economic value of the fish species, and it varies, also, with the characteristics of the species' life cycle, particularly with regard to the distance the species migrates upstream and its vulnerability to habitat disruption. And it varies with regard to the intensity of forestry/pastoral/industrial development in the species' catchment. The literature is extensive on the decline of species such as the sturgeons and salmons, large species of high economic value. But often much less widely publicised diadromous species like the Hawaiian goby, *Lentipes concolor*, and the New Zealand galaxiids, have suffered equivalent declines in abundance. A review of these occurrences would be both very prolonged and of no great value here.

Throughout the world, diverse problems, acting singly or together in various ways, have led to a reduction in the abundance of diadromous fishes, the futures of many species being in some doubt unless concerted efforts are made to reduce harmful impacts. Pavlov *et al.* (1985) believed that in Russia 'practically there is not a single large water body which has not been affected by the economic activity of man to some extent'. While this, fortunately, is not true on a global scale, it is nevertheless widely enough true to pose a major threat to the continued existence of a great many fisheries based on diadromous species, and, in a significant number of instances, the survival of the species themselves.

References

Aboussouan, A. (1969) Note sur les 'bichiques' de l'ile de la Reunion. *Recueil des Travaux Fasciaile Hors Series, Supplement - Station Marine d'Endoume 9*, 25–31

Aglen, A.J. (1980) International law and the United Nations Law of the Sea Conference in relation to Atlantic salmon. In A.E.J. Went (ed.), *Atlantic salmon — its future. Proceedings of the second international Atlantic salmon symposium*. Fishing News Books, Farnham, pp. 30–43

Akihito, Prince, Hayashi, M. and Yoshino, T. (1984) Suborder Gobioidea. In H. Masuda, K. Amaoka, C. Araya, and T. Yoshino, (eds), *The fishes of the Japanese Archipelago*. Tokai University Press, Tokyo, pp. 236–89

Al-Hassan, L.A.J. (1986) Variation in meristic characters of *Nematalosa nasus* (Bloch, 1795) from Iraqi and Kuwaiti waters. *Common strategies of anadromous and catadromous fishes — an international symposium*, 9–13 March, 1986, Boston, Massachusetts, USA, Program and Abstracts p. 46

Allen, G.R. (1982) *Inland fishes of Western Australia.* Western Australian Museum, Perth, 86 pp

Allen, K.R. (1949) The New Zealand grayling — a vanishing species. *Tuatara, 2* (1), 22–7

Alt, K.T. (1969) Taxonomy and ecology of the inconnu *Stenodus leucichthys* in Alaska. *Biological Papers, University of Alaska, 12*, 1–61

Anderson, G.J. and Bremer, A.E. (1976) *Salar: the story of the Atlantic salmon.* International Atlantic Salmon Federation, St Andrews. 74 pp

Anderson, W.W. (1957) Larval forms of the fresh-water mullet (*Agonostomus monticola*) from the open ocean off the Bahamas and South Atlantic coast of the United States. *Fisheries Bulletin of the United States Fish and Wildlife Service, 57* (120), 415–25

Anderson, W.W. (1958) Larval development, growth, and spawning of striped mullet (*Mugil cephalus*) along the South Atlantic coast of the United States. *Fisheries Bulletin of the United States Fish and Wildlife Service, 58* (144), 501–19

Annandale, N. and Hora, S.L. (1925) The freshwater fish from the Andaman Islands. *Records of the Indian Museum, 27* (2), 33–41

Anonymous (1954) 152 year old lake sturgeon caught in Ontario. *Commercial Fisheries Review, 16*(9), 28

Anonymous (1986) Freshwater fishery select committee report. *Tasmanian Inland Fisheries Commission Newsletter, 15* (2), 1–2.

Armstrong, R.H. and Morrow, J.E. (1980) The dolly varden, *Salvelinus malma.* In E.K. Balon (ed.), *Charrs - salmonid fishes of the genus* Salvelinus. *Perspectives in Vertebrate Science,* vol 1. Junk, The Hague, pp. 99–140

Azuma, M. (1981) On the origin of koayu, a landlocked form of amphidromous ayufish, *Plecoglossus altivelis. Verhandlungen International Verein fur Limnologie, 21*, 1291–6

Bagenal, T.B. and Braun, E. (1978) Eggs and early life history. In T.B. Bagenal (ed.), *Methods for the assessment of fish production in fresh waters.* International Biological Programme, Handbook No. 3. Blackwell, Oxford, pp.165–201

Baggerman, B. (1957) An experimental study on the timing of breeding and migration in the three-spined stickleback (*Gasterosteus aculeatus* Linnaeus). *Archives Neerlandaises de Zoologie 12*, 105–317

Baker, R.R. (1978) *The evolutionary ecology of animal migration.* Hodder and Stoughton, London, 1012 pp.

Bakshtansky, E.L. (1980) The introduction of pink salmon into the Kola Peninsula. In J.E. Thorpe, (ed.), *Salmon ranching*. Academic Press, London, pp. 245–60

Balon, E.K. (1968) Notes to the origin and evolution of trouts and salmons with special reference to the Danubian trouts. *Vestnik Ceskoslovenske Spolecnost Zoologicka, 32* (1), 1–21

Banister, K.E. (1986) Fishes of the Zaire system. In B.R. Davies and K.F. Walker (eds), *The ecology of river systems*. Junk, Dordrecht, pp. 215–24

Beamish, F.W.H. (1980) Biology of the North American anadromous sea lamprey, *Petromyzon marinus. Canadian Journal of Fisheries and Aquatic Sciences, 37* (11), 1924–43

Beamish, R.J. (1980) Adult biology of the river lamprey (*Lampetra ayresi*) and the Pacific lamprey (*Lampetra tridentata*) from the Pacific coast of Canada. *Canadian Journal of Fisheries and Aquatic Sciences, 37* (11), 1906–23

Beaumont, W.R.C. and Mann, R.H.K. (1984) The age, growth and diet of a freshwater population of the flounder *Platichthys flesus* (L.) in southern England. *Journal of Fish Biology, 25* (5), 607–16

Behnke, R.J. (1966) Relationships of the far eastern trout *Salmo mykiss* Walbaum. *Copeia, 1966* (2), 346–8

Behnke, R.J. (1972) The systematics of salmonid fishes of recently glaciated lakes. *Journal of the Fisheries Research Board of Canada 29*, 639–71

Behnke, R.J. (1980) A systematic revision of the genus *Salvelinus*. In E.K. Balon (ed.), *Charrs - salmonid fishes of the genus* Salvelinus. *Perspectives in Vertebrate Science*, vol l. Junk, the Hague, pp. 441–81

Behnke, R.J., Ting Pong Koh and Needham P.R. (1962) Status of the landlocked salmonid fishes of Formosa with a review of *Oncorhynchus masou* (Brevoort). *Copeia, 1962* (2), 400–7

Bell,M.A. (1979) Low-plate morph of the three-spine stickleback breeding in salt water. *Copeia, 1979* (3), 529–33

Bell, M.A. and Baumgartner, J.V. (1984) An unusual population of *Gasterosteus aculeatus* from Boston, Massachusetts. *Copeia, 1984*(1), 258–62

Benzie, V.L. (1968) A consideration of the whitebait stage of *Galaxias maculatus attenuatus* (Jenyns). *New Zealand Journal of Marine and Freshwater Research, 2* (3), 559–73

Berg, L.S. (1959) Vernal and hiemal races among anadromous fishes. *Journal of the Fisheries Research Board of Canada, 16* (4), 515–37

Berg, L.S. (1962) *Freshwater fishes of the U S S R and adjacent countries*. Israel Program for Scientific Translations, Jerusalem, 3 vols.

Berg, M. (1977) Pink salmon, *Oncorhynchus gorbuscha* (Walbaum), in Norway. *Report of the Institute for Freshwater Research, Drottningholm, 56*, 12–7

Berra, T.M. (1982) Life history of the Australian grayling *Prototroctes maraena* (Salmoniformes: Prototroctidae) in the Tambo River, Victoria. *Copeia, 1982*(4), 795–805

Best, E. (1929) Fishing methods and devices of the Maori. *Bulletin of the Dominion Museum, Wellington, 12*, 1–230

Bigelow, H.B. (1963) Genus *Salvelinus* Richardson,1836. In *Fishes of the western North Atlantic*. Part 3. *Soft-rayed bony fishes — Class Osteichthyes. Memoirs of the Sears Foundation for Marine Research* No. 1. Sears Foundation for Marine Research, New Haven, pp. 503–42

Bigelow, H.B. and Schroeder, W.C. (1948) Cyclostomes. In *Fishes of the Western North Atlantic*. Part 1. *Memoirs of the Sears Foundation for Marine Research* No. 1. Sears Foundation for Marine Research, New Haven, pp. 29–58

Bigelow, H.B. and Schroeder, W.C. (1953) Fishes of the Gulf of Maine. *Fisheries*

Bulletin of the United States Fish and Wildlife Service, 53(74), 1–577

Bigelow, H.B. and Schroeder, W.C. (1963) Family Osmeridae. In *Fishes of the Western North Atlantic. Part 3. Soft-rayed bony fishes — Class Osteichthyes. Memoirs of the Sears Foundation for Marine Research* No. 1. Sears Foundation for Marine Research, New Haven, pp. 553–97

Bishop, K.A. and Bell, J.D. (1978) Aspects of the biology of the Australian grayling *Prototroctes maraena* Gunther (Pisces: Prototroctidae). *Australian Journal of Marine and Freshwater Research, 29,* 743–61

Blaber, S.J. (1987) Factors affecting recruitment and survival of Mugilidae in estuaries and coastal waters of the Indo-West Pacific. *American Fisheries Society Symposium, 1,* 507–18

Black, R. and Wootton,R.J. (1982) Dispersion in a natural population of threespined sticklebacks. *Journal of Zoology,* London, *48,* 1133–5

Black, V.S. (1957) Excretion and osmoregulation. Vol. 1, In M.E.Brown, (ed.). *The physiology of fishes.* Academic Press, London, pp. 163–205

Blackburn, M. (1950) The Tasmanian whitebait, *Lovettia seali* (Johnston), and the whitebait fishery. *Australian Journal of Marine and Freshwater Research, 1* (2), 155–98

Blackett, R.F. (1973) Fecundity of resident and anadromous dolly varden (*Salvelinus malma*) in southeastern Alaska. *Journal of the Fisheries Research Board of Canada, 30* (4), 543–8

Blanc, M., Cadenat, J. and Stauch, A. (1968) Contribution a l'etude de l'Ichthyofaune de l'ile Ambon. *Bulletin de la Institude Fonde Africa Noire 30* (1), 238–56

Blanco, G.J. and Villadolid, D.V. (1939) Fish-fry industries of the Philippines. *Philippines Journal of Science, 69,* 69–100

Bok, A.H. (1979) The distribution and ecology of two mullet species in some fresh water rivers in the Eastern Cape, South Africa. *Journal of the Limnological Society of South Africa, 5* (2), 97–102

Bonetto, A.A. (1986) Fishes of the Parana System. In B.R. Davies and K.F. Walker, (eds), *The ecology of river systems.* Junk, Dordrecht, pp. 573–98

Boreman, J. and Austin, H.M. (1985) Production and harvest of anadromous striped bass stocks along the Atlantic coast. *Transactions of the American Fisheries Society, 114* (1), 3–7

Boulenger, G.A. (1905) The distribution of African freshwater fishes. *Nature,* 72, 413–21

Boulva, J. and Simard, A. (1968) Presence du *Salvelinus namaycush* (Pisces: Salmonidae) dans les eaux marines de l'Arctique occidental canadien. *Journal of the Fisheries Research Board of Canada, 25* (7), 1501–4

Breder, C.M. and Rosen, D.E. (1966) *Modes of reproduction in fishes.* Natural History Press, Garden City, 941 pp.

Briggs, J.C. (1953) The behaviour and reproduction of salmonid fishes in a small coastal stream. *California Department of Fish and Game Bulletin, 94,* 1–62

Bruton, M.N. (1986) Life history strategies of diadromous fishes in inland waters in southern Africa. *American Fisheries Society Symposium 1,* 104–21

Bruton, M.N. and Kok, H.M. (1980) The freshwater fishes of Maputaland. In M.N. Bruton and K.H. Cooper (eds.). *Studies on the ecology of Maputaland.* Rhodes University, Grahamstown, pp. 210–44

Burkov, A.I. and Solovkina, L.N. (1976) The main commercial and biological indicators of the omul, *Coregonus autumnalis,* from the North European zoogeographical region and the results of tagging. *Journal of Ichthyology, 16* (3), 327–31

Burnet, A. M. R. (1965) Observations on the spawning migrations of *Galaxias attenuatus* (Jenyns). *New Zealand Journal of Science, 8* (1), 79–87

REFERENCES

Burnet, A.M.R., Cranfield, H.J., and Benzie, V.L. (1969) The freshwater fishes. In G.A. Knox (ed.), *The natural history of Canterbury* . Reed, Wellington, pp. 498–508

Cadwallader, P.L. (1985) Freshwater fisheries production in Australia. *Australian Fisheries, 44* (9), 12–15

Cadwallader, P.L. and Backhouse, G.N. (1983) *A guide to the freshwater fish of Victoria*. Fisheries and Wildlife Division, Ministry of Conservation, Melbourne, 249 pp.

Cairns, D. (1941) Life history of the two species of freshwater eel in New Zealand. Part I. Taxonomy, age and growth, migration and distribution. *New Zealand Journal of Science and Technology, 23* (1), 53–72

Caldwell, D.K. (1966) *Marine and freshwater fishes of Jamaica*. Institute of Jamaica, Kingston, 120 pp.

Callemand, O. (1948) L'Anguille europenne (*Anguilla anguilla* L.). Les bases physiologiques de sa migration. *Annals of the Institute of Oceanography, Monaco, 21,* 361–440

Campos, H. (1969) Reproduccion del *Aplochiton taeniatus* Jenyns. *Boletin del Museo Nacional de Historia, Chile, 29,* 207–21

Campos, H. (1973) Migrations of *Galaxias maculatus* (Jenyns) (Galaxiidae, Pisces) in Valdivia Estuary, Chile. *Hydrobiologia, 43,* 301–12

Campos, H. (1974) Population studies of *Galaxias maculatus* (Jenyns) (Osteichthys: Galaxiidae) in Chile with reference to the number of vertebrae. *Studies of the Neotropical Fauna and Environment, 9,* 55–76

Campos, H. (1977) Osteichthyes. In *Biota acuatica de Sud-America Austral*. San Diego State University, San Diego, pp. 330–4

Campos, H. (1984) Gondwana and neotropical galaxiid fish biogeography. In T. Zaret, (ed.), *Evolutionary ecology of neotropical freshwater fishes*. Junk, The Hague, pp. 113–25

Castex, M.N. and Castello, H.P. (n.d.) *Potamotrygon leopoldi,* una nueva especie de raya de agua dulce par el Rio Xingu Brasil (Chondrichthyes:Potamotrygonidae). *Acta Scientifica, 10,* 1–16

Cavender, T.M. and Miller, R.R. (1972) *Smilodonichthys rastrosus,* a new Pliocene salmonid fish. *Bulletin of the Museum of Natural History, University of Oregon, 18,* 1–44

Ceskleba, D.G., Avelallemant, S. and Theulmer, T.F. (1985) Artificial spawning and rearing of lake sturgeon, *Acipenser fulvescens,* in Wild Rose State Fish Hatchery, Wisconsin, 1982–1983. *Environmental Biology of Fishes, 14* (1), 79–85

Chan, W.L. (1968) *Marine fishes of Hong Kong*. Part 1. Government Press, Hong Kong, 129 pp.

Chang, B.D. and Navas, W. (1984) Seasonal variations in growth, condition and gonads of *Dormitator latifrons* (Richardson) in the Chone River Basin, Ecuador. *Journal of Fish Biology, 24,* 637–48

Chervinski, J. (1975) Sea basses, *Dicentrarchus labrax* (Linne) and *D. punctatus* (Bloch) (Pisces: Serranidae), a control fish in fresh waters. *Aquaculture, 6* (3), 249–56

Childerhouse, R.J. and Trim, M. (1979) *Pacific salmon and steelhead trout*. Douglas and MacIntyre, Vancouver, 158 pp.

Chubb, C.F. and Potter, I.C. (1984) The reproductive biology and estuarine movements of the gizzard shad, *Nematalosa vlaminghi* (Munro). *Journal of Fish Biology, 25,* 527–43

Chubb, C.F., Potter, I.C., Grant, C.J., Lenanton, R.C.J. and Wallace, J. (1981) Age structure, growth rates and movements of sea mullet, *Mugil cephalus* L., and yellow-eyed mullet, *Aldrichetta forsteri* (Valenciennes), in the Swan-Avon River system, Western Australia. *Australian Journal of Marine and Freshwater Research, 32* (4),

605–28

Clark, A.H. (1905) Habits of West Indian whitebait. *American Naturalist, 39* (461), 335–7

Clarke, F.E. (1899) Notes on Galaxidae, more especially those of the western slopes: with descriptions of new species. *Transactions and Proceedings of the New Zealand Institute, 31,* 78–91

Coates, D. (n.d.) *The biology of tarpon or ox-eye herring,* Megalops cyprinoides *(Megalopidae) in the Sepik River.* Report Department of Primary Industry, Fisheries Research Survey Branch, Port Moresby (83–21), 19 pp.

Cohen, D. (1967) Optimization of seasonal migratory behaviour. *American Naturalist, 101*(917), 5–17

Cohen, D.M. (1970) How many recent fishes are there? *Proceedings of the California Academy of Science* (4), *38*(17), 341–6

Conte, F.P. and Wagner, H.H. (1965) Development of osmotic and ionic regulation in juvenile steelhead trout *Salmo gairdneri. Comparative Biochemistry and Physiology, 14,* 603–20

Contreras, S. and Escalante, M. (1984) Distribution and known impacts of exotic fishes in Mexico. In W.R. Courtenay and J.R. Stauffer (eds), *Distribution, biology and management of exotic fishes.* Johns Hopkins University Press, Baltimore, pp. 102–30

Coruya-Flores, I. (1980) A study of fish populations in the Esperitu Santo River estuary. MSc Thesis, University of Puerto Rico, Rio Piedras.

Couturier, C.Y., Clarke, L. and Sutterlin, A.M. (1986) Identification of spawning areas of two forms of Atlantic salmon (*Salmo salar* L.) inhabiting the same watershed. *Fisheries Research* 4(2), 131–44

Craig, P.C. and Poulin, V.A. (1975) Movements and growth of Arctic grayling (*Thymallus arcticus*) and juvenile Arctic char (*Salvelinus alpinus*) in a small Arctic stream, Alaska. *Journal of the Fisheries Research Board of Canada, 32,* 689–97

Craig-Bennett, A. (1931) The reproductive cycle of the three-spined stickleback *Gasterosteus aculeatus* L. *Philosophical Transactions of the Royal Society, London B, 219,* 197–279.

Craw, R.C. (1978) Two biogeographical frameworks: implications for the biogeography of New Zealand — a review. *Tuatara, 23* (2), 81–114

Craw, R.C. (1979) Generalized tracks and dispersal in biogeography: a response to R.M. McDowall. *Systematic Zoology, 28* (1), 99–107

Croizat, L., Nelson, G.J. and Rosen, D.E. (1974) Centers of origin and related concepts. *Systematic Zoology, 23* (2), 265–87

Crossman, E.J. (1984) Introduction of exotic fishes into Canada. In W.R. Courtenay and J.R. Stauffer (eds), *Distribution, biology and management of exotic fishes.* Johns Hopkins University Press, Baltimore, pp. 78–101

Cushing, D.H. (1969) Migration and abundance. In Perspectives in fisheries oceanography. *Bulletin of the Journal of the Society of Fisheries Oceanography (Special Number),* pp. 207–12

Dadswell, M.J. (1974) Distribution, ecology, and postglacial dispersal of certain crustaceans and fishes in eastern North America. *National Museum of Canada Publications in Zoology, 11,* 1–110

Dadswell, M.J. (1979) Biology and population characteristics of the shortnose sturgeon, *Acipenser brevirostrum* LeSueur 1818 (Osteichthyes: Acipenseridae), in the St John River estuary, New Brunswick, Canada. *Canadian Journal of Zoology, 57,* 2186–210

Dadswell, M.J. (1984) Status of the shortnose sturgeon, *Acipenser brevirostrum,* in Canada. *Canadian Field Naturalist, 98* (1), 75–9

Dadswell, M.J., Taubert, B.D., Squiers, T.S., Marchette, D. and Buckley, J. (1984) Synopsis of biological data on shortnose sturgeon, *Acipenser brevirostrum* LeSueur 1818. *FAO Fisheries Synopsis No. 140. NOAA Technical Report NMFS, 14,* 1–45

Dahl, G. (1971) *Los peces del norte de Colombia*. Ministerio de Agricultura, Instituto de Desarollo de los Recursos Naturales Renovables, Indernea, Bogota, 319 pp.

Dando, P.R. (1984) Reproduction in estuarine fish. In G.W. Potts and R.J. Wootton (eds), *Fish reproduction — strategies and tactics*. Academic Press, London, pp. 155–70

Dando, P.R. and Demir, N. (1985) On the spawning and nursery grounds of bass, *Dicentrarchus labrax*, in the Plymouth area. *Journal of the Marine Biological Association of the United Kingdom 65,* 159–68

Darlington, P.J. (1948) The geographical distribution of the cold-blooded vertebrates. *Quarterly Review of Biology 23* (1), 1–26

Darlington, P.J. (1957) *Zoogeography — the geographical distribution of animals*. Wiley, New York, 675 pp.

Darnell, R.M. (1962) Fishes of the Rio Tamesi and related coastal lagoons in east-central Mexico. *Publications of the Institute of Marine Science, University of Texas, 8* (2), 299–365

Davaine, P. and Beall, E. (1981a) Introductions de salmonides dans les terres australes et antarctiques Francaises. *Colloque sur les Ecosystemes Subantarctiques 1981. Paimont. CNFRA 51,* 289–99

Davaine, P. and Beall, E. (1981b) Acclimation de la trutte commune, *Salmo trutta* L., en mileue subantarctique (Iles Kerguelen): II. — Strategie adaptive. *Colloque sur les Ecosystemes Subantarctiques 1981. Paimont. CNFRA 51,* 399–411

David, L.R. (1946a) Some typical upper Eocene fish scales from California. *Contributions to Paleontology: Carnegie Institute, Washington, Publication, 551,* 47–79

David, L.R. (1946b) Upper Cretaceous fish remains from the western border of the San Joaquin Valley, California. *Contributions to Paleontology: Carnegie Institute, Washington, Publication, 551,* 83–112

Davies, B.R. and Walker, K.F. (eds) (1986) *The ecology of river systems* Junk, Dordrecht. 793 pp.

Davies, T.L.O. (1986) Migration patterns in barramundi, *Lates calcarifer* (Bloch) in van Diemen Gulf, Australia, with estimates of fishing mortality in specific areas. *Fisheries Research, 4,* 243–58

Davis, R.M. (1967.) Parasitism by newly-transformed anadromous sea lampreys on landlocked salmon and other fishes in a coastal Maine lake. *Transactions of the American Fisheries Society, 96,* 11–16

Day, J.H., Blaber, S.J.M. and Wallace, J.H. (1981) Estuarine fishes. In J.H. Day (ed.), *Estuarine ecology — with particular reference to Southern Africa*. Balkema, Rotterdam, pp. 197–221

Deelder, C.L. (1970) Synopsis of biological data on the eel, *Anguilla anguilla* (Linnaeus, 1758). *FAO, Fisheries Synopsis, 80,* 1.1–8.10

Dempson, J.B. and Green, J.M. (1985) Life history of anadromous Arctic char, *Salvelinus alpinus,* in the Fraser River, northern Labrador. *Canadian Journal of Zoology, 63,* 315–24

Deraniyagala, P.E.P. (1937) Two catadromous fishes new to Ceylon. *Ceylon Journal of Science (B), 20* (2), 181–4

De Silva, S.S. (1980) Biology of juvenile grey mullet: a short review. *Aquaculture, 19* (1), 21–36

Dingle, H. (1980) Ecology and evolution of migration. In S.A. Gauthreaux, (ed.), *Animal migration, orientation and navigation*. Academic Press, New York, 1–101

Di Persia, D.H. and Neiff, J.J. (1986) The Uruguay River system. In R.B. Davies and

K.F. Walker, (eds), *The ecology of river systems*. Junk, Dordrecht, pp. 599–621

Dorcey, A.H.J., Northcote, T.G. and Ward, D.V. (1978) Are the Fraser marshes essential to salmon? *Westwater Lectures, University of British Columbia, 1*, 1–29

Doroshov, I.S. (1985) The biology and culture of sturgeons, Acipenseriformes In J.E. Muir and R.J.Roberts, (eds), *Recent advances in aquaculture, 2*, 251–82

Dotu, Y. and Mito, S. (1955) Life history of the gobioid fish, *Sicydium japonicum* Tanaka. *Scientific Bulletin, Faculty of Agriculture, Kyushu University. 15*, 213–21

Dovel, W.L., Mihursky, J.A. and McErlean, A.J. (1969) Life history aspects of the hogchoker, *Trinectes maculatus*, in the Patuxent River estuary, Maryland. *Chesapeake Science, 10* (2), 104–19

Downes, T.W. (1918) Notes on eels and eel weirs (tuna and patuna). *Transactions and Proceedings of the New Zealand Institute, 50*, 296–316

Dudley, R.G., Mullis, A.W. and Terrell, J.W. (1977) Movements of adult striped bass (*Morone saxatilis*) in the Savannah River, Georgia. *Transactions of the American Fisheries Society, 106* (4), 314–22

Dunbar, M.J. and Hildebrand, H.H. (1952) Contribution to the study of the fishes of Ungava Bay. *Journal of the Fisheries Research Board of Canada, 9* (2), 83–128

Durbin, A.G., Nixon, S.W. and Oviatt, C.A. (1979) Effects of the spawning migration of the alewife, *Alosa pseudoharengus*, on freshwater ecosystems. *Ecology, 60* (1), 8–17

Dvinin, P.A. (1949) Lake coho, *Oncorhynchus kisutch* (Walbaum) morpha *relictus nova*. *Doklady Akademija Nauk SSSR, 69* (5), 695–7 (Fisheries Research Board of Canada, Translation No 225).

Dymond, J.R. and Vladykov, V.D. (1934) The distribution and relationships of the salmonoid fishes of North America and North Asia. *Proceedings of the Fifth Pacific Science Congress, 5*, 3741–50

Edge, T.A. (1984) Preliminary status of the Acadian whitefish, *Coregonus canadensis*, in southern Nova Scotia. *Canadian Field Naturalist, 98* (1), 86–90

Eggleston, D.A. (1972) New Zealand quinnat salmon — the marine phase and some problems. In C.J. Hardy (ed.), South Island Council of Acclimatisation Societies, Proceedings of the Salmon Fishery Symposium, 2–3 October, 1971, Ashburton (N Z). *New Zealand Marine Department Fisheries Technical Report, 83*, 68–87

Ego, K. (1954) *Life history of freshwater gobies*. Freshwater Game Fish Management Research Investigation Project. Job Completion Report, Hawaii F-4-R. 24 pp. (mimeo)

Eigenmann, C.H. (1928) The fresh-water fishes of Chile. *Memoirs of the National Academy of Sciences, 22* (1), 1–80

Eiras, J.C. (1983) Some aspects of the biology of a landlocked population of anadromous shad *Alosa alosa* L. *Publicaciones Instituto de Zoologia, Dr Augusto Noble, 180*, 1–16

Ekman, S. (1953) *The zoogeography of the sea*. Sidgwick and Jackson, London, 417 pp.

Eldon, G.A. and Greager, A.J. (1983) Fishes of the Rakaia Lagoon. *New Zealand Ministry of Agriculture and Fisheries, Fisheries Environmental Report, 30*, 1–65

Englert, T.L., Lawler, J.P., Aydin, F.N. and Vachtsevanos, G. (1976) A model of striped bass population dynamics in the Hudson River. In M. Wiley, (ed.), *Estuarine processes. Volume 1. Uses, stresses and adaptation to the estuary*. Academic Press, New York, pp. 137–50

Erdman, D.S. (1961) Notes on the biology of the gobiid fish *Sicydium plumieri* in Puerto Rico. *Bulletin of Marine Science of the Gulf and Caribbean, 11*, 448–56

Erdman, (1974) Common names of fishes in Puerto Rico. *Commercial Fisheries Laboratory, Agriculture and Fisheries, Contribution Departamento de Agricultura Cabo Rojo,6* (2), 1–50

REFERENCES

Erdman, D.S. (1984) Exotic fishes in Puerto Rico. In W.R. Courtenay, and J.R. Stauffer (eds), *Distribution, biology and management of exotic fishes*. Johns Hopkins University Press, Baltimore, pp. 162–76

Erdman, D.S. (1986) The green stream goby, *Sicydium plumieri* in Puerto Rico. *Tropical Fish Hobbyist, 34* (6/630), 70–4

Evans, D.H. (1984) The roles of gill permeability and transport mechanisms in euryhalinity. In W.S. Hoar and D.J. Randall (eds), *Fish physiology Vol. 10: Gills. Part B. Ion and water exchange*, Academic Press, London, pp. 239–83

Fang, P.W. (1934) Study on the fishes referring to Salangidae of China. *Sinensia (Nanking), 4* (9), 231–68

FAO. (1974) *Yearbook of fisheries statistics — catches and landings, 1973*. Vol. 40. FAO, Rome. 586 pp.

FAO. (1978) *Yearbook of fisheries statistics — catches and landings, 1977*. FAO, Rome. 343 pp.

FAO. (1984) *Yearbook of fishery statistics — catches and landings, 1983*. Vol. 56. FAO, Rome. 393 pp.

FAO. (1986) Development of aquaculture in Europe. *Fifteenth FAO Regional Conference for Europe, Istanbul, Turkey, 28 April–2 May, 1986, FRC/86/4*, 1–31

Fink, W.L. (1984) Basal euteleosts: relationships. In G.H. Moser (ed.), Ontogeny and systematics of fishes. *American Society of Ichthyologists and Herpetologists Special Publication, 1*, 202–6

Fink, W.L. and Weitzman, S.H. (1982) Relationships of the stomiiform fishes (Teleostei), with a description of *Diplophos*. *Bulletin of the Museum of Comparative Zoology, Harvard University, 150*(2), 31–93

Finlay, H.J. (1972) Report on the examination of the scales of quinnat salmon (*Oncorhynchus tshawytscha* (Walbaum)) for the determination of age and growth-rate. *New Zealand Ministry of Agriculture and Fisheries, Fisheries Technical Report, 66*, 1–27

Flain, M. (1981) A drift to the north. *Freshwater Catch (NZ), 13*, 9–10

Flain, M. (1982) Quinnat salmon — a study. *Freshwater Catch (NZ), 16*, 6–7

Flain, M. (1983) N Z only success with transplants. *Freshwater Catch (NZ), 19*, 11–3

Foerster, R.E. (1947) Experiments to develop sea-run from land-locked sockeye salmon (*Oncorhynchus nerka kennerlyi*). *Journal of the Fisheries Research Board of Canada, 7* (2), 88–93

Foerster, R.E. (1955) The Pacific salmon (genus *Oncorhynchus*) of the Canadian Pacific coast, with particular reference to their occurrence in or near fresh water. *Bulletin of the International North Pacific Fisheries Commission, 1*, 5–576

Foerster, R.E. (1968) The sockeye salmon *Oncorhynchus nerka*. *Bulletin of the Fisheries Research Board of Canada, 162*, 1–422

Fontaine, M. (1975) Physiological mechanisms in the migrations of marine and amphihaline fish. In F.S. Russell and M. Young (eds), *Advances in marine biology 13*, 241–335

Fontaine, M. and Callemand, O. (1941) Sur l'hydrotropisme des civelles. *Bulletin of the Institute of Oceanography, Monaco, 811*, 1–6

Foster, N.R. (1969) Factors in the origins of fish migrations. *Underwater Naturalist, 6* (1), 27–31, 48

Foster, N.R. (1985) Lake trout reproductive behaviour: influence of chemosensory cues from the young-of-the-year by-products. *Transactions of the American Fisheries Society 114* (6), 795–803

Frank, S. (1969) *The pictorial encyclopedia of fishes*. Hamlyn, London, 552 pp.

Frost, W. (1963) The homing of charr, *Salvelinus willughbii* (Gunther) in Windermere. *Animal Behaviour, 11* (1), 74–82

Fukui, S. (1979) On the rock-climbing behaviour of the goby *Sicyopterus japonicus*. *Japanese Journal of Ichthyology, 26* (1), 84–8

Fulton, W. (1984) Whitebait wipeout. *Tasmanian Inland Fisheries Commission Newsletter, 13*(3),6

Fulton, W. (1986) The Tasmanian mudfish *Galaxias cleaveri* Scott. *Fishes of Sahul, 4* (1), 150–1

Ganapati, S.V. (1973) Ecological problems of man-made lakes of South India. *Archiv fur Hydrobiologie, 71* (3), 363–80

Gerking, S.D. (1959) The restricted movements of fish populations. *Biological Reviews, 34*, 221–42

Gery, J. (1969) The freshwater fish of South America. In E.J. Fittkau, J. Illies, H. Kunge, G.H. Schwabe and H. Sioli (eds), *Biogeography and ecology in South America.* Junk, The Hague, pp. 828–48

Giamas, M.T.D., Santos, L.E. and Vermulem, H. (1983) Influencia de fatores climaticos sobre reproducao da manjuba, *Anchoviella lepidentostole* (Fowler, 1911) (Teleostei: Engraulidae). *Boletin de Instituto Pesca, 10* (1), 95–100

Gilbert, C.H. and O'Malley, H. (1921) Investigation of the salmon fisheries of the Yukon River. *United States Bureau of Fisheries. Appendix 6. Report of the United States Fisheries Commission for 1921 (Document 909)*, pp. 128–54

Gilbert, C.R. and Kelso, D.P. (1971) Fishes of the Tortuguero area, Caribbean Costa Rica. *Bulletin of the Florida State Museum, Biological Science, 16* (1), 1–54

Gilmore, R.G. (1977) Notes on the opossum pipefish, *Oostethus lineatus*, from the Indian River lagoon and vicinity, Florida. *Copeia, 1977* (4), 781–3

Glova, G.J. and McCart, P.J. (1974) Life history of the Arctic char (*Salvelinus alpinus*) in the Firth River, Yukon Territory. In P.J. McCart (ed.), *Life histories of anadromous and freshwater fishes in the western Arctic. Canadian Arctic Gas Studies Ltd, Biological Report Series, 20* (3), 1–50

Gorham, S.W. and McAllister, D.E. (1974), The shortnose sturgeon, *Acipenser brevirostrum*, in the Saint John River, New Brunswick, Canada, a rare and possibly endangered species. *Syllogeus, 5*, 1–18

Gosline, W.A. (1960) Mode of life, functional morphology, and the classification of modern teleostean fishes. *Systematic Zoology, 8*, 160–4

Goto, A. (1981) Life history and distribution of a river sculpin, *Cottus hangiongensis. Bulletin of the Faculty of Fisheries Hokkaido University, 32*, 10–21 (In Japanese)

Goto, A. (1987) Life history variation in males of the river sculpin *Cottus hangiongensis* along the course of a river. *Environmental Biology of Fishes, 19* (2), 81–91.

Gould, S.J. (1983) What, if anything else, is a zebra? *Natural History, 90* (7), 6–12

Graham, D.H. (1953) *A treasury of New Zealand fishes.* Reed, Wellington. 404 pp.

Grainger, E.H. (1953) On the age, growth, migration, reproductive potential and feeding habits of the Arctic char (*Salvelinus alpinus*) of Frobisher Bay, Baffin Island. *Journal of the Fisheries Research Board of Canada, 10* (6), 326–70

Grant, C.J. and Spain, A.V. (1975) Reproduction, growth and size allometry of *Mugil cephalus* Linnaeus (Pisces: Mugilidae) from north Queensland inshore waters. *Australian Journal of Zoology, 23* (2), 181–201

Griffin, R.K. (1987) Life history distribution and seasonal migration of the barramundi (*Lates calcarifer*) in the Daly river northern territory Australia. *American Fisheries Society Symposium, 1*, 358–63

Gross, M. (1984) Sunfish, salmon, and the evolution of alternative reproductive strategies and tactics in fishes. In G.W Potts and J.R. Wootton (eds), *Fish reproduction — strategies and tactics.* Academic Press, London, 410 pp.

Gross, M. (1985) Disruptive selection for alternative life histories in salmon. *Nature 313*, 47–8

Gross, M. (1987) The evolution of diadromy in fishes. *American Fisheries Society Symposium 1, 14–25*

Guelpen, L. van, and Davis, C.C. (1979) Seasonal movements of the winter flounder, *Pseudopleuronectes americanus,* in two contrasting inshore locations in Newfoundland. *Transactions of the American Fisheries Society, 108,* 26–37

Gunter, G. (1938) Notes on invasion of fresh water by fishes of the Gulf of Mexico, with special reference to the Mississippi–Atchafalaya River system. *Copeia, 1938 (2), 69–72*

Gunter, G. (1956) A revised list of the euryhaline fishes of North and Middle America. *American Midland Naturalist, 56,* 345–54

Haedrich, R.L. (1983) Estuarine fishes. In B.H. Ketchum, (ed.), Estuaries and enclosed seas. *Ecosystems of the World, 26,* 183–207

Hagen, D.W. (1967) Isolating mechanisms in threespine sticklebacks (*Gasterosteus*). *Journal of the Fisheries Research Board of Canada 24* (8), 1637–92

Hall, S.L. (1985) The prehistoric exploitation of freshwater fish from the Koonap River. *Ichthos* (10), 10

Hall, T.S. (1905) The distribution of the fresh-water eel in Australia and its means of dispersal. *Victorian Naturalist, 22* (5), 80–3

Hamada, K. (1961) Taxonomic and ecological studies of the genus *Hypomesus* in Japan. *Memoirs of the Faculty of Fisheries, Hokkaido University, 9* (1), 1–56

Hanavan, M.G. and Skud, B.E. (1954) Survival of pink salmon spawn in an intertidal area with special reference to the influence of crowding. *Fisheries Bulletin of the United States Fish and Wildlife Service, 56* (95), 167–76

Hanson, J.A. and Smith, H.D. (1967) Mate selection in a population of sockeye salmon (*Oncorhynchus nerka*) of mixed age-groups. *Journal of the Fisheries Research Board of Canada, 24* (9), 1955–77

Harden-Jones, F.R. (1968) *Fish migration* Arnold, London. 325 pp.

Harden-Jones, F.R. (1981) Fish migration: strategy and tactics. *Society for Experimental Biology, Seminar Series, 13,* 139–65

Hardisty, M.W. and Potter, I.C. (1971a) The behaviour, ecology and growth of larval lampreys. In M.W. Hardisty and I.C. Potter (eds), *The biology of lampreys,* Vol. 1. Academic Press, London, pp. 85–125

Hardisty, M.W. and Potter, I.C. (1971b) The general biology of adult lampreys. In M.W. Hardisty and I.C. Potter (eds), *The biology of lampreys.* Vol. 1. Academic Press, London, pp. 127–206

Hardisty, M.W. and Potter, I.C. (1971c) Paired species. In M.W. Hardisty and I.C. Potter, (eds), *The biology of lampreys.* Vol. 1. Academic Press, London, pp. 249–77

Harrington, D. and Beumer, J.D. (1980) Eels support a valuable fishery in Victoria. *Australian Fisheries, 39* (8), 23

Harris, J.H. (1984) Impoundment of coastal drainages of south-eastern Australia and a review of its relevance to fish migration. *Australian Zoologist, 21* (2–3), 235–50

Harrison, A.C. (1963) *Freshwater fish and fishing in Africa.* Nelson, Johannesburg

Hart, J.L. (1973) Pacific fishes of Canada. *Bulletin of the Fisheries Research Board of Canada, 180,* 1–740

Hartley, P.H.T. (1940) The saltash tuck-net fishery and the ecology of some estuarine fishes. *Journal of the Marine Biological Association of the United Kingdom, 24* (1), 1–68

Haslar, A.D. (1966) *Underwater guideposts — homing of salmon.* University of Wisconsin, Madison, 155 pp.

Haslar, A.D. (1971) Orientation and fish migration. In W.S.Hoar and D.J. Randall (eds), *Fish physiology.* Vol. VI. Academic Press, London, pp. 429–510

Havey, K.A. (1961) Restoration of anadromous alewives at Long Pond, Maine.

Transactions of the American Fisheries Society, 90 (3), 281–6

Hayes, J.W. (1987) Competition for spawning space between brown (*Salmo trutta*) and rainbow trout (*S. gairdneri*) in a lake inlet spawning tributary, New Zealand. *Canadian Journal of Fisheries and Aquatic Sciences, 44* (1), 40–7

Haynes, J.M., Gray, R.H. and Montgomery, J.C. (1978) Seasonal movements of the white sturgeon (*Acipenser transmontanus*) in the mid-Columbia River. *Transactions of the American Fisheries Society, 107* (2), 275–80

Heape, W. (1931) *Emigration, migration and nomadism.* Heffer, Cambridge, 369 pp.

Herre, A.W.C.T. (1927) Gobies of the Philippine and the China Sea. *Manila Bureau of Science Monograph, 23,* 1–352

Herre, A.W.C.T. (1958) Marine fishes in Philippines rivers and lakes. *Philippines Journal of Science, 87,* 65–88

Hickling, C.F. (1970) A contribution to the natural history of the English grey mullets (Pisces, Mugilidae). *Journal of the Marine Biological Association of the United Kingdom, 50,* 609–33

Hildebrand, S.F. (1939) The Panama Canal as a passageway for fishes with lists and remarks on the fishes and invertebrates observed. *Zoologica, New York, 24* (3), 15–40

Hildebrand, S.F. (1963) Family Clupeidae. In *Fishes of the Western North Atlantic. Part 3. Soft-rayed bony fishes —Class Osteichthyes. Memoirs of the Sears Foundation for Marine Research,* No. 1. Sears Foundation for Marine Research, New Haven, pp. 257–454

Hildebrand, S.F. and Schroeder, W.C. (1927) Fishes of Chesapeake Bay. *Bulletin of the United States Bureau of Fisheries, 43* (1024), 1–366

Hoar, W.S. (1976) Smolt transformation: evolution, behaviour and physiology. *Journal of the Fisheries Research Board of Canada, 33* (5),1234–52

Hobbs, D.F. (1947) Migrating eels in Lake Ellesmere. *Transactions of the Royal Society of New Zealand, 77* (5), 228–32

Hoese, D.F., Larson, H.K. and Llewellyn, L.C. (1980) Family Eleotridae — gudgeons. In R.M. McDowall (ed.), *Freshwater fishes of southeastern Australia.* Reed, Sydney, pp. 169–85

Hoover, E.E. (1936) Contributions to the life history of the chinook and landlocked salmon in New Hampshire. *Copeia, 1936* (4), 193–8

Hogan, A.E. and Nicholson, J.C. (1987) Sperm motility of sooty grunter, *Hephaeustus fuliginosus* (Macleay) and jungle perch, *Kuhlia rupestris* (lacepede) in differenct salinities. *Australian Journal of Marine and Freshwater Research 38(4),* 523–8

Hopkins, C.L. (1979a) Age-growth characteristics of *Galaxias fasciatus* (Salmoniformes: Galaxiidae). *New Zealand Journal of Marine and Freshwater Research, 13* (1), 39–46

Hopkins, C.L. (1979b) Reproduction in *Galaxias fasciatus* Gray (Salmoniformes: Galaxiidae). *New Zealand Journal of Marine and Freshwater Research, 13* (2), 225-30

Hopkins, C.L. (1985) Pacific salmon in Chile. *Freshwater Catch (NZ), 28,* 18

Horton, D. (1984) Dispersal and speciation: Pleistocene biogeography and the modern Australian biota. In M. Aichen and G. Clayton (eds), *Vertebrate zoogeography and evolution in Australasia (animals in space and time).* Hesperian, Carslisle, pp. 113–18

Hubbard, N. (1979) The Wanganui smelt fishery. *Freshwater Catch (N.Z.), 4,* 10–11

Hubbs, C.L. and Potter, I.C. (1971) Distribution, phylogeny and taxonomy. In M.W. Hardisty and I.C. Potter (eds), *The biology of lampreys.* Vol. 1. Academic Press, London, pp. 1–65

Huff, J.A. (1975) Life history of the Gulf of Mexico sturgeon, *Acipenser oxyrhynchus desotoi,* in Suwannee River, Florida. *Florida Marine Research Publications, 16,* 1-32

271

Humphries, P. (1986) Spawning habits of the spotted mountain trout. *Tasmanian Inland Fisheries Commission Newsletter, 15* (3), 4

Huntsman, A.G. and Dymond, J.R. (1940) Pacific salmon not established in Atlantic waters. *Science, 91* (2367), 447–9

Hutchings, J.A. and Morris, D.W. (1986) How are anadromous life histories linked to freshwater residency? *Common strategies of anadromous and catadromous fishes - an international symposium, March 9-13, 1986, Boston, Massachusetts, USA Programs and Abstracts,* p. 56.

Hutchings, J.A. and Myers, R.A. (1985) Mating between anadromous and nonanadromous Atlantic salmon, *Salmo salar. Canadian Journal of Zoology, 63,* 2219–21

Hynes, H.B.N. (1970) *The ecology of running waters.* Liverpool University Press, Liverpool, 555 pp.

Islam, B.N. and Talbot, G.B. (1968) Fluvial migration, spawning and fecundity of Indus River hilsa, *Hilsa ilisha. Transactions of the American Fisheries Society, 97* (4), 350–5

Jackson, P.B.N. (1960) On the desirability or otherwise of introducing fish to waters that are foreign to them. *CCT/CSA Publication, 63,* 157–64

Jackson, P.B.N. (1986) Fishes of the Zambesi System. In B.P.Davies and K.F. Walker (eds) *The ecology of river systems.* Junk, Dordrecht pp. 269–88

Jager, T., Nellen, W., Schoffer, W. and Shodjai, F. (1981) Influence of salinity and temperature on early life stages of *Coregonus albula, C. lavaretus, R. rutilus, and L. lota. Proces-Verbeaux Reunion Conseil International Exploration del Mer, 178,* 345–8

Jarvik. E. (1968) Aspects of vertebrate phylogeny. In T. Orvig (ed.), *Current problems of lower vertebrate phylogeny. Nobel Symposium* vol. 4. Interscience, New York, pp. 497–527

Jayaram, K.C. (1974) Ecology and distribution of freshwater fishes, amphibians and reptiles. In M.S. Mani (ed.), *Ecology and biogeography in India. Monographiae Biologicae* Vol. 23, Junk, The Hague, pp. 517–84

Jehangeer, I. (1986) Freshwater fish and fisheries of Mauritius. *IPFC Expert Consultation on Inland Fisheries of the larger Indo-Pacific Islands, Bangkok, Thailand,* 4–9 August, 1986. 10 pp. (mimeo)

Jellyman, D.J. (1977a) Juvenile New Zealand eels found in Japan. *Catch (NZ), 4* (5), 10

Jellyman, D.J. (1977b) Summer upstream migration of juvenile freshwater eels in New Zealand. *New Zealand Journal of Marine and Freshwater Research, 11* (1), 61–71

Jellyman, D.J. (1979) Upstream migration of glass-eels (*Anguilla* spp.) in the Waikato River. *New Zealand Journal of Marine and Freshwater Research, 13* (1), 13–22

Jellyman, D.J. (1984) Eels surmount climbing challenge. *Freshwater Catch (NZ), 25,* 8–9

Jellyman, D.J. (1987) A review of the marine life history of the Australasian temperate species of *Anguilla. American Fisheries Society Symposium, 1,* 276–85.

Jellyman, D.J. and Todd, P.R. (1982) New Zealand eels: their biology and fishery. *New Zealand Ministry of Agriculture and Fisheries, Fisheries Research Division Information Leaflet, 11,* 1–19

Jhingram, V.G. (1975) *Fish and fisheries of India.* Hindustan Publishing Company, Delhi, 954 pp.

Johnson, D.W. and McLendron, E.L. (1970) Differential distribution of the striped mullet, *Mugil cephalus* Linnaeus. *California Fish and Game, 56,* 138–9

Johnson, L. (1964) Marine-glacial relics of the Canadian Arctic islands. *Systematic Zoology, 13,* 76–91

Johnson, L. (1980) The Arctic char, *Salvelinus alpinus.* In E.K. Balon (ed), Charrs salmonid fishes of the genus *Salvelinus. Perspectives in Vertebrate Science,* Vol. 1.

Junk, The Hague, pp. 15–98

Johnston, J. M. (1981) Life histories of anadromous cutthroat with emphasis on migratory behavior. In E. L. Brannon and E. O. Salo (eds.), *Proceedings of the salmon and trout migratory behavior symposium.* School of Fisheries, University of Washington, Seattle, 123–7

Jonsson, B. (1982) Diadromous and resident trout *Salmo trutta*: is their difference due to genetics? *Oikos, 38,* 297–300

Jonsson, B. (1985) Life history patterns of freshwater resident and sea-run migrant brown trout in Norway. *Transactions of the American Fisheries Society, 114,* 182–94

Jordan, D.S. (1907) *American nature series — Fishes.* Holt, New York

Joyner, T. (1980a) Salmon ranching in South America. In J.E. Thorpe (ed.), *Salmon ranching.* Academic Press, London, pp. 261–76

Joyner, T. (1980b) Ocean ranching — first returns in Chile project. *Fish Farming International, 7* (1), 29

Joyner, T., Mahnken, C.V.W. and Clark, R.C. (1974) Salmon — future harvest from the Antarctic Ocean? *Marine Fisheries Review, 36* (5), 20–8

Jubb, R.A. (1965) Freshwater fishes of the Cape Province. *Annals of the Cape Province Museums (Natural History), 4,* 1–72

Jubb, R.A. (1967) *Freshwater fishes of southern Africa.* Balkema, Cape Town, 248 pp.

Kafuku, T. and Ikenoue, H. (1983) Modern methods of aquaculture in Japan. *Developments in Aquaculture and Fisheries Science* No. 11. Elsevier, Amsterdam, 216 pp.

Kami, H. (1986) Rehabilitation of the Masso Reservoir to a recreational fishery. *IPFC Expert Consultation on Inland Fisheries of the Larger Indo-Pacific islands, Bangkok, Thailand,* 4-9 August, 1986. 3 pp.

Kawanabe, H. (1969) The significance of social structure in production of the 'Ayu', *Plecoglossus altivelis.* In T.G. Northcote, (ed.), *Symposium on salmon and trout in streams.* H.R. MacMillan Lectures in Fisheries. University of British Columbia, Vanvouver, pp. 243–51

Kawanabe, H., Saito, Y.T., Sunaga, T., Maki, I. and Azuma, M. (1968) Ecology and biological production of Lake Naka-umi and adjacent regions. 4. Distribution of fishes and their foods. *Special Publications of the Seto Marine Biological Laboratory, 2,* (2), 45–73

Kedney, G.I., Bopule, V. and Fitzgerald, G.J. (1987) The comparative reproductive ecology of a population of threespined sticklebacks, (*Gasterosteus aculeatus* Linnaeus, forma *trachurus*) breeding in freshwater and saltwater habitats. *American Fisheries Society Symposium, 1,* 151–61

Kendall, W.C. (1935) The fishes of New England. II. The salmon family. Part 2. The salmons. *Memoirs of the Boston Society for Natural History, 9* (1), 1–166

Kendeigh, S.C. (1961) *Animal ecology.* Prentice-Hall, Englewood Cliffs, 468 pp.

Kessler, K.T. (1877) Fishes of the Aralo-Caspio-Pontine region. *Trudy Aralo-Caspian Expedition,* (4), 360 pp. (In Russian)

Kesteven, G.L. (1953) Further results of tagging sea mullet, *Mugil cephalus* Linnaeus, on the eastern Australian coast. *Australian Journal of Marine and Freshwater Research, 4* (2), 251–306

Kissil, G.W. (1974) Spawning of the anadromous alewife, *Alosa pseudoharengus,* in Bride Lake, Connecticut. *Transactions of the American Fisheries Society, 103* (2), 312–7

Kohlhorst, D.W. (1976) Sturgeon spawning in the Sacramento River in 1973, as determined by distribution of larvae. *California Fish and Game, 62* (1), 32–40

Korringa, P. (1967) Estuarine fisheries in Europe as affected by man's multiple activities. In G.H. Lauff (ed.). Estuaries. *American Association for the Advancement of*

Science Publication No 93, Washington, pp. 658–63

Koski, R.T. (1978) Age, growth, and maturity of the hogchoker, *Trinectes maculatus*, in the Hudson River, New York. *Transactions of the American Fisheries Society, 107* (3), 449–53

Koumans, F.P. (1953) *The fishes of the Indo-Australian archipelago. X. Gobioidea.* Brill, Leiden, 423 pp.

Kovtun, I.F., Bondarenko, T.S. and Rekov, Yu. I. (1984) Characteristics of the Azov sturgeon fishery. *Journal of Ichthyology, 24* (6), 119–24

Krejsa, R.J. (1967) The systematics of the prickly sculpin, *Cottus asper* Richardson, a polytypic species. Part 1. Synonymy, nomenclatural history, and distribution. *Pacific Science, 21* (2), 241–51

Kutty, M.N. and Mohamed, M.P. (1975) Metabolic adaptation of mullet *Rhinomugil corsula* (Hamilton) with special reference to energy utilization. *Aquaculture, 5*, 253–70

Kwain, A. (1987) Pink salmon of the Great Lakes. *American Fisheries Society Symposium, 1*, 57–65

Lagler, K.F., Bardach, J.E. and Miller, R.R. (1962) *Ichthyology — the study of fishes.* Wiley, New York, 545 pp.

Lake, J.S. (1971) *Freshwater fishes and rivers of Australia.* Nelson, Melbourne, 61 pp.

Lake, J.S. (1978) *Australian freshwater fishes — an illustrated field guide.* Nelson, Melbourne, 160 pp.

Landsborough-Thompson, A. (1926) *Problems of bird migration.* Witherby, London

Lasserre, P. and Gallis, J-L. (1975) Osmoregulation and differential penetration of two grey mullets, *Chelon labrosus* (Risso) and *Liza ramada* (Risso), in estuarine fish ponds. *Aquaculture, 5* (4), 323–44

Lear, W.H. (1975) Evaluation of the transplant of Pacific pink salmon (*Oncorhynchus gorbuscha*) from British Columbia to Newfoundland. *Journal of the Fisheries Research Board of Canada, 32* (12), 2343–56

Lear, W.H. (1980) The pink salmon transplant experiment in Newfoundland. In J.S. Thorpe (ed.), *Salmon ranching.* Academic Press, London, pp. 213–43

Lee, D.S., Gilbert, C.R., Hocutt, C.H., Jenkins, R.E., McAllister, D.E. and Stauffer, J.R. (1980) *Atlas of North American freshwater fishes.* North Carolina State Museum of Natural History, 854 pp.

Lee, G.(1937) Oral gestation in the marine catfish, *Galeichthys felis. Copeia, 1937* (1), 49–56

Legendre, P. and Legendre, V. (1984) Postglacial dispersal of freshwater fishes in the Quebec Peninsula. *Canadian Journal of Fisheries and Aquatic Science, 41* (12), 1781–802

Legendre, P., Schreck, C.B. and Behnke, R.J. (1972) Taximetric analysis of selected groups of western North American *Salmo* with regard to phylogenetic divergence. *Systematic Zoology, 21*(3), 292–307

Leggett, W.C. (1973) The migrations of the shad. *Scientific American, 228* (3), 92–8

Leggett, W.C. and Carscadden, J.E. (1978) Latitudinal variation and reproductive characteristics of American shad (*Alosa sapidissima*): evidence for population specific life history strategies in fish. *Journal of the Fisheries Research Board of Canada, 35*, 1469–78

Lenteigne, J., and McAllister, D.E. (1983) The pygmy smelt, *Osmerus spectrum* Cope 1870, a forgotten sibling species of eastern North American fish. *Syllogeus, 45*, 1–32

Lesack, L.F.W. (1986) Estimates of catch and potential yield for the riverine artisanal fishery in the Gambia, West Africa. *Journal of Fish Biology, 28*, 679–700

Levenge, R. (1980) A general review of the state of the salmon fisheries of the north Atlantic. In A.E.J. Went (ed), *Atlantic salmon — its future. Proceedings of the Second International Atlantic Salmon Symposium.* Fishing News Books, Farnham, pp. 6-17

Lewis, A.D. and Hogan, A.E. (1987) The enigmatic jungle perch — recent research provides some answers. *SPC Newsletter 40,* 24–31

Lewis, A.D. and Pring, C.K. (1986) Freshwater and brackishwater fish and fisheries of Fiji. *Country review prepared for the expert consultation on inland fisheries of the larger Indo-Pacific islands, Bangkok, Thailand,* 4–9 August, 1986, 14 pp.

Lindbergh, J.M. and Brown, P. (1982) Continuing experiments in salmon ocean ranching in southern Chile. *International Council for the Exploration of the Sea,* CM 1982/M21, 5 pp.

Lindberg, G.U. and Legeza, M. (1969) Fishes of the sea of Japan and the adjacent areas of the Sea of Okhotsk and the Yellow Sea. Part 2. Teleostomi. XII Acipenseriformes — XVIII Polynemiformes. *Keys to the fauna of the USSR published by the Zoological Institute of the Academy of Sciences of the USSR, 84,* 1–389

Lindbergh, J.M., Noble, R.E. and Blackburn, K. (1981) First returns of Pacific salmon in Chile. *International Council for the Exploration of the Sea,* CM 1981/F27, 9 pp.

Lindroth, A. (1957) A study of the whitefish (*Coregonus*) of the Sundsvall Bay district. *Report of the Institute for Freshwater Research, Drottningholm, 38,* 70–108

Llewellyn, L.C. and MacDonald, M.C. (1980) Family Percichthyidae — Australian freshwater basses and cods. In R.M. McDowall (ed.), *Freshwater fishes of southeastern Australia,* Reed, Sydney, pp. 142–7

Loesch, J.G. (1987) Life history aspects of anadromous alewife (*Alosa pseudoharengus*) and blueback herring (*A. aestivalis*) in freshwater lakes. *American Fisheries Society Symposium, 1,* 89–103

Loesch, J.G. and Lund, W.A. (1977) A contribution to the life history of the blueback herring, *Alosa aestivalis. Transactions of the American Fisheries Society 106* (6), 583–9

Loftus, W.F., Kushlan, J.A. and Voorhees, S.A. (1984) Status of the mountain mullet in southern Florida. *Florida Scientist, 47* (4), 256–63

Lowe-McConnell, R.H. (1986) Fishes of the Amazon system. In B.P. Davies and K.F. Walker (eds), *The ecology of river systems.* Junk, Dordrecht, pp. 339–51

Lynch, D.D. (1965) Changes in Tasmanian fishery. *Australian Fisheries Newsletter, 24* (4), 13, 15

McAllister, D.E. (1963) A revision of the smelt family, Osmeridae. *Bulletin of the National Museum of Canada, 191,* 1–53,

McAllister, D.E. (1970) Rare or endangered Canadian fishes. *Canadian Field Naturalist, 84* (1), 5–26

McAllister, D.E. (1984) Osmeridae. In P.J.P. Whitehead, M-L. Bauchot, J-G Hureau, J. Nielsen and E.Tortonese, (eds), *Fishes of the northeast Atlantic and Mediterranean.* Vol. 1. Unesco, Rome, pp. 399–400

McAllister, D.E. and Lindsey, C.C. (1959) Systematics of the freshwater sculpins (*Cottus*) of British Columbia. *Bulletin of the National Museums of Canada, 172,* 66–89

McAllister, D.E., Parker, B.J. and McKee, P.M. (1985) Rare, endangered and extinct fishes in Canada. *Syllogeus, 54,* 1–192

McCart, P. and Craig, P. (1973) Life history of two isolated populations of Arctic char (*Salvelinus alpinus*) in spring fed tributaries of the Canning River, Alaska. *Journal of the Fisheries Research Board of Canada, 30,* 1215–20

McCleave, J.D., Fried, S.M. and Towt, A.K. (1977) Daily movements of shortnose sturgeon, *Acipenser brevirostrum,* in a Maine estuary. *Copeia, 1977* (1), 149–57

McCleave, J.D., Arnold, G.P., Dodson., J.S. and Neill, W.H. (1984) *Mechanisms in migration of fishes.* Plenum, New York, 374 pp.

McCleave, J.D., Kleckner, R.C. and Castonguay, M. (1987) Reproductive sympatry of

American and European eels and implications for migration and taxonomy. *American Fisheries Society Symposium, 1*, 286–97

McCormick, S. D., Naiman, R. J., and Montgomery, E. T. (1985) Physiological smolt characteristics of anadromous and non-anadromous brook trout (*Salvelinus fontinalis*) and Atlantic salmon (*Salmo salar*). *Canadian Journal of Fisheries and Aquatic Sciences, 42* (3), 529–38

MacCrimmon, H.R. (1971) World distribution of rainbow trout (*Salmo gairdneri*). *Journal of the Fisheries Research Board of Canada, 28* (5), 663–704

MacCrimmon, H.R. and Campbell, J.S. (1969) World distribution of brook trout, *Salvelinus fontinalis*. *Journal of the Fisheries Research Board of Canada, 26* (7), 1699–725

MacCrimmon, H.R. and Gots, B.L. (1979) World distribution of Atlantic salmon, *Salmo salar*. *Journal of the Fisheries Research Board of Canada, 36*, 422–57

MacCrimmon, H.L. and Gots, B.L. (1980) Fisheries for charrs. In E.K. Balon (ed.), Charrs — salmonid fishes of the genus *Salvelinus. Perspectives in Vertebrate Science*, Vol. 1. Junk, The Hague, pp. 797–839

MacCrimmon, H.R. and Marshall, T.C. (1968) World distribution of brown trout, *Salmo trutta. Journal of the Fisheries Research Board of Canada, 25* (12), 2527–48

McDowall, R.M. (1965) Studies on the biology of the redfinned bully *Gobiomorphus huttoni* (Ogilby). II. Breeding and life history. *Transactions of the Royal Society of New Zealand, Zoology, 5* (14), 177–96

McDowall, R.M. (1968a) The application of the terms anadromous and catadromous to the Southern Hemisphere salmonoid fishes. *Copeia, 1968 (1)*, 176–8

McDowall, R.M. (1968b) *Galaxias maculatus* (Jenyns) — the New Zealand whitebait. *New Zealand Marine Department Fisheries Research Bulletin, 2*, 1–84

McDowall, R.M. (1969) Relationships of galaxioid fishes with a further discussion of salmoniform classification. *Copeia, 1969* (4), 796–824

McDowall, R.M. (1970) The galaxiid fishes of New Zealand. *Bulletin of the Museum of Comparative Zoology, Harvard University, 139* (7), 341–431

McDowall, R.M. (1971a) Fishes of the family Aplochitonidae. *Journal of the Royal Society of New Zealand, 1* (1), 31–52

McDowall, R.M. (1971b) The galaxiid fishes of South America. *Zoological Journal of the Linnean Society, 50* (1), 33–73

McDowall, R.M. (1972) The species problem in freshwater fishes and the taxonomy of diadromous and lacustrine populations of *Galaxias maculatus* (Jenyns). *Journal of the Royal Society of New Zealand, 2* (3), 325–67

McDowall, R.M. (1973) Relationships and taxonomy of the New Zealand torrentfish *Cheimarrichthys fosteri* Haast (Pisces: Mugiloididae). *Journal of the Royal Society of New Zealand, 3* (2), 199–217

McDowall, R.M. (1975) A revision of the New Zealand species of *Gobiomorphus* (Pisces: Eleotridae). *National Museum of New Zealand Records, 1* (1), 1–32

McDowall, R.M. (1976) Fishes of the family Prototroctidae (Salmoniformes). *Australian Journal of Marine and Freshwater Research, 27* (4), 641–59

McDowall, R.M. (1978a) *New Zealand freshwater fishes — a guide and natural history.* Heinemann Educational Books, Auckland, 230 pp.

McDowall, R.M. (1978b) Generalized tracks and dispersal in biogeography. *Systematic Zoology, 27* (1), 88–104

McDowall, R.M. (1979) Fishes of the family Retropinnidae (Pisces:Salmoniformes): a taxonomic revision and synopsis. *Journal of the Royal Society of New Zealand, 9* (1), 85–121

McDowall, R.M. (1983) Nothing new under the sun. *Freshwater Catch (NZ), 19*, 15–17

McDowall, R.M. (1984a) Southern hemisphere freshwater salmoniforms: development and relationships. In G.H. Moser (ed.), *Ontogeny and systematics of fishes. American Society of Ichthyologists and Herpetologists, Special Publication, 1*, 150–3

McDowall, R.M. (1984b) *Trout in New Zealand waters*. Wetland Press, Wellington, 120 pp.

McDowall, R.M. (1984c) *The New Zealand whitebait book*. Reed, Wellington, 210 pp.

McDowall, R.M. (1987a) The occurrence and distribution of diadromy in fishes. *American Fisheries Society Symposium, 1*, 1–13

McDowall, R.M. (1987b) Zoogeography — the native fish. In A.B. Viner (ed.). *New Zealand inland waters*. Department of Scientific and Industrial Research, Wellington, pp. 291–306

McDowall, R.M. and Eldon, G.A. (1980) The ecology of whitebait migrations (Galaxiidae: *Galaxias* spp.). *New Zealand Ministry of Agriculture and Fisheries, Fisheries Research Bulletin, 20*, 1–171

McDowall, R.M. and Frankenberg, R.S. (1981) The galaxiid fishes of Australia (Pisces: Galaxiidae). *Records of the Australian Museum, 33* (10) , 443–605

McDowall, R.M., Robertson, D.A. and Saito, R. (1975) Occurrence of galaxiid larvae and juveniles in the sea. *New Zealand Journal of Marine and Freshwater Research, 9* (1), 1–9

McEnroe, M. and Cech, J.J. (1985) Osmoregulation in juvenile and adult white sturgeon, *Acipenser transmontanus. Environmental Biology of Fishes, 14* (1), 23–30

McHugh, J.L. (1967) Estuarine nekton. In G. Lauff (ed.). Estuaries. *American Association for the Advancement of Science Publication No.* 83. Washington, pp. 581–620

Maciolek, J.A. (1977) Taxonomic status, biology and distribution of Hawaiian *Lentipes*, a diadromous goby. *Pacific Science, 31* (4), 355–62

Maciolek, J.A. (1984) Exotic fishes in Hawaii and other islands of Oceania. In W.R. Courtenay and J.R. Stauffer (eds), *Distribution, biology and management of exotic fishes*. Johns Hopkins University Press, Baltimore, pp. 131–61

McKenzie, R.A. (1959) Marine and freshwater fishes of the Miramichi River and estuary, New Brunswick. *Journal of the Fisheries Research Board of Canada, 16* (6), 807–29

McKenzie, R.A. (1964) Smelt life history and fishery in the Miramichi River, New Brunswick. *Bulletin of the Fisheries Research Board of Canada, 144*, 1–77

McKeown, B.A. (1984) *Fish migration*. Croom-Helm, London, 224 pp.

McKernan, D.L. (1980) The future of the Atlantic salmon — an international issue. In A.E.J. Went (ed.), *Atlantic salmon — its future. Proceedings of the Second International Atlantic Salmon Symposium*. Fishing News Books, Farnham pp. 18–29

McLane, W.M. (1955) The fishes of the St Johns River system. PhD thesis, University of Florida, Gainesville

McPhail, J.D. and Lindsey, C.C. (1970) Freshwater fishes of northwestern Canada and Alaska. *Fisheries Research Board of Canada Bulletin, 173*, 1–381

Maitland, P.S. (1979) The status and conservation of rare freshwater fishes in the British Isles. *Proceedings of the British Freshwater Fish Conference, 1*, 237–48

Maitland, P.S. (1980a) Review of the ecology of lampreys in northern Europe. *Canadian Journal of Fisheries and Aquatic Sciences, 37* (11), 1944–52

Maitland, P.S. (1980b) Scarring of whitefish (*Coregonus lavaretus*) by European river lamprey (*Lampetra fluviatilis*) in Loch Lomond, Scotland. *Canadian Journal of Fisheries and Aquatic Sciences, 37* (11), 1981–88

Maitland, P.S. (1985) Criteria for the selection of important sites for freshwater fish in the British Isles. *Biological Conservation, 31*, 335–53

Manacop, P.R. (1953) The life history and habits of the goby *Sicyopterus extraneus* Herre (Anga) Gobiidae, with an account of the goby-fry fishery of Cagayan River,

Oriental Misamis. *The Philippine Journal of Fisheries, 2* (1), 1–58

Mansueti, R.J. (1964) Eggs, larvae and young of the white perch, *Roccus americanus,* with comments on its ecology in the estuary. *Chesapeake Science,* 5, 3–45

Manter, H.W. (1955) The zoogeography of trematodes of marine fishes. *Experimental Parasitology, 4,* 62–86

Margolis, L. (1965) Parasites as an auxillary source of information about the biology of Pacific salmon (Genus *Oncorhynchus*). *Journal of the Fisheries Research Board of Canada, 22* (6), 1387–95

Masuda, H., Amaoka, K., Araga, C., Uyeno, T. and Yoshino, T. (1984) *The fishes of the Japanese archipelago.* Tokai University Press, Tokyo, 437 pp.

Mathison, O.A. and Berg, M. (1968) Growth rates of the char *Salvelinus alpinus* (L.) in the Vardnes River, Tromso, northern Norway. *Report of the Institute for Freshwater Research, Drottningholm, 48,* 177–86

Matsui, S. (1986) Studies on the ecology and migration of the ice goby *Leucopsarion petersi* Hilgendorf. *Scientific Bulletin of the Faculty of Agriculture, Kyushu University, 40,* (2-3), 135–74

Matsumiya, Y., Mitani, T. and Tanaka, M. (1982) Changes in distribution pattern and condition coefficient of the juvenile Japanese sea bass with the Chikugo River ascending. *Bulletin of the Japanese Society of Scientific Fisheries, 48* (2), 129–38

Meek, A. (1916) *The migrations of fishes.* Arnold, London, 427 pp.

Melvin, G.D., Dadswell, M.J. and Martin, J.D. (1986) Fidelity of American shad, *Alosa sapidissima* (Clupeidae), to its river of previous spawning. *Canadian Journal of Fisheries and Aquatic Sciences, 43,* 640–6

Meredyth-Young, J.L. and Pullan, S.G. (1977) Fisheries survey of Lake Chalice, Marlborough Acclimatisation District, South Island. *New Zealand Ministry of Agriculture and Fisheries, Fisheries Technical Report, 150,* 1–21

Merrick, J.R. and Schmida, G.E. (1984) *Australian freshwater fishes — biology and management.* Merrick, North Ryde, 409 pp.

Michaelis, F.B. (1985a) Rare or threatened species from inland waters of Tasmania, Australia. *Records of the Queen Victoria Museum, Launceston, 87,* 1–13

Michaelis, F.B. (1985b) Threatened fish — a report on the threatened fish of inland waters in Australia. *Australian National Parks and Wildlife Service, Report Series, 3,* 1–45

Miller, L.W. (1972) Migrations of sturgeon tagged in the Sacramento-San Joaquin estuaries. *California Fish and Game, 58* (2), 102–6

Miller, P.J. (1973a) The species of *Pseudaphya* (Teleostei: Gobiidae) and the evolution of the aphyiine gobies. *Journal of Fish Biology, 5,* 353–65

Miller, P.J. (1973b) The osteology and adaptive features of *Rhyacichthys aspro* (Teleostei: Gobioidei) and the classification of gobioid fishes. *Journal of Zoology, London, 171,* 397–434

Miller, P.J. (1984) The tokology of gobioid fishes. In G.W. Potts and J.R. Wootton (eds), *Fish reproduction — strategies and tactics.* Academic Press, New York, pp. 119–53

Miller, R.J. and Brannon, E.L. (1982) The origin and development of life history patterns in Pacific salmonids. In E.L. Brannon and E.O. Salo (eds), *Proceedings of the salmon and trout migratory behaviour symposium.* School of Fisheries, University of Washington, Seattle, pp. 296–309

Miller, R.R. (1959) Origins and affinities of the freshwater fish fauna of western North America. In C.L. Hubbs (ed.), *Zoogeography. American Association for the Advancement of Science Symposium Vol.* 51, pp. 187–222

Miller, R.R. (1960) Systematics and biology of the gizzard shad (*Dorosoma cepedianum*) and related fishes. *Fisheries Bulletin of the United States Fish and Wildlife Ser-*

vice, 60 (173), 371–92

Miller, R.R. (1966) Geographical distribution of Central American freshwater fishes. *Copeia, 1966* (4), 773–802

Miller, R.R. (1972a) Classification of the native trouts of Arizona with the description of a new species, *Salmo apache. Copeia, 1972* (3), 401–22

Miller, R.R. (1972b) Threatened freshwater fishes of the United States. *Transactions of the American Fisheries Society, 101* (2), 239–52

Miller, R.R. (1977) *International Union for the Conservation of Nature and Natural Resources Red Data Book. Vol. 4: Pisces — freshwater fishes.* IUCN, Morges

Miller, R.R. and Hubbs, C.L. (1969) Systematics of *Gasterosteus aculeatus* with particular reference to intergradation and introgression along the Pacific coast of North America: a commentary on a recent contribution. *Copeia, 1969* (1), 52–69

Mills, D.C. (1971) *Salmon and trout: the resource, its ecology, conservation and management.* Oliver and Boyd, Edinburgh, 351 pp.

Mirza, M.R. (1975) Freshwater fishes and zoogeography of Pakistan. *Bijdragen tot de Dierkunde, 45* (2), 143–80

Mitchel, N.C. (1965) The lower Bain fisheries. *Ulster Folklore, 11*, 1–32

Mitchell, C.P. (1986) Effects of introducing grass carp on populations of two species of small native fishes in a small lake. *New Zealand Journal of Marine and Freshwater Research, 20*, 219–30

Mitchell, C.P. and Penlington, B.P. (1982) Spawning of *Galaxias fasciatus* Gray (Salmoniformes: Galaxiidae). *New Zealand Journal of Marine and Freshwater Research, 16*, 131–3

Mizuno, N. (1960) Study on a freshwater goby, *Rhinogobius similis* Gill, with a proposition on the relationships between landlocking and speciation of some freshwater gobies of Japan. *Memoirs of the College of Science, University of Kyoto, Series B, 27* (2), 97–113

Mizuno, N. (1963a) Distributions of *Cottus japonicus* Okada (Cottidae) and *Tukogobius flumineus* Mizuno (Gobiidae), with special reference to their peculiarities in both landlocking and the speciation from their amphidromous ancestors. *Memoirs of the Osaka Gakugei University* (11), 129–61 (In Japanese, English summary)

Mizuno, N. (1963b) Distributions of two gobiid fishes, *Rhinogobius brunneus (=R. similis)* and *Tukogobius flumineus.* I. Distribution of them in and near stagnant water. *Japanese Journal of Ecology, 13* (6), 242–7 (In Japanese, English summary)

Mizuno, N. (1980) Recent trends in investigations of freshwater gobies in Japan. *Proceedings of the First Workshop for the Promotion of Limnology in Developing Countries.* Kyoto, Japan, pp. 25–30

Montgomery, W.L., McCormick, S.D., Naiman, R.J., Whoriskey, F.J. and Black, G.A. (1983) Spring migratory synchrony of salmonid, catastomid and cyprinid fishes in Riviere a la Truite, Quebec. *Canadian Journal of Zoology, 61*, 2495–502

Montilla, J. (1931) The ipon fisheries of northern Luzon. *Philippines Journal of Science, 45*, 61–75

Moore, J.W. (1975a) Reproductive biology of anadromous Arctic char, *Salvelinus alpinus* (L.) in the Cumberland Sound area of Baffin Island. *Journal of Fish Biology, 7*, 143–51

Moore, J.W. (1975b) Distribution, movements and mortality of anadromous Arctic char, *Salvelinus alpinus* L., in the Cumberland Sound area of Baffin Island. *Journal of Fish Biology, 7*, 339–48

Moore, R. (1982) Spawning and early life history of barramundi, *Lates calcarifer* (Bloch), in Papua New Guinea. *Australian Journal of Marine and Freshwater Research, 33* (4), 647–61

Moore, R. and Reynolds, L.F. (1982) Migration patterns of barramundi, *Lates calcarifer*

(Bloch), in Papua New Guinea. *Australian Journal of Marine and Freshwater Research, 33* (4), 671–682

Moriarty, C. (1978) *Eels — a natural and unnatural history.* David and Charles, Newton Abbott, 192 pp.

Moreau, J. (1986) Freshwater and brackishwater fishes and fisheries in Madagascar — present status and potential, a review. *IPFC expert consultation on inland fishes of the larger Indo-Pacific islands, Bangkok, Thailand,* 4–9 August, 1986. 42 pp.

Morin, R., Dodson, J. and Power, G. (1980) Estuarine fish communities of the eastern James-Hudson Bay coast. *Environmental Biology of Fishes, 5* (2), 135–41

Morin, R., Dodson, J.J., and Power, G. (1981) The migrations of anadromous cisco (*Coregonus artedii*) and lake whitefish (*C. clupeaformis*) in estuaries of eastern James Bay. *Canadian Journal of Zoology, 59,* 1600–7

Morrisey, N. (19850 The commercial fishery for barramundi (*Lates calcarifer*) in Western Australia. *Department of Fisheries and Wildlife Western Australia Report, 68,* 1–32

Morrow, J.W. (1980) *Freshwater fish of Alaska.* Alaska Northwest, Anchorage, 248 pp.

Mottley, C.M. (1934) The origin and relations of the rainbow trout. *Transactions of the American Fisheries Society, 64,* 323–7

Moule, T. (1842) *Heraldry of fish — notice of the principal families bearing fish in their arms.* Van Voorst, London.

Moyle, P.B. and Cech, J.J. (1982) *Fishes: an introduction to ichthyology.* Prentice-Hall, Englewood Cliffs, 593 pp.

Mullan, J.W. (1958) The sea-run or 'salter' brook trout (*Salvelinus fontinalis*) fishery of the coastal streams of Cape Cod, Massachusetts. *Bulletin of the Massachusetts Division of Fisheries and Game, 17,* 1–25

Mullem, P. van, and Vlugt, J. C. van der (1964) On the age, growth, and migration of the anadromous stickleback *Gasterosteus aculeatus* L. investigated in a mixed population. *Archives Neerlandaises de Zoologie, 16* (1), 111–39

Munro, I.S.R. (1967) *The fishes of New Guinea.* Department of Agriculture, Stock and Fisheries, Port Moresby, 650 pp.

Myers, G.S. (1949a) Salt tolerance of fresh-water fish groups in relation to zoogeographical problems. *Bijdragen tot de Dierkunde, 28,* 315–22

Nagasawa, A. and Aguilera, P. (1985) Photographic samples data of the Pacific salmon after hatchery release in Aysen Region, Chile, 1982-1984. *Servicio Nacional de Pesca/Japan International Cooperation Agency. Introduction into Aysen, Chile, of Pacific salmon, Information Brief, 15,* 1–35

Neave, F. (1944) Racial characteristics and migratory habits in *Salmo gairdneri. Journal of the Fisheries Research Board of Canada, 6* (3), 245–51

Neave, F. (1949) Game fish populations of the Cowichan River. *Bulletin of the Fisheries Research Board of Canada, 84,* 1–32

Neave, F. (1966a) Salmon of the North Pacific Ocean. Part III. A review of the life history of North Pacific salmon. 5. Pink salmon in British Columbia. *Bulletin of the North Pacific International Fisheries Commission, 18,* 71–9

Neave, F. (1966b) Salmon of the North Pacific Ocean. Part III. A review of the life history of North Pacific salmon. 6. Chum salmon in British Columbia. *Bulletin of the North Pacific International Salmon Fisheries Commission, 18,* 81–5

Nelson, G.J. and Rothman, M.N. (1973) The species of gizzard shads (Dorosomatinae) with particular reference to the Indo-Pacific region. *Bulletin of the American Museum of Natural History, 150* (2), 131–206

Nelson, J.S. (1976) *Fishes of the world.* Wiley, New York, 416 pp.

Netboy, A. (1958) *Salmon of the Pacific northwest — fish vs dams.* Binfords and Mort, Portland, 122 pp.

Netboy, A. (1974). *The salmon — the fight for survival*. Houghton, Miflin, Boston, 613 pp.

Netboy, A. (1980) *Salmon - the world's most harrassed fish*. Deutsch, London, 304 pp.

Nicholls, J.T. (1928) Fish from the White Nile collected by the Taylor expedition of 1927: a discussion of the fresh-water fauna of Africa. *American Museum Novitates, 319*, 1–7

Nikolskii, G.V. (1961) *Special ichthyology*. Israel Programme for Scientific Translations, Jerusalem, 538 pp.

Nikolsky, G.V. (1963) *The ecology of fishes*. Academic Press, London 352 pp.

Nikolsky G.V. and Reshetnikov, Yu. S. (1970) Systematics of coregonid fishes in the USSR.: Infraspecific variability and difficulties in taxonomy. In C.C. Lindsey and C.S. Woods (eds), *Biology of coregonid fishes*. University of Manitoba, Winnipeg, pp. 251–66

Nilsson, N.A. (1975) Skyddsvarda fiskar. *Sveriges Natur Arsbok, 66*, 141–50

Nomura, H. (1962) Manjuba or anchovy fishery of southern Brazil. *Commercial Fishery Review, 24* (7), 54–5

Norden, C.R. (1970) Evolution and distribution of the genus *Prosopium*. In C.C. Lindsey and C.S. Woods (eds), *Biology of coregonid fishes*. University of Manitoba Press, Winnipeg, pp. 67–80

Nordeng, H. (1961) On the biology of the char (*Salvelinus alpinus* Linnaeus) in Salangen, Norway. 1. Age and spawning frequency determined from scales and otoliths. *Nytt Magazin for Zoologi, 10*, 67–123

Nordeng, H. (1983) Solution to the 'char problem' based on Arctic char (*Salvelinus alpinus*) in Norway. *Journal of the Fisheries Research Board of Canada, 40* (9), 1372–87

Nordlie, F.G. (1981) Feeding and reproductive biology of eleotrid fishes in a tropical estuary. *Journal of Fish Biology, 18* (1), 97–110

Nordlie, F.G., Szelistowski, N.A. and Nordlie, W.C. (1982) Ontogenesis of osmotic regulation in the striped mullet, *Mugil cephalus* L. *Journal of Fish Biology, 20* (1), 79–86

Norman, J.R. and Greenwood, P.H. (1963) *A history of fishes*. Benn, London, 398 pp.

Northcote, T.G. (1967) The relation of movements and migrations to production in freshwater fishes. In S.D. Gerking (ed.), *The biological basis of freshwater fish production*. Wiley, New York, pp. 315–44

Northcote, T.G. (1969) Patterns and mechanisms in the lakeward migratory behaviour of juvenile trout. In T.G. Northcote (ed.), *Symposium on salmon and trout in streams*. H.R. McMillan Lectures in Fisheries, University of British Columbia, pp. 183–203

Northcote, T.G. (1979) Migratory strategies and production in fresh water. In S.D. Gerking (ed.), *Ecology of freshwater fish production*. Blackwell, Oxford, pp.326–59

Northcote, T.G. and Ward, F.J. (1985) Lake resident and migratory smelt, *Retropinna retropinna* (Richardson), of the lower Waikato River system, New Zealand. *Journal of Fish Biology, 27* (1), 113–29

Odell, T.T. (1934) The life history and ecological relationships of the alewife (*Pomobolos pseudoharengus* (Wilson)) in Seneca Lake, New York. *Transactions of the American Fisheries Society, 64*, 118–26

Okada, Y. (1955) *Fishes of Japan*. Maruzen, Tokyo, 434 pp.

Okada, Y. (1960) *Studies on the freshwater fishes of Japan*. Prefectural University of Mie, Tsu, 860 pp.

Okazaki, T. (1984) Genetic divergence and its zoogeographical implications in closely related species *Salmo gairdneri* and *Salmo mykiss*. *Japanese Journal of Ichthyology, 31* (3), 297–308

REFERENCES

Ommanney, F.D. (1964) *The fishes*. Time-Life International, New York, 192 pp.

Osbeck, P. (1762) *A voyage to China and the East Indies ... with a voyage to Suratte by O. Toreen ... and an Account of Chinese history by Capt. C.G. Eckeberg*. Translated from the German by J.R. Forster to which are added a faunula and flora sinensis'. London. 2 vols.

Ots, J-P. and Eldon, G.A. (1975) Downstream movement of fry of *Galaxias fasciatus* Gray. *New Zealand Journal of Marine and Freshwater Research, 9* (1), 97–9

Pantulu, V.R. (1963) Studies on the age and growth, fecundity and spawning of *Osteogeneiosus militaris* (Linn.). *Journal du Conseil, 28* (1), 295–315

Pantulu, V.R. (1986) Fish of the lower Mekong Basin. In B.P. Davies and K.F. Walker (eds), *The ecology of river systems*. Junk, Dordrecht, pp. 721–41

Parker, P. 1980. Family Scorpaenidae — scorpionfishes. In R.M. McDowall (ed.), *Freshwater fishes of southeastern Australia*. Reed, Sydney, pp. 138–9

Parry, G. (1960) The development of salinity tolerance in the salmon, *Salmo salar* (L.) and some related species. *Journal of Experimental Biology, 37*, 425–34

Pavlov, D.S., Reshetnikov, Yu. S., Shatunovskiy, M.I. and Shilin, N.I. (1985) Rare and disappearing fishes in the USSR and the principles of their inclusion in the 'Red Book'. *Journal of Ichthyology, 25* (1), 88–99

Paxton, J.R. and Hoese, D.F. (1985) The Japanese sea bass, *Lateolabrax japonicus* (Pisces, Percichthyidae) an apparent marine introduction into eastern Australia. *Japanese Journal of Ichthyology, 31* (4), 369–72

Payne, A.I. (1976) The relative abundance and feeding habits of the grey mullet species occurring in an estuary in Sierra Leone, West Africa. *Marine Biology, 35*, 277–86

Pearcy, W.G. and Richards, S.W. (1962) Distribution and ecology of fishes of the Mystic River estuary, Connecticut. *Ecology, 43*, 248–59

Peck, J.W. (1970) Straying and reproduction of coho salmon, *Oncorhynchus kisutch*, planted in a Lake Superior tributary. *Transactions of the American Fisheries Society, 99* (3), 591–5

Peterson, R.H. Johansen, P.H. and Metcalfe, J.C. (1980) Observations on the early life stages of Atlantic tomcod, *Microgadus tomcod*. *Fisheries Bulletin of the United States Fish and Wildlife Service, 78* (1), 147–58

Petr, T. (1985) Technical report on the possibilities of Sepik River fish stock enhancement (Papua New Guinea). *FAO Fisheries Travel Report and Aide Memoire, 2505*, 1–22

Phillipps, W.J. (1923) Life history of the New Zealand grayling, *Prototroctes oxyrhynchus*. *New Zealand Journal of Science and Technology, 6* (2), 115–17

Phillipps, W.J. and Hodgkinson, E.R. (1922) Further notes on the edible fishes of New Zealand. *New Zealand Journal of Science and Technology, 5* (2), 91–7

Pillay, S.R. and Rosa, H. (1963) Synopsis of biological data on hilsa, *Hilsa ilisha* (Hamilton) 1822. *FAO Fisheries Biology Synopsis, 25*, 1:1–6:8

Pillay, T.V.R. (1967) Estuarine fisheries of the Indian Ocean coastal zone. In G.H. Lauff (ed.), *Estuaries. American Association for the Advancement of Science, Publication 83*. Washington, pp. 647–57

Podlesnyy, A.V. (1968) The fundamental difference between migratory and non-migratory teleost fishes. *Problems in Ichthyology, 8*, 165–8

Pollard, D.A. (1971) The biology of a landlocked form of the normally catadromous salmoniform fish *Galaxias maculatus* (Jenyns). I. Life cycle and origin. *Australian Journal of Marine and Freshwater Research, 22* (1), 91–123

Potter, I.C. (1970) The life cycles and ecology of Australian lampreys of the genus *Mordacia*. *Journal of Zoology, London, 161*, 487–511

Potter, I.C. (1980) The Petromyzoniformes with particular reference to paired species. *Canadian Journal of Fisheries and Aquatic Sciences, 37* (11), 1595–615

282

Potter, I.C. and Strahan, R. (1968) The taxonomy of lampreys *Geotria* and *Mordacia* and their distribution in Australia. *Proceedings of the Linnaean Society, London, 179* (2), 229–40

Potter, I.C., Prince, P.A. and Croxall, J.P. (1979) Data on the adult marine and migratory phases in the life cycles of the Southern Hemisphere lamprey, *Geotria australis* Gray. *Environmental Biology of Fishes, 4* (1), 65–9

Power, G. (1969) The salmon of Ungava Bay. *Arctic Institute of North America, Technical Paper, 22*, 1–72

Power, G. (1980) The brook char, *Salvelinus fontinalis*. In E.K. Balon (ed.), *Charrs - salmonid fishes of the genus* Salvelinus. *Perspectives in Vertebrate Science* vol. 1. Junk, The Hague, pp. 141–203

Power, G., Power, M.V. Dumas, R. and Gordon, A. (1987) marine migration of Atlantic salmon *(Salmo salar)* in Ungava Bay. *American Fisheries Society Symposium, 1*, 364–76

Prince, J.D. and Potter, I.C. (1983) Life cycle duration, growth and spawning times of five species of Atherinidae (Teleostei) found in a Western Australian estuary. *Australian Journal of Marine and Freshwater Research, 34*, 287–301

Pritchett, R.S. (1975) Eel fisheries. In V.R.P. Sinha and J.W. Jones (eds), *The European freshwater eel*. Liverpool University Press, Liverpool, pp. 100–23

Reed, W. (1967) *Fish and fisheries of northern Nigeria*. Ministry of Agriculture, Zaria, 226 pp.

Regan, C.T. (1911) *The freshwater fishes of the British Isles*. Methuen, London, 287 pp.

Reintjes, J.W. and Pacheco, A.L. (1966) The relation of menhaden to estuaries. In R.F. Smith, A.H. Swartz and W.H. Mossman (eds), *A symposium on estuarine fisheries*. *American Fisheries Society Special Publication* No. 3, pp. 50–8

Richkus, W.A. (1975) Migratory behaviour and growth of juvenile anadromous alewives, *Alosa pseudoharengus*, in a Rhode Island drainage. *Transactions of the American Fisheries Society, 104* (3), 483–93

Ricker, W.E. (1940) On the origin of kokanee, a fresh-water type of sockeye salmon. *Transactions of the Royal Society of Canada, (3),34* (5), 121–35

Ricker, W.E. (1954) Pacific salmon for Atlantic waters? *Canadian Fish Culturalist, 16* (6), 6–14

Ricker, W.E. (1959) Additional observations concerning residual sockeye and kokanee *(Oncorhynchus nerka)*. *Journal of the Fisheries Research Board of Canada, 16*, 897–903

Ricker, W.E. (1972) Hereditary and environmental factors affecting certain salmonid populations. In R.C. Simon and P.A. Larkin (eds), *The stock concept and management of Pacific salmon. H.R. McMillan Lectures in Fisheries. University of British Columbia, Vancouver*, pp. 11–160

Rimmer, M.A. (1985) Growth, feeding and condition of the fork-tailed catfish *Arius graeffei* Kner and Steindachner (Pisces: Ariidae) from the Clarence River, New South Wales. *Australian Journal of Marine and Freshwater Research, 36* (1), 33–9

Rimmer, M.A. and Merrick, J.R. (1983) A review of reproduction and development in the fork-tailed catfishes (Ariidae). *Proceedings of the Linnean Society of New South Wales, 107* (1), 41–50

Roberts, T.R. (19750) Geographical distribution of African freshwater fishes. *Zoological Journal of the Linnaean Society, 57*, 249–319

Roberts, T.R. (1978) An ichthyological survey of the Fly River in Papua New Guinea with descriptions of new species. *Smithsonian Contributions to Zoology, 281*, 1–72

Roberts, T.R. (1981) Sundasalangidae, a new family of minute freshwater salmoniform fishes from southeast Asia. *Proceedings of the California Academy of Sciences, 42* (9), 295–302

Roberts, T.R. (1984) Skeletal anatomy and classification of the neotenic Asian salmoniform superfamily Salangoidea (icefishes or noodlefishes). *Proceedings of the California Academy of Sciences, 43* (13), 179–220

Robertson, O.H. (1957) Survival of preociously mature king salmon male parr (*Oncorhnychus tshawytscha* juveniles) after spawning. *California Fish and Game. 43* (2), 119–30

Rogan, P.L. (1(1981) The Australian chinook salmon fishery. In C.L. Hopkins (ed.), *Proceedings of the salmon symposium. New Zealand Ministry of Agriculture and Fisheries, Fisheries Research Division Occasional Publication, 30,* 78–92

Romer, A.S. (1966) *Vertebrate paleontology,* 3rd edn. University of Chicago, Chicago, 464 pp.

Rosen D.E. (1974) The phylogeny and zoogeography of salmoniform fishes and the relationships of *Lepidogalaxias salamandroides. Bulletin of the American Museum of Natural History, 153* (2), 265–326

Roule, L. (1933) *Fishes — their journeys and migrations.* Routledge, London, 270 pp.

Rounsefell, G.A. (1958) Anadromy in North American Salmonidae. *Fisheries Bulletin of the United States Fish and Wildlife Service, 58* (131), 171–85

Rounsefell, G.A. (1962) Relationships among North American Salmonidae. *Fisheries Bulletin of the United States Fish and Wildlife Service, 62* (209), 1–49

Rulifson, R.A., Huish, M.T. and Thoeson, R.W. (1982a) *Anadromous fish in the southeastern United States and recommendations for development of a management plan.* United States Fish and Wildlife Service, Fishery Resource Region 4, Atlanta. 525 pp.

Rulifson, R.A., Huish, M.T., and Thoeson, R.W. (1982b) Status of anadromous fishes in southeastern U.S. estuaries. In V.S. Kennedy (ed.), *Estuarine comparisons.* Academic Press, New York, pp. 413–25

Rutherford, A.J. (1901) Notes on the Salmonidae in their new home in the south Pacific. *Transactions and Proceedings of the New Zealand Institute, 33,* 240–49

Sagar, P.M. (1986) Salmonids in the Southern Ocean. *Freshwater Catch (NZ), 30,* 11–13

Sakuda, H.M. (1986) *Hawaii freshwater and estuarine fish and fisheries. IPFC expert consultation on inland fisheries of the larger Indo-Pacific islands, Bangkok, Thailand,* 4–9 August, 1986. 17 pp.

Sano, S. (1966) Salmon of the North Pacific Ocean. Part III. A review of the life history of North Pacific salmon. 3. Chum salmon in the far east. *Bulletin of the North Pacific Fisheries Commission, 18,* 41–57

Sarojini, K.K. (1957) Biology and fisheries of the grey mullets of Bengal. I. Biology of *Mugil parsia* Hamilton with notes on its fishery. *Indian Journal of Fisheries, 4* (1), 160–206

Sarojini, K.K. (1958) Biology and fisheries of the grey mullets of Bengal. II. Biology of *Mugil cunnesius* Valenciennes. *Indian Journal of Fisheries, 5* (1), 56–76

Schaffer, W.M. and Elson, P.F. (1975) The adaptive significance of variations in life history among local populations of Atlantic salmon in North America. *Ecology, 56,* 577–90

Schmidt, P.J. (1936) *Migrations of fishes.* State Publishing House of Biological and Medical Literature, Moscow. (In Russian)

Schreck, C.B. and Behnke, R.J. (1971) Trouts of the upper Kern River Basin, California, with reference to systematics and evolution of western North American *Salmo. Journal of the Fisheries Research Board of Canada, 28,* 987–98

Schultz, L.P. (1943) Fishes of the Phoenix and Samoan Islands collected in 1939 during the expedition of the USS 'Bushnell'. *Bulletin of the United States National Museum, 180,* 1–316

Scott, D. (1984) Origin of the New Zealand sockeye salmon, *Oncorhynchus nerka* (Walbaum). *Journal of the Royal Society of New Zealand, 14* (3), 245–9

Scott, D. (1985) Migration and the transequatorial establishment of salmonids. *Proceedings of the International Association of Theoretical and Applied Limnology, 22* (4), 2684–90

Scott, W.B. and Crossman, E.J. (1973) Freshwater fishes of Canada. *Bulletin of the Fisheries Research Board of Canada, 184,* 1–966

Seale, A. (1932) *Agonostomus hancocki* Seale, sp. nov. *Proceedings of the California Academy of Sciences,* (4), *20* (10), 467–9

Senta, T. (1973a) Spawning ground of the salmonoid fish, *Salangichthys microdon,* in Takahashi River, Okayama Prefecture *Japanese Journal of Ichthyology, 20* (1), 25–8

Senta, T. (1973b) On the salmonoid fish, *Salangichthys microdon,* in spawning season, in Takahashi River, Okayama Prefecture. *Japanese Journal of Ichthyology, 20* (1), 29–35

Setzler, E.M., Boynton, W.R., Wood, K.V., Zion, H.H., Lubbers, L., Mountford, N.K., Frere, P., Tucker, L. and Mihursky, J.A. (1980) Synopsis of biological data on striped bass *Morone saxatilis* (Walbaum). *NOAA Technical Report, National Marine Fisheries Service Circular 433,* 1–69 (FAO Synopsis 121)

Shiraishi, Y. (1970) The migration of fishes in the Mekong River. In *Proceedings of the International Biological Programme, Section PF (Freshwater Productivity) Regional Meeting of inland water biologists in South-east Asia, Kuala Lumpur and Malacca, Malaysia, 5–11 May, 1969.* Unesco Field Science Office for South-east Asia, Djakarta, pp. 135–40

Shireman, J.V. (1975) Gonadal development of striped mullet (*Mugil cephalus*) in fresh water. *Progressive Fish Culturalist, 37* (4), 205–8

Shubnikov, D.A. (1976) Types of migrations of diadromous and semidiadromous fishes. *Journal of Ichthyology, 16* (4), 531–5

Sibley, C.G. and Ahlquist, J.E. (1981) The phylogeny and relationships of the ratite birds as indicated by DNA–DNA hybridisation. In G.G.E. Scudder and J.L. Reveal (eds), *Evolution today. Proceedings of the Second International Congress on Systematics and Evolutionary Biology,* pp. 301–35

Sinha, V.R.P. and Jones, J.W. (1975) *The European freshwater eel.* Liverpool University Press, Liverpool, 146 pp.

Sjoberg, K. (1980) Ecology of the European river lamprey (*Lampetra fluviatilis*) in northern Sweden. *Canadian Journal of Fisheries and Aquatic Sciences, 37* (11), 1974–80

Skead, C.J. (1959) The climbing of juvenile eels. *Piscator, 46,* 74–86

Skelton, P.M. (1977) South African red data book — fishes. *South African Scientific Programme Report* No. 14

Skelton, P.M. (1983) Perspectives on the conservation of threatened fishes in southern Africa. *The Naturalist, 27,* 3–12

Skelton, P.M. (1986) Fishes of the Orange-Vaal System. In B.P. Davies and K.F. Walker (eds), *The ecology of river systems.* Junk, Dordrecht, pp. 147–61

Slaney, P.A. and Northcote, T.G. (1974) Effects of prey abundance on density and territorial behaviour of young rainbow trout (*Salmo gairdneri*) in laboratory stream channels. *Journal of the Fisheries Research Board of Canada, 31,* 1201–9

Sloane, R.D. (1982) The Tasmanian eel fishery — some facts and figures. *Australian Fisheries, 41* (12), 14–7

Sloane, R.D. (1984a) Upstream migration by young pigmented freshwater eels (*Anguilla australis australis* Richardson) in Tasmania. *Australian Journal of Marine and Freshwater Research, 35* (1), 61–73

Sloane, R.D. (1984b) The upstream movement of fish in the Plenty River, Tasmania. *Papers and Proceedings of the Royal Society of Tasmania, 118*, 163–71

Smith, J.L.B. (1935) The fishes of the family Mugilidae in South Africa. *Annals of the South African Museum, 30* (5), 587–644

Smith, J.L.B. (1937) Freshwater fishes of the eastern Cape Province. In *A guide to the vertebrate fauna of the eastern Cape Province of South Africa*. Albany Museum, Grahamstown, pp. 119–41

Smith, M.W. and Saunders, J.W. (1958) Movements of brook trout, *Salvelinus fontinalis* (Mitchill), between and within fresh and salt water. *Journal of the Fisheries Research Board of Canada, 15*, 1403–49

Smith, R.J.F. (1985) *The control of fish migration*. Springer-Verlag, Berlin, 243 pp.

Smith, S.H. (1957) Evolution and distribution of the coregonids. *Journal of the Fisheries Research Board of Canada, 14* (4), 599–604

Smith, S.H. (1968a) Species succession and fishery exploitation in the Great Lakes. *Journal of the Fisheries Research Board of Canada, 25* (4), 667–93

Smith, S.H. (1968b) That litle pest the alewife. *Limnos, 1* (2), 12–20

Smith, T.I.J. (1985) The fishery, biology and management of Atlantic sturgeon, *Acipenser oxyrhynchus*, in North America. *Environmental Biology of Fishes, 14* (1), 61–72

Sokolov, L.I. and Malyutin, V.S. (1977) Features of the population structure and characteristics of spawners of the Siberian sturgeon, *Acipenser baieri*, in the spawning grounds of the Lena River. *Journal of Ichthyology, 10*, 210–8

Sprules, W.M. (1952) The Arctic char of the west coast of Hudson Bay. *Journal of the Fisheries Research Board of Canada, 9* (1), 1–15

Stanford, J.A. and Ward, J.V. (1986) Fish of the Colorado. In B.P. Davies and K.F. Walker (eds), *The ecology of river systems*. Junk, Dordrecht, pp. 385–402

Stearns, S.C. (1977) The evolution of life history traits — a critique of the theory and a review of the data. *Annual Review of Ecology and Systematics, 8*, 145–71

Stewart, L. (1980) A history of migratory salmon acclimatisation experiments in parts of the Southern Hemisphere and the possible effects of oceanic currents and gyres upon their outcome. *Advances in Marine Biology, 17*, 397–466

Stokell, G. (1941) *Wildlife control — defects in present scheme exposed, some constructive suggestions*. Publ. Author, 21 pp.

Stokell, G. (1962) Pacific salmon in New Zealand. *Transactions of the Royal Society of New Zealand, Zoology, 2* (21), 181–90

Suttkus, R.D. (1963) Order Lepisostei. In H.B. Bigelow (ed.), *Fishes of the Western North Atlantic. Part 3. Memoirs of the Sears Foundation for Marine Research* No. 1. Sears Foundation for Marine Research, New Haven, pp. 61–88

Svardson, G. (1949) The coregonid problem. I. Some general aspects of the problem. *Report of the Institute for Freshwater Research, Drottningholm, 29*, 89–101

Svardson, G. (1952) The coregonid problem. IV. The significance of scales and gill rakers. *Report of the Institute for Freshwater Research, Drottningholm, 33*, 204–32

Svardson, G. (1979) Speciation of Scandinavian *Coregonus*. *Report of the Institute for Freshwater Research, Drottningholm, 57*, 1–95

Svetovidov, A.N. (1963) *Fauna of the USSR — Fishes* - Vol. II, No. 1. *Clupeidae*. Israel Programme for Scientific Translation, Jerusalem, 428 pp.

Sykes, J.E. (1956) Shad fishery of the Ogeechee River, Georgia, in 1954. *United States Fish and Wildlife Service Special Scientific Report, Fisheries, 191*, 1–11

Talbot, G.B. (1966) Estuarine environmental requirements and limiting factors for striped bass. In R.F. Smith, A.H. Swartz and W.H. Massman (eds), *A symposium on estuarine fishes. American Fisheries Society Special Publication, 3*, 37–49

Talbot, G. B. and Sykes, J. E. (1958) Atlantic coast migrations of American shad.

Fisheries Bulletin of the United States Fish and Wildlife Service, 58 (142), 473–90

Tamura, E. and Honma, Y. (1969) Histological changes in the organs and tissues of the gobioid fishes throughout the life-span. I. Hypothalamo-hypophysial neurosecretory system of the ice goby, *Leucopsarion petersi* Hilgendorf. *Journal of the Japanese Society for Scientific Fisheries, 35* (9), 875–84

Taubert, B.D. (1980) Reproduction of shortnose sturgeon (*Acipenser brevirostrum*) in Holyoke Pool, Connecticut River, Massachusetts. *Copeia, 1980* (1), 114–7

Taubert, B.D. and Dadswell, M.J. (1980) Description of some larval shortnose sturgeon (*Acipenser brevirostrum*) from the Holyoke Pool, Connecticut River, Massachusetts, USA and the St Johns River, New Brunswick, Canada. *Canadian Journal of Zoology, 58*, 1125–8

Taylor, E.H. (1919) Ipon fisheries of Abra River. *Philippines Journal of Science, 14*, 127–30

Tchernavin, V. (1939) The origin of salmon — is its ancestry marine or fresh water? *Salmon and Trout Magazine, 95*, 120–40

Tesch, F.W. (1977) *The eel — biology and management of anguillid eels.* Chapman and Hall, London, 434 pp.

Teugels, G.G., Janssens, L.J.M., Bogaert, J. and Dumalin, M. (1985) Sur une collection de poissons de riviere des Comores. *Cybium, 9* (1), 41–56

Thomson, J.M. (1966) The grey mullets. In H. Barnes (ed.), *Oceanography and marine biology — an annual review* .Vol. 4 Allen and Unwin, London, pp. 301–55

Thomson, J.M. and Luther, G. (1984) Family Mugilidae. In W. Fischer and G. Bianchi (eds), *FAO Species identification sheets for fishery purposes — Western Indian Ocean (Fishing Area 51).* FAO, Rome, 6 vols

Thorpe, J.E. (1982) Migration in salmonids with special reference to juvenile movements in fresh water. In E.S. Brannon and E.O. Salo (eds), *Proceedings of the salmon and trout migratory behaviour symposium.* School of Fisheries, University of Washington, Seattle, pp. 86–97

Thorpe, J.E. 1987. Smolting versus residency: developmental conflicts in salmonids. *American Fisheries Society Symposium, 1*, 224–52

Thorson, T.B. (1976) The status of the Lake Nicaragua shark: an updated appraisal. In T.B. Thorson (ed.), *Investigations of the ichthyofauna of Nicaraguan lakes.* School of Life Sciences, University of Nebraska, Lincoln, pp. 561–74

Thorson, T.B. (1982) Life history implications of a tagging study of the largetoothed sawfish, *Pristis perotteti*, in the Lake Nicaragua–Rio San Juan system. *Environmental Biology of Fishes, 7* (3), 207–28

Tinker, S.W. (1978) *Fishes of Hawaii — a handbook of the marine fishes of Hawaii and the central Pacific Ocean.* Hawaii Service Inc., Honolulu, 532 pp.

Todd, P.R. (1979) Wanganui lamprey fishery. *Freshwater Catch (NZ), 2*, 19–20

Todd, P.R. (1980) Size and age of migrating New Zealand freshwater eels (*Anguilla* spp.). *New Zealand Journal of Marine and Freshwater Research, 14*, 283–93

Tomihama, M.T. (1972) The biology of *Sicydium stimpsoni,* a freshwater goby endemic to Hawaii. BSc Honours Thesis, University of Hawaii, Honolulu, 127 pp.

Torricelli, P., Tongiorgi, P, and Almansi, P. (1982) Migration of grey mullet fry into the Arno River: seasonal appearance, daily activity and feeding rhythms. *Fisheries Research, 1*, 219–34

Trewevas, E. (1953) Sea trout and brown trout. *Salmon and Trout Magazine, 139*, 199–215

Tuunainen, P., Ikonen, E. and Auvinen, H. (1980) Lampreys and lamprey fisheries in Finland. *Canadian Journal of Fisheries and Aquatic Sciences, 37* (11), 1953–9

Tyus, H.M. (1974) Movements and spawning of anadromous alewives, *Alosa pseudoharengus* (Wilson) at Lake Mattamuskeet, North Carolina. *Transactions of the*

American Fisheries Society, 103 (2), 392–6

Uchihashi, K., Iitaka, Y., Morinaga, T., and Kikkawa, J. (1985) Acclimatization of king salmon, *Oncorhynchus tshawytscha*, into New Zealand. *Memoirs of the Faculty of Agriculture, Kinki University, 17*, 221–43

Unwin, M.J. (1981) Aspects of the juvenile salmon outmigration from the Glenariffe Stream. In C.L. Hopkins (ed.), *Proceedings of the salmon symposium. New Zealand Ministry of Agriculture and Fisheries, Fisheries Research Division Occasional Publication, 30*, 15–18

Ustyugov, A.F. (1972) The ecological and morphological characteristics of the Siberian cisco [*Coregonus albula sardinella* (Valenciennes)] from the Yenisey Basin. *Journal of Ichthyology, 12*, 745–66

Vainola, R. (1986) Sibling species and phylogenetic relationships of *Mysis relicta* (Crustacea: Mysidacea). *Annales Zoologica Fennica, 23*, 207–21

Valdez, R.A. and Helm, W.T. (1971) Ecology of the threespine stickleback, *Gasterosteus aculeatus*, on Amchitka Island, Alaska. *Bioscience, 21* (12), 641–5

Valtonen, T. (1980) European river lamprey (*Lampetra fluviatilis*) fishing and lamprey populations in some rivers running into Bothnian Bay, Finland. *Canadian Journal of Fisheries and Aquatic Sciences, 37* (11), 1967–73

Vladykov, V.D. (1950) Movements of Quebec shad (*Alosa sapidissima*) as demonstrated by tagging. *Canadian Naturalist, 77* (5-6), 121–35

Vladykov, V.D. (1963) A review of salmonid genera and their broad geographical distribution. *Transactions of the Royal Society of Canada* (4), *1* (3), 459–504

Vladykov, V.D. and Greeley, J.R. (1963) Order Acipenseroidei. In *Fishes of the Western North Atlantic. Part 3. Soft rayed bony fishes — Class Osteichthys. Memoirs of the Sears Foundation for Marine Research* No. 1. Sears Foundation for Marine Research, New Haven, pp. 24–60

Vladykov, V.D. and Kott, E. (1979) Satellite species among the holarctic lampreys (Petromyzoniformes). *Canadian Journal of Zoology, 57* (4), 860–7

Wagner, H.H., Conte, F.P. and Fessler, J.L. (1969) Development of osmotic and ionic regulation in two races of chinook salmon *Oncorhynchus tshawytscha. Comparative Biochemistry and Physiology, 29*, 325–41

Wakiya, Y. and Takahasi, W. (1937) Study on fishes of the family Salangidae. *Journal of the College of Agriculture, Tokyo University, 14* (4), 265–96

Wal, E.J. van der. (1985) Effects of temperature and salinity on the hatch rate and survival of Australian bass (*Macquaria novemaculeata*) eggs and yolk-sac larvae. *Aquaculture, 47* (2+3), 239–44

Walford, L. and Wicklund, R. (1973) Contribution to a world-wide inventory of exotic marine and anadromous organisms. *FAO Fisheries Technical Paper, 121*, 1–49

Walker, B.W. (1952) A guide to the grunion. *California Fish and Game, 38* (3), 409-20

Walters, V. (1955) Fishes of western Arctic America and eastern Arctic Siberia. *Bulletin of the American Museum of Natural History, 106*, 261–368

Warburg, C.H. (1956) Commercial and sport shad fisheries of the Edisto River, South Carolina, 1955. *United States Fish and Wildlife Service Special Scientific Report, Fisheries, 187*, 1–9

Warburg,C.H. (1957) Neuse River shad investigations, 1953. *United States Fish and Wildlife Service Special Scientific Report, Fisheries, 206*, 1–13

Webster, D. (1952) Early history of the Atlantic salmon in New York. *New York Fish and Game Journal, 29* (1), 26–44

Weeks, H.J. (1985) Ecology of the threespined stickleback (*Gasterosteus aculeatus*): reproduction in rocky tide pools. PhD Thesis, Cornell University, Ithaca, 136 pp.

Welcomme, R.L. (1979) *Fisheries ecology of floodplain rivers.* Longman, London, 317 pp.

Welcomme, R.L. (1981) Register of international transfers of inland fish species. *FAO Fisheries Technical Paper, 213*, 1–120

Welcomme, R.L. (1985) River fisheries. *FAO Technical Paper* 262: 1–330.

Welcomme, R.L. (1986) The fish of the Niger system. In B.P. Davies and K.F. Walker (eds), *The ecology of river systems*. Junk, Dordrecht, pp. 25–48

Wheeler, A. (1969) *The fishes of the British Isles and north west Europe.* McMillan, London, 613 pp.

Wheeler, A. (1985) *The world encyclopedia of fishes*. McDonald, London, 368 pp.

Whitehead, P.J.P. (1974) Clupeidae. In W. Fischer and P.J.P. Whitehead (eds), *FAO species identification sheets for fishery purposes. Eastern Indian Ocean (fishing area 57) and western central Pacific (Fishing area 71)*. FAO, Rome, 4 vols

Whitehead, P.J.P. (1981) Clupeidae. In W. Fischer, G. Bianchi and W.B. Scott (eds), *FAO species identification sheets for fishery purposes — eastern central Atlantic (Fishing Areas 34, 41)*. FAO/Department of Fisheries and Oceans, Ottawa, 6 vols

Whitehead, P.J.P. (1985) Clupeoid fishes of the world (suborder Clupeoidei) — an annotated and illustrated catalogue of the herrings, sardines, pilchards, sprats, shads, anchovies and wolf-herrings. Part I. Chirocentridae, Clupeidae and Pristgasteridae. *FAO Fisheries Synopsis*, (125), 7 (1), 1–303

Wilder, D.G. (1952) A comparitive study of anadromous and freshwater populations of brook trout (*Salvelinus fontinalis* (Mitchill)). *Journal of the Fisheries Research Board of Canada, 9* (4), 169–203

Williams, G.R. and Given, D.R. (1981) *The red data book of New Zealand — rare and endangered species of endemic terrestrial vertebrates and vascular plants*. Nature Conservation Council, Wellington, 175 pp.

Withler, I.C. (1966) Variability in life history characteristics of steelhead trout (*Salmo gairdneri*) along the Pacific coast of North America. *Journal of the Fisheries Research Board of Canada, 23* (3), 365–94

Wootton, J.R. (1976) *The biology of sticklebacks*. Academic Press, London, 387 pp.

Wurman, C. (1985) Chilean salmon: from dreams to reality. *Infofish Marketing Digest, 1985* (4), 13–16

Wynne-Edwards, V.C. (1952) Freshwater vertebrates of the Arctic and sub-Arctic. *Bulletin of the Fisheries Research Board of Canada, 94*, 5–24

Yerger, R. (1978) River goby. In C.R. Gilbert (ed.), *Rare and endangered biota of Florida*. Vol. 4. *Fishes*. University Press, Florida, pp. 46–7

Zama, A. and Cardenas, E. (1984) *Introduction into Aysen Chile of Pacific salmon. No. 9. Descriptive catalogue of marine and freshwater fishes from the Aysen region, southern Chile, with zoogeographical notes on the fish fauna*. Servicio Nacional de Pesca, Ministerio de Economia, Fomenta y Reconstructions, Chile, and Japan International Cooperative Agency, 75 pp.

Appendix

Listing of Diadromous Fish Species According to Type of Diadromy with Latitudinal Range

ANADROMOUS FISHES

Family Petromozontidae

Petromyzon marinus	58°N – 28°N
Lampetra japonica	71°N – 34°N
Lampetra eyresi	58°N – 38°N
Lampetra fluviatilis	59°N – 42°N
Lampetra tridentata	58°N – 32°N
Caspiomyzon wagneri	46°N – 37°N

Family Geotriidae

Geotria australis	34°S – 55°S

Family Mordaciidae

Mordacia mordax	35°S – 44°S
Mordacia lacipida	34°S – 64°S

Family Carcarhinidae

Probably no strictly diadromous species

Family Pristidae

Probably no strictly diadromous species

Family Acipensidedae

Acipenser oxyrhynchus	54°N – 28°N
Acipenser brevirostrum	45°N – 34°N
Acipenser medirostris	59°N – 30°N
Acipenser nudiventris	46°N – 37°N
Acipenser ruthensis	46°N – 37°N
Acipenser stellatus	46°N – 40°N
Acipenser transmontanus	60°N – 36°N
Acipenser baieri	70°N – 68°N
Acipenser sturio	71°N – 43°N

Huso huso	56°N – 46°N
Huso dauricus	52°N – 52°N

Family Lepisosteidae

Probably no strictly diadromous species

Family Salmonidae

Salmo salar	68°N – 41°N
Salmo trutta	68°N – 43°N
Salmo clarkii	61°N – 40°N
Salmo gairdnerii	61°N – 32°N
Salmo mykiss	60°N – 50°N
Salvelinus alpinus	82°N – 45°N
Salvelinus fontinalis	60°N – 41°N
Salvelinus malma	71°N – 48°N
Salvelinus leucomaenis	60°N – 42°N
Hucho perryi	50°N – 42°N
Stenodus leucichthys	73°N – 60°N
Oncorhynchus masou	65°N – 58°N
Oncorhynchus nerka	69°N – 38°N
Oncorhynchus kisutch	68°N – 37°N
Oncorhynchus tshawytscha	68°N – 37°N
Oncorhynchus keta	71°N – 33°N
Oncorhynchus gorbuscha	71°N – 35°N
Coregonus autumnalis	73°N – 65°N
Coregonus laurettae	69°N – 58°N
Coregonus nasus	73°N – 60°N
Coregonus lavaretus	73°N – 65°N
Coregonus oxyrinchus	57°N – 51°N
Coregonus muskun	73°N – 69°N
Coregonus canadensis	44°N – 43°N
Coregonus albula	73°N – 58°N
Coregonus artedii	69°N – 44°N
Coregunus clupeaformis	71°N – 44°N

Family Osmeridae

Osmerus mordax	73°N – 41°N
Osmerus eperlanus	68°N – 38°N
Spirinchus lanceolatus	45°N – 42°N
Spirinchus thaleichthys	61°N – 38°N
Thaleichthys pacificus	60°N – 36°N
Hypomesus transpacificus	73°N – 38°N

Family Salangidae

Salangichthys microdon	52°N – 32°N
Salanx ariakensis	33°N – 31°N
Salanx ishikawae	39°N – 33°N
Salanx cuvieri	39°N – 38°N
Salanx acuticeps	39°N – 23°N
Protosalanx hyalocranium	35°N – 31°N
Neosalanx jordani	40°N – 35°N
Neosalanx hubbsi	40°N – 23°N
Neosalanx andersoni	40°N – 24°N
Neosalanx reganius	33°N – 32°N
Hemisalanx prognathus	40°N – 31°N

Family Aplochitonidae

Lovettia sealii	41°S – 44°S

Family Retropinnidae

Retropinna retropinna	34°S – 47°S
Retropinna tasmanica	41°S – 44°S
Stokellia anisodon	43°S – 47°S

Family Clupeidae

Alosa pseudoharengus	51°N – 34°N
Alosa sapidissima	51°N – 28°N
Alosa mediocris	44°N – 26°N
Alosa chrysochloris	30°N – 25°N
Alosa aestivalis	47°N – 26°N
Alosa alabamae	30°N – 26°N
Alosa alosa	60°N – 20°N
Alosa fallax	65°N – 28°N
Alosa kessleri	46°N – 37°N
Alosa pontica	46°N – 41°N
Alosa caspia	46°N – 37°N
Nematalosa vlaminghi	17°S – 32°S
Nematalosa nasus	34°N – 12°N
Clupeonella cultiventris	46°N – 37°N
Pellonula leonensis	17°N – 5°S
Pellonula vorax	6°N – 13°S
Dorosoma cepedianum	49°N – 21°N
Dorosoma petenense	39°N – 22°N
Tenualosa ilisha	51°N – 30°N
Tenualosa toli	22°N – 10°S

Tenualosa reevesi	28°N – 5°N
Hilsa macrura	11°N – 9°S
Hilsa keelee	26°N – 30°S
Clupanodon thrissa	25°N – 10°N
Anodontasoma chacunda	30°N – 20°S
Herkotslichthys gotoi	18°N – 4°S
Herkotslichthys koningsbergeri	13°S – 25°S

Family Pristigasteridae

Ilisha megaloptera	23°N – 9°S
Pellona ditcheli	24°N – 30°S

Family Engraulidae

Anchoviella lepidentostole	6°N – 20°S

Family Ariidae

Arius madagascariensis	5°N – 25°S

Gasterosteidae

Gasterosteus aculeatus	71°N – 29°N
Pungitius pungitius	73°N – 37°N

Family Syngnathidae

Microphis brachyurus	35°N – 32°S

Family Belonidae

Probably no strictly diadromous species

Family Gadidae

Microgadus tomcod	54°N – 36°N

Family Percichthyidae

Morone saxatilis	46°N – 34°N
Morone americanus	52°N – 29°N

Family Gobiidae

Batanga lebretonis	? ?
Leucopsarion petersi	45°N – 30°N
Awaous tajasica	31°N – 5°S

Awaous stamineus	22°N – 9°S
Tasmanogobius lordi	41°S – 44°S

CATADROMOUS FISHES

Family Anguillidae

Anguilla anguilla	70°N – 28°N
Anguilla rostrata	60°N – 9°N
Anguilla japonica	42°N – 22°N
Anguilla mossambica	25°N – 35°S
Anguilla celebensis	24°N – 11°S
Anguilla marmoratus	24°N – 33°S
Anguilla bengalensis	23°N – 33°S
Anguilla bicolor	22°N – 27°S
Anguilla obscura	2°N – 27°S
Anguilla borneensis	2°N – 2°S
Anguilla megastoma	1°S – 25°S
Anguilla interioris	3°S – 9°S
Anguilla reinhardtii	10°S – 38°S
Anguilla australis	18°S – 47°S
Anguilla dieffenbachii	34°S – 47°S

Family Megalopidae

Probably no strictly diadromous species

Family Galaxiidae

Galaxias maculatus	33°S – 55°S

Family Clupeidae

Potamalosa richmondia	25°S – 44°S
Ethmalosa fimbriata	19°N – 23°S

Family Engraulidae

Thryssa scratchleyi

Family Atherinidae

Probably no strictly diadromous species

294

Family Percichthyidae

Macquaria novemaculeata	26°S – 39°S
Lateolabrax japonicum	37°N – 21°N

Family Centropomidae

Lates calcarifer	23°N – 26°S

Family Mugilidae

Mugil cephalus	42°N – 42°S
Mugil robustus	
Agonostomus monticola	34°N – 28°N
Agonostomus telfairi	4°S – 25°S
Joturus pichardi	20°N – 8°N
Myxus petardi	26°S – 33°S
Myxus capensis	27°S – 34°S
Myxus parsia	25°N – 6°N
Myxus cunnesius	24°N – 30°S
Liza dumerilii	15°N – 32°S
Liza falcipinnis	14°N – 8°S
Liza ramada	60°N – 15°N
Liza aurata	25°S – 45°S
Liza abu	
Trachynotus eurynotus	28°S – 34°S

Family Bovichthyidae

Pseudaphritis urvillii	37°S – 44°S

Family Scorpaenidae

Notesthes robusta	26°S – 36°S

Family Kuhliidae

Kuhlia rupestris	18°N – 32°S
Kuhlia marginata	41°N – 34°N
Kuhlia sandvicensis	22°N – 19°N

Family Lutjanidae

Lutjanus goldei	5°S – 10°S

Family Theraponidae

Mesopristes kneri	22°S – 24°S

Family Cottidae

Cottus asper	58°N – 34°N
Cottus aleuticus	58°N – 35°N
Cottus kazika	41°N – 31°N
Cottus hongiongensis	48°N – 36°N

Family Gobiidae

Gobiomorus dormitor	27°N – 4°N
Awaous tajasica	31°N – 5°S
Mugiolgobius abei	36°N – 30°N
Redigobius bikolanus	35°N – 6°N

Family Pleuronectidae

Rhombosolea retiaria	34°S – 47°S
Platichthys stellatus	70°N – 34°N
Pleuronectes flesus	60°N – 35°N

Family Soleidae

Trinectes maculatus	70°N – 43°N

AMPHIDROMOUS FISHES

Family Plecoglossidae

Plecoglossus altivelis	44°N – 23°N

Family Galaxiidae

Galaxias truttaceus	37°S – 44°S
Galaxias brevipinnis	34°S – 53°S
Galaxias fasciatus	34°S – 47°S
Galaxias argenteus	34°S – 47°S
Galaxias postvectis	34°S – 47°S
Galaxias platei	40°S – 54°S
Galaxias cleaveri	38°S – 44°S

Family Aplochitonidae

Aplochiton zebra	40°S – 54°S
Aplochiton taeniatus	40°S – 55°S

Family Prototroctidae

Prototroctes maraena	34°S – 44°S
Prototroctes oxyrhynchus	34°S – 47°S

Family Clupeidae

Esculaosa thoracata	14°N – 8°N

Family Mugiloididae

Cheimarrichthys fosteri	34°S – 47°S

Family Cottidae

Leptocottus armatus	60°N – 33°N
Cottus japonicus	46°N – 31°N

Family Eleotridae

Gobiomorphus huttoni	34°S – 47°S
Gobiomorphus hubbsi	34°S – 47°S
Gobiormorphus gobioides	34°S – 47°S
Gobiomorphus cotidianus	34°S – 47°S
Gobiomorphus australis	30°S – 37°S
Eleotris melanosoma	45°N – 22°S
Eleotris sandwichensis	22°N – 19°N
Eleotris fusca	45°N – 27°S
Eleotris oxycephala	37°N – 27°N

Family Gobiidae

Sicyopterus extraneus	
Sicypoterus micrurus	18°N – 22°S
Sicypoterus gymnauchen	10°N – 18°S
Sicyopterus lachrymosus	18°N – 8°S
Sicyopterus fuliag	18°N – 11°S
Sicyopterus taeniurus	13°S – 19°S
Sicyopterus japonicum	37°N – 22°N
Sicypoterus lagocephalus	21°S
Sicypoterus macrostetholepis	10°N – 10°S
Sicydium stimpsoni	19°N – 16°N
Sicydium plumieri	20°N – 13°N
Sicyopus lepurus	
Sicyopus zosterophorum	10°N – 9°S
Stiphodon elegans	13°N
Stiphodon stevensoni	

Lentipes concolor	22°N – 19°N
Lentipes armatus	
Acanthogobius flavimanus	45°N – 23°N
Rhinogobius brunneus	45°N – 23°N
Rhinogobius giurinus	45°N – 22°N
Chonophorus ocellaris	
Chonophorus guamensis	13°N – 13°N
Ophiocara aporos	18°N – 25°S
Glossogobius giurus	45°N – 32°S
Glossogobius celebensis	
Dormitator latifrons	
Dormitator maculatus	32°N – 5°S
Tridentiger brevispinis	45°N – 31°N
Tridentiger obscurum	43°N – 31°N
Chaenogobius urotaenia	55°N – 31°N
Chaenogobius castaenas	45°N – 22°N
Chaenogobius sp. 1	
Chaenogobius sp. 2	45°N – 35°N
Luciogobius pallidus	
Luciogobius guttatus	45°N – 22°N

Family Rhyacichthyidae

Rhyacichthys aspro	24°N – 13°S

Index

Numerals in **bold face** indicate
principal family entries

Acanthogobius flavimanus 106, 204,
   236
Acanthopagrus barda 26
Acanthopterygii 31, 133
Acclimation 150, 153
Acclimatization 186
Acentrogobius 108
Achirus 93
Acipenser baieri 39-40, 241
Acipenser brevirostrum 38-40, 112,
   212, 251, 254
Acipenser fulvescens 38, 40
Acipenser guldenstaedtii 38, 212
Acipenser medirostris 38-9, 212, 257
Acipenser nudiventris 38-9, 257
Acipenser oxyrhynchus 38-9 212, 252,
   255, 258
Acipenser ruthensis 39-40, 189
Acipenser stellatus 39, 213
Acipenser sturio 38-40, 212-13, 257
Acipenser transmontanus 38-40, 212,
   258
Acipenseridae 6-7, 17, 19, 30, 33, **37**,
   115-16, 122, 131-2, 133, 140, 137,
   142, 145, 147-8, 167, 172, 187,
   206, 211
Agnatha 133
Agonostomus 138
Agonostomus catalai 88
Agonostomus hancocki 88
Agonostomus monticola 87, 138, 147,
   172
Agonostomus telfairi 88
Aholeholes, *see* Kuhliidae
Albulidae 30
Aldrichetta forsteri 89, 138
Alepisauridae 31
Alepocephalidae 31
Alewife, *see Alosa pseudoharengus*
Alimental migration 22
Alligator gar, *see Lepisosteus osseus*
Allosmerus elongata 57
Alosa 63, 138
Alosa aestivalis 64-5, 230
Alosa alabamae 65, 259

Alosa alosa 65, 68, 182, 229, 259
Alosa brashnikova 231
Alosa caspia 66, 189, 231
Alosa chrysochloris 65, 259
Alosa fallax 65, 146, 229, 259
Alosa kessleri 65, 231, 241
Alosa mediocris 65, 230
Alosa pontica 66
Alosa pseudoharengus 6-7, 61, 63,
   146, 209, 229-30
Alosa sapidissima 7, 64, 113, 146,
   187, 189-90, 197-8, 229-30, 259
American mountain mullet, *see*
   Agonostomus monticola
American shad, *see Alosa sapidissima*
American striped bass, *see Morone*
   saxatilis
Amiidae 30
Ammocoete 34, 161
Amphibiotic migration 22
Amphidromy - definition 20, 33
Amphihaline 23, 71, 149
Anabantidae 32
Anadromy - definition 20, 33-4
Anchoviella lepidentostole 68, 232
Anchovies, *see* Engraulidae
Anodontosoma chacunda 67
Anguilla 8, 14, 27, **77**, 153, 177, 213,
   243
Anguilla anguilla 168, 188, 214
Anguilla australis 188, 215, 244
Anguilla dieffenbachii 215, 242
Anguilla reinhardtii 215, 244
Anguilla mossambica 4
Anguilla rostrata 168, 258
Anguillidae 6, 8, 24, 27-30, 33, **77**,
   114-16, 119, 122, 131-2, 133-5,
   137, 139, 140, 142, 145, 147-8,
   156, 161-2, 165, 187, 247,
Anguilliformes 30, 133
Aplochiton marinus 99
Aplochiton taeniatus 99-100
Aplochiton zebra 100, 229
Aplochitonidae 25, 28, 31, 33, **60, 99**,
   115-16, 120, 131, 140, 142, 147,
   188, 228
Aquaculture 212, 216, 223
Arctic char, *see Salvelinus alpinus*